P9-EEE-555

Praise for *Nuclear Roulette*

"It will be an auspicious start to our new century if we can encourage a revitalized movement to stop all nuclear production and immediately close down every nuclear facility—military and civilian. Then we can dedicate our skills and resources to finding true solutions to the real challenges of our time: evolving a sustainable, energy-wise, and peaceful society."

—**ERNEST CALLENBACH AND JERRY MANDER**,
from the foreword

"If ever there was a book that people need to read at this moment in the history of the world it is *Nuclear Roulette*. Comprehensively referenced, it is not only an encyclopedia of the nuclear age related specifically to nuclear power, it is a potent warning of the almost incomprehensible dangers that lie ahead, as well as the damage that has already contaminated portions of our beloved planet beyond repair. I highly recommend this wonderful book to all who care about our children, future generations, and the thirty million other species that cohabit this earth with us."

—**DR. HELEN CALDICOTT**, pediatrician,
founding president of Physicians for Social Responsibility
and author of *Nuclear Madness* and *Nuclear Power Is Not the Answer*

"A thoroughly brilliant work. Extraordinary! Gar Smith cuts through the lies of the nuclear promoters to document the deadliness of atomic power."

—**KARL GROSSMAN**, professor of journalism,
State University of New York College at Old Westbury,
and author of *Cover Up: What You Are Not Supposed to Know About Nuclear Power*

"This powerful solartopian screed leaves no doubt that the experiment with atomic energy is the most dangerous and expensive technological failure in human history. Gar Smith writes with extraordinary power and clarity on an industry whose failure threatens the future of our species—ecologically, economically, and in terms of biological survival. *Nuclear Roulette* is a strong signpost pointing straight to a green-powered world, where we get our energy cheaper, safer, cleaner, community-owned, and quicker. Take this book with you next time you're out shutting a nuke or opening a wind farm."

—**HARVEY WASSERMAN**, author of *SOLARTOPIA! Our Green-Powered Earth, A.D. 2030*, and editor of www.nukefree.org.

"*Nuclear Roulette* by Gar Smith is a timely and necessary book. Nuclear energy is unaffordable by every measure—by the measure of financial costs, of safety, and of the destruction of democracy. We are witnessing this in India where the US-India nuclear agreement is imposing a 'nuclear renaissance' by creating a police state in Koodankulam and in Jaitapur. *Nuclear Roulette* should be in the hands of everyone who cares for life and freedom."

—**DR. VANDANA SHIVA**, founder of Navdanya Research Foundation for Science, Technology and Ecology

"*Nuclear Roulette* is an act of love and reason for Mother Earth. We've had five decades of poisonous decision making in the face of millennia of life. Now is the time to safeguard our generations yet to come. Gar Smith's powerful writing tells the stories that inform our good work."

—**WINONA LADUKE**, indigenous-rights activist and author of *All Our Relations* and *Recovering the Sacred*

NUCLEAR ROULETTE

NUCLEAR ROULETTE

THE TRUTH ABOUT THE MOST DANGEROUS ENERGY SOURCE ON EARTH

GAR SMITH

Foreword by Jerry Mander and Ernest Callenbach

CHELSEA GREEN PUBLISHING
WHITE RIVER JUNCTION, VERMONT

Nuclear Roulette is a project of the International Forum on Globalization.
1009 General Kennedy Avenue, #2, San Francisco, CA 94129, www.ifg.org

Project Manager: Hillary Gregory
Project Editor: Brianne Goodspeed
Copy Editor: Nancy W. Ringer
Proofreader: Susan Barnett
Indexer: Lee Lawton
Designer: Melissa Jacobson

Printed in the United States of America
First printing September, 2012
10 9 8 7 6 5 4 3 2 1   12 13 14 15 16

Our Commitment to Green Publishing
Chelsea Green sees publishing as a tool for cultural change and ecological stewardship. We strive to align our book manufacturing practices with our editorial mission and to reduce the impact of our business enterprise in the environment. We print our books and catalogs on chlorine-free recycled paper, using vegetable-based inks whenever possible. This book may cost slightly more because it was printed on paper that contains recycled fiber, and we hope you'll agree that it's worth it. Chelsea Green is a member of the Green Press Initiative (www.greenpressinitiative.org), a nonprofit coalition of publishers, manufacturers, and authors working to protect the world's endangered forests and conserve natural resources. *Nuclear Roulette* was printed on FSC®-certified paper supplied by Thomson-Shore that contains at least 30 percent postconsumer recycled fiber.

Library of Congress Cataloging-in-Publication Data
Smith, Gar.
Nuclear roulette : the truth about the most dangerous energy source on earth / Gar Smith ; foreword by Jerry Mander and Ernest Callenbach.
p. cm.
Includes bibliographical references and index.
ISBN 978-1-60358-477-7 (hardcover) — ISBN 978-1-60358-434-0 (pbk.) — ISBN 978-1-60358-435-7 (ebook)
1. Nuclear energy. 2. Nuclear power plants—Risk assessment. 3. Nuclear power plants—Accidents. 4. Nuclear power plants—Natural disaster effects I. Title.

TK9145.S556 2012
621.48'3—dc23

2012027407

Chelsea Green Publishing
85 North Main Street, Suite 120
White River Junction, VT 05001
(802) 295-6300
www.chelseagreen.com

Dedicated to Ernest "Chick" Callenbach

Author, visionary, and friend. A genial provocateur who faced every challenge with a smile and inspiration.

(April 3, 1929–April 16, 2012)

Cover Image Background

The 1,130-MW Trojan plant in Oregon, which began operating in 1976, was plagued with construction problems. In 1992, after a cracked steam tube released radioactive gases into the atmosphere, the plant was forced to shut down—after only 16 years of its expected 36-year life. Once the largest reactor in the United States, the $450 million plant became the first US commercial reactor to be decommissioned. The 1,000-ton reactor vessel was shipped by barge to the Hanford Nuclear Reservation in Washington State and buried in a 45-foot-deep pit. The 500-foot-tall cooling tower was brought down by dynamite in 2006. Removing the plant cost nearly as much as building it; Portland General Electric customers are still paying for the costs of decommissioning the plant. All 800 spent fuel rods are still stored in a pool on the reactor site, since there is no place to send them. And the site has now been turned into a 75-acre lakeside park.

Contents

Foreword ix
Preface xv
Introduction xviii

PART ONE
FOURTEEN ARGUMENTS AGAINST NUCLEAR POWER

1. The Impossibility of Speedy Deployment 1
2. Unaffordable Costs 9
3. Inefficient and Unreliable 23
4. Catastrophic Dangers: From Three Mile Island to Fukushima 31
5. Aging Reactors 49
6. Environmental Pollution: Water, Air, and Land 56
7. Damage to Indigenous Peoples 65
8. Earthquake Risks 69
9. Climate Change: Droughts, Floods and Solar Flares 80
10. Radioactive Wastes 87
11. The Decommissioning Dilemma 95
12. Proliferation Dangers 102
13. Reactors, War, and Terrorism 109
14. "New, Improved" Reactors Won't Save Us 116

PART TWO
COVER-UPS AND CONSEQUENCES

15. Japan: Living with the Consequences 123
16. Why You Can't Believe the Official Story 130
17. The Regulatory-Industrial Complex 137
18. Five of the Worst US Reactors 147
19. Near Misses and Unbelievable Mishaps 168

PART THREE
THE PATH FORWARD: BETTER OPTIONS EXIST

Introduction 179
20. The Path Ahead 185
21. Efficiency: The "Fifth Fuel" 188
22. The Power of Renewables 193
23. Public Policy Reforms 207
24. Powering Down: The Conservation Imperative 221

Conclusion 227
Acknowledgments 231
Notes 233
Index 269

FOREWORD

False Solutions

THIS BOOK is based on the fifth report in the International Forum on Globalization's series on false solutions to the global climate and energy crises. Both the June 2011 report and this expanded book-length investigation focus on the appalling attempt to revive and celebrate nuclear power as a grand "green," climate-friendly energy system that can successfully replace fossil fuels and continue to sustain our industrial society at its present level. Author Gar Smith systematically refutes all the nuclear industry arguments, including one of its most critical assumptions—that the deadly radiation from nuclear wastes can be successfully sequestered for the 250,000 years it will remain dangerous to all life, an assertion bordering on insane.

Nuclear technology was originally devised as a tool to fabricate weapons of mass destruction. The goal was to create a nuclear arsenal that could deliver an unprecedented level of death and devastation (with accompanying pollution to air, water, and biological systems), well beyond anything that had ever before been achieved. At Hiroshima and Nagasaki, atomic power proved it could excel at destroying buildings and lives. But after World War II, nuclear's corporate advocates tried to rebrand it as a benign, efficient, sustainable source of energy that would be "too cheap to meter." Today, the atom's corporate boosters have begun to tout nuclear reactors as the best "green" solution to the world's energy and climate crisis.

In the pages that follow, Gar Smith demolishes these claims. If the nuclear dragon can be slain with the weapons of logic and information, this document should prove fatal. Nuclear's inherent problems include its extravagant and noncompetitive costs, the absurdly long time spans required to deploy it, and the technology's little-noted but very important net energy deficiencies. On this latter point, all full-life-cycle studies—measuring total energy expended on everything from mining, processing, and shipping uranium to plant construction, operation, and ultimate decommissioning—conclude that nuclear energy requires about as much energy *input* as the *output* it may ultimately provide. There are no bargains here. Sustainable alternatives, including wind, solar, hydro, and geothermal, have far better *net energy* ratios.

But that's the least of it. The overriding problem with nuclear energy, which dwarfs all others, is that nuclear production demands that society deal with all the spectacularly dangerous emanations and waste products intrinsic to that production. The risks lie far beyond the horizon of our imaginations. Nuclear accidents can make large regions uninhabitable. They can bankrupt nations. Even a nuclear reactor's routine, day-by-day leaks pose generational hazards that should make us shudder. To run the industry safely would mean isolating its waste products for at least tens of thousands of years—not to mention controlling the risks of weapons proliferation. Are we somehow exempt from history? Few civilizations last more than a couple of hundred years. The Romans lasted about 700 years and did very well for a while. But they made some mistaken assumptions about their own permanence that are very similar to those we are making now. The situation would be comical if it weren't so deadly. What hubris. What madness!

How exactly will the nuclear industry's vast tonnage of radioactive waste be safely isolated from future generations? What will shield our grandchildren—and theirs? This report goes to great lengths to demonstrate that no successful containment system has yet been invented to package this stuff successfully beyond a very short period. Everywhere, efforts to store these wastes underground have been resisted vigorously by local communities—and with good reason.

The problem of long-term storage lasting eons begins to suggest that we contemplate some kind of permanent signage we might deploy around these danger zones. But what language should we use? Will English still be spoken after a quarter million years, in say, the year 25,2012 AD? And even then, the radioactive danger will still be present. Our legacy of poisoned earth will be, effectively, eternal.

The only thing more bizarre than the demand that we accept these circumstances is that the governments of numerous highly developed, well-educated countries, including our own, have agreed to let this technology operate. They expect the public to agree to these grim prospects. They tell us we must let this technology proliferate; we must blind ourselves to its risks; we must continue to contribute our tax dollars to subsidize the unknowable costs of maintaining the industry and cleaning up the devastation it leaves behind. But it's not only governments, desperate for energy, that accept these conditions. Much of the public, as well, remains entranced by the energy industry's reassuring public relations and advertising mantras. We need to shake off this state of denial. The situation is so grave that we should more properly be camped out night and day in front of legislative and regulatory bodies demanding the permanent end to any and every expression of this continued nuclear menace, be it civilian or military.

Snake Eyes

At Japan's Fukushima Daiichi nuclear power complex—a facility knowingly constructed in a seismically active site—the reactors' spent fuel rods (which are more radioactive than the working core itself) were stored aboveground in open pools of water. If that arrangement sounds somehow primitive—more like an afterthought than a serious design solution—please remember that most of Japan's nuclear power plants were designed by General Electric, which also built 23 similar Mark 1 reactors in the United States. Many of these US reactors share the same aboveground storage systems that were devastated by the earthquake and tsunami that swept Fukushima in 2011.

Three of Fukushima Daiichi's six reactors experienced meltdowns and a huge region of north-central Japan has been left uninhabitable. Some argue that thousands of square miles should be declared uninhabitable. Radioactive isotopes now spill wildly into the air, infecting the jet stream and bleeding into the ocean. The future risks from these radioactive emissions remain incalculable. So, instead of safeguarding the future for 250,000-plus years, the containment lasted barely 40 years.

How do such things happen? It's not as if the Japanese government and corporations were not warned. In Japan, an active citizens movement comprising tens of thousands of people had been demanding the shutdown of the nuclear industry for nearly three decades. This anti-nuclear movement predicted exactly what happened at Fukushima, but both government and the media criminally ignored its warnings. In a dramatic turnaround, the first anniversary of the Fukushima disaster saw nearly every one of Japan's 54 reactors shut down—and government officials calling for a swift transition to renewable energy and conservation. Unfortunately, when it comes to nuclear power, "shutdown" doesn't actually shut things down, since reactor cores remain hot and dangerous even when not generating electricity. But Japan still has done more than the United States to address the terrifying dangers of nuclear power.

How could President Obama, so soon after the Fukushima meltdowns, propose an *increase* in Washington's already immense subsidies for nuclear power? The sad truth is, the US government's primary commitment is not to sustain life on earth but, rather, to sustain an energy-desperate economic system that is based on the insane notion that economic growth can continue forever. But we live on a finite planet. The depletion of our resources—fossil fuels, fresh water, forests, arable soils, and key minerals—is already well advanced. Faced with escalating examples of climate change—tornados, floods, droughts, and melting glaciers—the preposterousness of our fundamental assumptions could not be more obvious.

It was all just a really bad gamble. We have been rolling the dice since the industrial revolution—or maybe since the Enlightenment—when we decided our technologic superiority would get us anything we want and would solve all problems. A world without limits. We could control Nature and stand in for God. But then the dice rolled snake eyes.

Atoms for Peace?

The writing of this report was well under way months before the quake and tsunami added the name "Fukushima" to the growing list of nuclear disasters. Gar Smith has added a host of new material on the implications of Fukushima for the planet. In addition to assessing the costs, inefficiencies, delays, and dangers of nuclear power, this report documents the industry's multifold impacts on water, air, and land, whether from routine operations, from hundreds of "near misses" that have received scandalously little notice, and from the full-blown disasters at Three Mile Island, Chernobyl, and Fukushima. Gar also investigates a number of critical but overlooked issues ranging from aging equipment to regulatory failure, the impacts of extreme weather, and the damage caused to generations of indigenous peoples (especially in the United States and Australia) who mined uranium and raised families alongside the so-called low-level radioactive wastes that continue to poison their soil and their homes.

This report also discusses what may be the least-talked-about problem: the industry's contribution to weapons proliferation. The military-industrial complex is also a political-corporate alliance. A courtship that began in the earliest days of the Atomic Age quickly blossomed into a marriage of military and civilian atomic enterprises. Westinghouse and General Electric have been called the "Coke and Pepsi" of nuclear power. Westinghouse, the world's leading manufacturer of nuclear reactors, is also a major Pentagon weapons contractor. Similarly, General Electric derives its wealth from the sales of both civilian reactors and military weapons.

When World War II ended, the United States faced two postwar dilemmas: how to maintain the Pentagon's huge infrastructure of nuclear scientists, uranium enrichment facilities, and research labs and how to profit from America's massively expanded wartime uranium production. There was considerable fear that, without a war to sustain jobs, the country would soon slide back into a Great Depression, so finding ways to pursue "peaceful uses for nuclear energy" seemed like a good idea. The transition to a "civilian nuclear program" was accelerated by President Eisenhower's Atoms for Peace campaign, which was little more than what we now call a "corporate stimulus" program. The program promoted "peaceful uses of the atom" by introducing nuclear reactors to the United States and other countries. Among the first countries to receive US nuclear reactors were Iran and Pakistan.

Then there was Eisenhower's Project Plowshare, which proposed a series of nuclear explosions to replace the Panama Canal with a "sea-level canal" across Central America. His Project Chariot would have used five A-bombs to create a harbor near Cape Thompson, Alaska. Other plans for "nuclear landscaping" involved using "peaceful" A-bombs to blast tunnels and create reservoirs for dams. Although none of these plans were ever carried out, these make-work projects kept the Pentagon's nuclear scientists and atomic labs occupied during peacetime, while the "civilian" energy program assured General Electric, Westinghouse, and other corporations a steady supply of tax dollars while they were waiting for the next war—which came along soon enough.

From Irresolvable Dilemmas to Better Options

Part 1 of this book offers "Fourteen Arguments against a Nuclear Renaissance," In part 2, "Cover-ups and Consequences," Gar takes a close look at the culture of political and industrial collusion that puts profits before protection, and he illustrates the consequences by profiling some of the most dangerous nuclear plants in the United States. Part 3, "The Path Forward: Better Options Exist," surveys a vast and expanding landscape of renewable energy and public policy options. These include an optimistic perspective on ways to "power down" an overheated industrial economy to a sustainable level by rapidly deploying an array of alternative energy ideas that are now gaining attention. Alternative sources of energy are now outracing not only nuclear pipe dreams but also the outdated technologies of coal and oil.

No sane utility executive has ever contemplated building a nuclear plant without huge government subsidies. Still, with about 440 aging nuclear power plants in the world (including more than a hundred in the United States) and plans for at least 60 new plants in the works, despite the unsolved problems of nuclear waste storage and radiation, how much hope is really appropriate?

The Japanese government is now turning away from its former gung-ho nuclear ambitions and accelerating its transition toward renewables. The Chinese also are reexamining their nuclear plans and pushing harder to accelerate their wind and solar projects. It seems we are at a tipping point, where the powerful and wealthy are being forced to realize that nuclear is not viable as an energy solution. We simply cannot afford more Three Mile Islands, Chernobyls, and Fukushimas.

But does this mean that the corporate world is at last finally ready to back off from its commitment to endless growth—an absurd drive that no combination of renewables can possibly sustain? We shall see whether governments can finally get real about the central challenge of this century: learning to live in new ways that we have the power to sustain. Thousands of groups worldwide are already working hard on "economic transitions" that involve realistic assessments of our global dilemma, the possibilities for increased efficiencies, alternative technologies, and opportunities to replace the emphasis on globalization with a reinforcement of local sustainable systems in all crucial areas, from food production to energy to transport and banking. But first, we need to let go of our fantasies of limitless production and consumption.

It will be an auspicious start to our new century if we can encourage a revitalized movement to stop all nuclear production and immediately close down every nuclear facility—military and civilian. Then we can dedicate our skills and resources to finding true solutions to the real challenges of our time: evolving a sustainable, energy-wise, and peaceful society.

JERRY MANDER AND ERNEST CALLENBACH

PREFACE

Lessons from Fukushima

WITHIN WEEKS of the thirty-second anniversary of the Three Mile Island meltdown and the twenty-fifth anniversary of the Chernobyl explosion, an earthquake-driven tsunami swamped the 40-year-old Fukushima reactor complex on the east coast of Japan, unleashing a nuclear calamity that forced massive evacuations, contaminated local crops, dumped radioactive water into the sea, and spread radiation around the world. Hydrogen explosions destroyed buildings at reactor units 1, 2, 3, and 4. A month later, Japanese officials raised the severity rating of the accident to level 7—the highest international rating, on par with the Chernobyl disaster. The devastation of the Sendai tsunami was immediately apparent: 25,000 killed and entire cities washed out to sea. Fukushima's devastation, while not as visible, will be longer lasting.

The timeline for catastrophe has shifted. Where our ancestors used to fear "once-in-a-lifetime" floods, droughts, and earthquakes, our children now face epochal, intergenerational traumas. These include the chemicals that haunt Bhopal, the poisoned waters in the Gulf of Mexico, the defoliated landscapes of Vietnam, and the deformed bodies of children in Serbia, Iraq, and Afghanistan, who were born in countries irradiated by the Pentagon's "depleted uranium" weapons.

Nuclear mishaps, minor and major, can create legacies that outlive even the great-great-grandchildren of the engineers who designed the nuclear plants. Our short-term failings now have extremely long-term consequences, as with each passing year our planet grows increasingly more radioactive. "Hot particles" released at Chernobyl are still being detected on radiation monitors around the world. Fukushima's isotopes have further raised the level of "global glowing." We already may have reached a tipping point for irreversible climate change. Is there a tipping point for radioactivity in the environment?

Biologists report that bird and insect populations have been falling in the Fukushima region since the disaster. Comparing populations of 14 matched species of birds in Fukushima and Chernobyl, researchers found that in areas with similar radiation levels, bird counts in Fukushima (recent exposure) were half what they were in Chernobyl (long-term exposure). Writing in the journal *Environmental Pollution*,[1] Timothy Mousseau, a biological sciences professor at the University of South Carolina, theorizes that the difference may arise from the fact that "evolution has already been at work near Chernobyl, killing off individual birds that cannot cope with the background radiation and allowing the genes of those that have some tolerance to be passed on." In Fukushima, the birds have not yet had a chance to adapt to the Radiocene—a troubling, new global epoch defined by rising levels of ambient radioactivity.

A similar inadvertent culling may already be under way among the human populations exposed to slowly rising levels of ambient radioactivity. If this is happening, we have created a technology that may now be slowly transforming us genetically, with the goal of producing a new, more radiation-resistant species.

Meanwhile, more than a year after the Fukushima meltdowns, the situation in Japan is still not under control. Nuclear engineers and government officials have predicted that efforts to prevent future catastrophic damage at the broken reactor complex could take decades. Like the devastated landscapes downwind of Chernobyl, parts of northeastern Japan may remain a no-man's-land for centuries. The cleanup and containment of Fukushima's radioactive ruins will extend over generations.

Fukushima was supposedly designed to withstand the greatest credible threats. In the aftermath of the quake and tsunami, officials at TEPCO, Japan's electric utility, were left facing problems "beyond the design capacity" of the plant. In the United States, 23 Fukushima-style General Electric Mark 1 reactors sit at 16 sites in Alabama, Georgia, Illinois, Iowa, Massachusetts, Minnesota, Missouri, New Jersey, New York, North Carolina, Pennsylvania, and Vermont. To what capacity is their design capable of handling disaster? Can we reasonably predict the scope of disaster likely to befall them?

The lesson is clear: if Japan, one of the world's most technologically sophisticated countries, cannot assure the safety of nuclear power, it is time to consider the only rational course of action—reactor shutdowns followed by an immediate global decommissioning.

Dr. Natalia Mironova, a Russian nuclear engineer who risked her life as a Chernobyl "liquidator," offered this sobering reflection during a visit to San Francisco in April 2011: "Chernobyl was one reactor releasing radioactive materials for two weeks in a relatively rural setting with low population. Fukushima is at least four [active] reactors, plus their spent fuel pools, venting radionuclides . . . in a populated area. Fukushima is not just 'as bad as' Chernobyl. It is eight times worse already, with no end in sight."[2]

With new reactors under construction in 15 countries, we should take Mironova's warning to heart. Before another ton of concrete is poured, before another bolt is secured, before another nail is driven, we should insist on four basic demands:

1. end government subsidies that make it possible to build new reactors;
2. require owners to accept full financial responsibility for covering all potential damages;
3. rescind the extended operating license for any reactor that has reached the end of its 40-year design lifetime; and
4. shut down all currently operating reactors pending a solution to the problem of storing radioactive wastes.

Holding the nuclear lobby financially accountable would pull the plug on this dangerous technology. The money saved could then be put to better use promoting the transition to sustainable, zero-growth economies powered by clean, safe, affordable, renewable energy.

It is not as if we were not warned. On February 2, 1976, three former General Electric nuclear engineers—Dale Bridenbaugh, Richard Hubbard, and Gregory Minor—appeared before the US Joint Committee on Atomic Energy to explain why they had decided to abandon their high-paying careers. While they had been initially attracted by "the promise of a virtually limitless source of safe, clean and economic energy for this and future generations," the engineers explained that they could no longer continue working for General Electric "because we could no longer justify devoting our life energies to the continued development and expansion of nuclear fission power—a system we believe to be so dangerous that it now threatens the very existence of life on this planet."

"Nuclear power can never be safe," Greenpeace International concluded in its one-year anniversary report on the TEPCO meltdowns. Once this truth is acknowledged, Greenpeace argued, it becomes incumbent on everyone to "put the nuclear regime under close public scrutiny" and prepare to end the Atomic Age, once and for all.

GAR SMITH

INTRODUCTION

The NRC's New Reactors: Renaissance or Recrudescence?

ON FEBRUARY 9, 2012, the US Nuclear Regulatory Commission (NRC) did something it hadn't done in 30 years: it granted a license to build and operate a new nuclear power plant. Pro-nuclear lobbyists hailed the decision as an enormous milestone. After all, the last order to build a new reactor had been filed in 1977 (a year before the Three Mile Island meltdown), and the previous 40 years had seen the cancellation of more than 120 planned nuclear units.[1] In February 2010, President Barack Obama had helped pave the way by offering an $8.3 billion loan guarantee to the Georgia-based Southern Company (a private energy firm with nearly $16 billion of operating revenues) for the construction of two Toshiba-Westinghouse AP1000 reactors at the Vogtle nuclear facility in Burke, Georgia. The Southern Company confirmed its acceptance of the loans on June 18, 2010 and announced that the first of its newly licensed Vogtle reactors was expected to begin operation in 2016, with the second unit firing up in 2017.

Despite President Obama's promise to provide an unprecedented level of openness in government, a shroud of secrecy covered the Department of Energy's multibillion-dollar loan guarantee process. Although taxpayers will bear the costs of a loan default, the public has no role in determining who gets these handouts. The DOE has not explained how it plans to process and approve the loans, and the applicants and proposed projects remain secret. On March 17, 2010, after the DOE routinely ignored numerous Freedom of Information Act requests from various public interest groups, Public Citizen and a coalition of environmental groups wrote to Energy Secretary Steven Chu to protest the DOE's "continuing refusal to disclose even the most basic information about the program."[2]

But did Obama's billions really signal a federal affirmation of the nuclear industry's long-promised "nuclear renaissance"? Not according to NRC chairman Gregory Jaczko, who cast the lone dissenting vote against the new Vogtle reactors. "I cannot support issuing this license as if Fukushima never happened," he stated.[3] Jaczko was particularly concerned that there was no guarantee that the Vogtle project's AP1000 reactors would incorporate essential "Fukushima enhancements" designed to guard against catastrophic damage. Southern Company CEO Thomas Fanning called the NRC's decision a "monumental accomplishment"[4] but failed to explain why his company would not promise to incorporate post-Fukushima protections into the reactors' design.

Jaczko wasn't the only one who feared the Vogtle project might prove to be more a millstone than a milestone. The NRC majority felt compelled to include a special "condition" on the license that called for inspection and testing of "important components of the new reactors' passive cooling system."[5] Congressman Edward Markey (D-Mass.) charged that the NRC had "abdicated its duty to protect public health and safety, just to make construction faster and cheaper for the nuclear industry,"[6] and Alison Fisher, an energy expert with Public Citizen, lamented, "It is inexplicable that we've chosen this moment in history to expand the use of a failed and dangerous technology. While Germany, Italy, and Switzerland responded to the [Fukushima] catastrophe by announcing plans to eliminate nuclear power from their energy sector, US nuclear regulators are doing the exact opposite. The US is approving new reactors . . . before new safety regulations that were recommended by a task force established after the meltdown at Fukushima have been implemented."[7]

Far from heralding a renaissance, the AP1000 reactors are emblematic of the irreducible problems that inevitably accompany nuclear power—a technology that consistently has proven to be costly, unwieldy, unpredictable, and unsafe. The NRC's ruling was more a recrudescence than a renaissance. "Recrudescence" (a grand old word that deserves to be dusted off for the present occasion) has two contradictory meanings: 1) "the process of renewal or rebirth" and 2) "the return of an illness after a period of remission" or "a sudden, violent, spontaneous occurrence (usually of some undesirable condition)."

A Rigged System

To anyone living near the Vogtle plant, the NRC's decision must have looked like a foregone conclusion, since work on the two new reactors—Units 3 and 4—had been under way for years. Vast swaths of land already had been cleared, huge trenches had been gouged into the earth, tons of concrete had been poured, and miles of pipe had been laid. Even the mammoth metal containment vessel was under construction. How is it possible that a nuclear plant can be built before it is even approved? We have former NRC commissioner Jeffrey Merrifield to thank for this. Before he left the NRC to join the

Shaw Group (a company that builds nuclear reactors), Merryfield managed to tweak the NRC's definition of "construction" to allow "pre-construction" of new nuclear facilities even *before* a construction license has been granted.[8]

As nuclear critic Karl Grossman has observed, "The nuclear power program in the US was set up rigged—to allow the federal government to push atomic energy" over the concern of local residents and ratepayers.[9] In the case of the two Vogtle reactors, Georgia's Southern Company started charging its customers a "construction work in progress" fee long before the NRC acted on its licensing request. The day before the NRC decision, the *Atlanta Journal-Constitution* reported that the utility's customers "already are paying down the project's financing costs through a fee that will increase to $8.74 a month by 2015."[10] Meanwhile, US taxpayers are on the hook to subsidize the plant's construction, thanks to generous government loans created to bail out a business that has *never* been self-supporting.

Pro-nuclear forces claimed another victory on March 30, 2012, when the NRC approved the construction of two more Toshiba-Westinghouse AP1000 reactors at the Virgil C. Summer site in South Carolina. The two 1,117 MW reactors will join a 1,000 MW Westinghouse pressurized water reactor already operating at the site. Again NRC chairman Jaczko cast the lone dissenting vote. "I continue to believe," Jaczko explained, "that we should require all Fukushima-related safety enhancements are implemented before these new reactors begin operating."[11] (And, again, the fact that a thousand construction workers had already been laboring at the site for months suggested the NRC's deliberations were merely a formality.)

The NRC majority accepted Westinghouse's argument that damage to the reactor or spent fuel from a loss-of-cooling accident need not be addressed because an extended power blackout was unlikely to occur. At the NRC licensing hearing, nuclear expert Arnie Gundersen testified that this reasoning was "a blatant manipulation of a safety code designed to protect public health and safety." (At the same time the NRC was approving the Summer plants, the Japanese energy giant TEPCO was announcing some alarming news to the people of Japan: Fukushima's damaged Unit 2 reactor had been releasing much more radioactivity than previously believed. Radiation levels had reportedly reached 73 sieverts—ten times the lethal dose.)

Still, the new US plants were not necessarily a "done deal." As Beyond Nuclear's Kevin Kamps noted, "An NRC license does not guarantee ultimate project success. Atomic reactors have been NRC-licensed and then nearly, or even entirely, constructed, and still blocked from operating." Kamps pointed to several plants in Michigan, New York, and Indiana that had never opened, despite billions of dollars paid by ratepayers. Kamps also noted that a default on the government's $8.3 billion loan guarantee for Vogtle's Unit 3 and Unit 4 reactors "would be 15 times worse than the Solyndra solar energy manufacturer default that cost US taxpayers $550 million."[12]

Sure enough, despite the NRC's approval, the Vogtle and Summer projects soon ran into difficulties. The Southern Company objected to the terms the White House Office of Management and Budget had imposed on the $8.3 billion loan to build the Vogtle reactors, and Georgia ratepayers were up in arms about the company's reactor-related rate hikes. The Southern Alliance for Clean Energy (SACE), believing good science and sound fiscal reasoning should guide nuclear policy, sued the NRC over its decision to grant a license to construct the Vogtle reactors and challenged the White House loan guarantee for "socializing the risk and privatizing the profits for the big power companies."[13] (The Congressional Budget Office has predicted the risk of default is well over 50 percent.) Meanwhile, South Carolinians were digging in to challenge "pre-construction" rate hikes imposed to finance construction of the two proposed Summer reactors.

Perhaps the biggest impediment for SACE and other anti-nuclear activists continues to be the pay-to-play politics that dominate most modern industrial economies. Politicians are more beholden to corporate lobbyists and super PACs than they are to homeowners and Joe Six-Packs. This troubling anti-democratic dynamic is in play from the council chambers of small Nebraska towns right up to the carpeted inner sanctum of the Oval Office. Even in the shadow of Fukushima, while other countries were abandoning the sinking nuclear battleship and climbing aboard a waiting fleet of renewable life rafts, Washington's otherwise squabbling Democrats and Republicans still managed to agree that nuclear energy must continue to be supported, no matter the costs, no matter the risks.

President Barack Obama's implacable allegiance to all things nuclear (from mini-reactors to a new generation of nuclear bombs) comes as no surprise to folks who recall that the president's chief advisors, David Axelrod and Rahm Emanuel, share links to the Exelon Corporation, a major player in the nuclear game. Over the first four years of Obama's presidency, Exelon and its employees donated generously to the president's campaign. As *Democracy Now!* host Amy Goodman once ruefully observed, "As long as our politicians dance to the tune of their donors, the threat of nuclear disaster will never be far off."[14]

Renaissance or Retreat?

It looked like a nuclear renaissance back in the fall of 2007, when the first of 15 companies filed applications with the NRC to build 25 new US reactors.

But most of these ambitious plans were scaled back due to a combination of plummeting natural gas prices and lower energy demands as the US economy sputtered. (In 2011, a 1,000 MW plant powered by natural gas could be licensed and built in a few years at a cost of $1 billion or less.) Instead of 25 reactors, the industry now hopes for five—two units at Vogtle, two at Summer in South Carolina, and one government-built reactor at the Tennessee Valley Authority (TVA) site at Watts Bar, Tennessee. (In February 2012, the TVA reported that its Watts Bar project was behind schedule and expected to "significantly exceed" its $2.5 billion projected cost.)[15]

While 2007 saw 5 applications for 8 reactors (in Alabama, Maryland, South Carolina, Texas, and Virginia) and 2008 set a record with 11 applications to build a whopping 16 reactors (in Florida, Georgia, Louisiana, Michigan, Mississippi, Missouri, New York, North Carolina, Pennsyvlania, South Carolina, and Texas), the seeming "renaissance" soon went into retreat. In 2009 there was only one license request—for two AP1000 reactors at New York's Turkey Point. According to NRC records (updated on October 6, 2011), since 2010, only three applications had been submitted for new reactors.

Nuclear power has been taking body blows in the global arena as well. In 2011 nuclear capacity dropped by 10 gigawatts to 366.6 GW, accounting for only 5 percent of the world's commercial energy production. (Bear in mind that, even at its peak years in 2001 and 2002, nuclear managed to meet only 6 percent of the globe's energy needs.)[16] In June 2011, a global survey of 19,000 people in 24 countries conducted for Reuters found that support for nuclear energy had fallen from 54 percent to 38 percent (lower than the support for coal). Three-quarters of the respondents reported that they saw the nuclear option as "a limited and soon obsolete form of energy."[17] The preferred energy options were solar (97 percent) and wind (93 percent).

According to the Worldwatch Institute, the world's total number of working reactors dropped from 441 to 433 in 2011, and nuclear programs were suspended or curtailed in Austria, China, Japan, Italy, the Philippines, Switzerland, Taiwan, the United Kingdom, and Venezuela. While 65 reactors are under construction worldwide, an essential (and often omitted) fact is that 20 of these reactors have been "under construction" for more than 20 years. "Although construction on 16 new reactors began in 2010," Worldwatch notes, "that number fell to just two in 2011, with India and Pakistan each starting construction on a plant."[18] There's your "nuclear renaissance."

India's approval of plans to complete two Russian-designed reactors ("under construction" since 1988) came on March 19, 2012, and was met by large protests. Three million residents of Tamil Nadu live within 18 miles of the Kudankulam power plants. After more than 10,000 protesters gathered to express opposition, the government threatened to arrest protest leaders on charges of terrorism and waging war against the state.

Only two other large countries—China and the United States—continue to be fired up over nuclear power. In addition to its 27 existing plants, China began construction of 10 new plants in 2010, representing an increase of 27 gigawatts. Unlike the United States, though, China halted construction of 25 reactors in response to the Fukushima disaster. In the United States, 2010 was the year the Obama administration announced its $8.3 billion loan construction guarantee to help build two new reactors in Georgia. The president also proposed adding another $36 billion in nuclear loan guarantees to the 2011 federal budget (on top of the existing $18.5 billion allotment for loan guarantees). After a groundswell of grassroots resistance, Congress rejected the plan but, in February 2011, Obama inserted the same $36 billion request in his 2012 budget proposal. Once again, public protests stopped the bailout. The White House appears to be getting the message: for the first time in three years, the 2013 federal budget included no proposals for new nuclear loan guarantees.[19]

Turning Away from the Atom

Nuclear power is in retreat around the world. Since 2010, France, Germany, Japan, and the United Kingdom have unplugged 11.5 GW of nuclear capacity. In 2011 alone, Germany (the world's fourth-largest national economy) eliminated 8 GW of nuclear power. By January 2012, Japan had shut down all but four of its 54 nuclear reactors and was getting by with only 20 percent of its previous nuclear capacity. With local authorities electing to ban restarts of off-line reactors in their prefectures, the last remaining active Japanese reactor (on the northern island of Hokkaido) was shut down for safety checks in May 2012.[20]

In France, the Greens and Socialists have joined forces to campaign for the closure of 24 of the country's 58 nuclear reactors by 2025. The Socialists are aiming for a 50 percent reduction in nuclear power by 2025; the Greens want a complete ban and are insisting on the immediate closure of the country's oldest plant at Fessenheim.[21] Meanwhile, the completion date for the first new reactor to be built in France in 15 years has been pushed back to 2016, while the cost has nearly doubled from $4.4 billion to an estimated $8 billion.[22] And the French nuclear giant AREVA announced it was abandoning its ambitious plans for building new reactors across the United States, including a controversial plan to build two reactors in Fresno, California.

On January 3, 2012, the French Nuclear Safety Authority (ASN) dropped a devastating document on the front desk of the country's national energy conglomerate, Électricité de France (EDF). The report ordered EDF to immediately invest an estimated 10 billion euros to address safety issues at 58 French reactors. Clearly inspired by the devastation wrought in Fukushima, ASN president André Claude told the press, "If EDF estimates that what we are asking for is so expensive that it no longer makes it worthwhile to operate one facility, it can decide to shut that facility."[23]

Looking ahead to a nuclear-free future, French ecology minister Nathalie Kosciusko-Morizet told the *Financial Times* that the country's new objective "is to rebalance the energy mix in favor of renewables" beginning with a bid to install five new wind farms off the French coast.[24] The European Union's commissioner for climate action, Connie Hedegaard, followed up by championing wind power. "Some people tend to believe that nuclear power is very, very cheap," Hedegaard told *The Guardian*, "but offshore wind is cheaper than nuclear."[25] A February 2012 report by the European Wind Energy Association projected that by 2020, nuclear power will cost 102 euros/MWh, while the cost of offshore wind power will reach 75 euros/MWh and onshore generation will fall to 58 euros/MWh.[26]

The downsides of nuclear investment became clear to the citizens of France on January 31, 2012, when they awoke to the news that the anticipated cost of decommissioning the country's aging reactors and storing 30-plus years of radioactive waste would approach 79 billion euros ($103.5 billion). Dismantling EDF's 58 reactors is expected to cost 18 billion euros, storing the waste will require an additional 28.4 billion euros, and annual maintenance charges will double from 1.5 billion euros to 3.7 billion euros by 2025.[27]

On July 14, 2011, a month before he resigned, Japanese prime minister Naoto Kan addressed the country in a national TV broadcast. "We will aim to bring about a society that can exist without nuclear power," Kan declared. "The risk of nuclear energy is too high. It involves technology that cannot be controlled according to our conventional concept of safety." The message was not lost on Japan's energy conglomerates. Three major nuclear suppliers—Toshiba, Mitsubishi, and Hitachi—announced plans to shift their attention to renewable energy programs (solar, wind, and geothermal) and "energy-smart" communities.

In September 2011, the upper house of the Swiss Parliament followed a 101–54 vote of the lower house and agreed to eliminate all of the country's nuclear plants over the next 20 years, relying instead on hydroelectric power and renewable energy.

On September 19, 2011, Siemens (the German engineering multinational that constructed Germany's entire fleet of 17 reactors) announced that it was getting out of the nuclear power business. Siemens also announced that it would cancel its nuclear joint venture with Rosatom, Russia's state atomic energy corporation. Siemens CEO Peter Loescher hailed the government's decision to shutter all nuclear plants by 2020 and called the plan to move to a "renewable energy economy" by 2050 the "project of the century."[28]

There is evidence to back up Loescher's optimism. Following the Fukushima disaster, Germany shut down 40 percent of its nuclear power, which prompted dire predictions that winter would bring nationwide blackouts. The first months of 2012 tested that prediction with a bone-chilling blast of arctic weather that sent snow falling as far south as Athens and Morocco. Back in Germany, however, a fleet of wind turbines was easily contributing 9 percent of the country's total power needs and churning out enough extra electricity that Germany wound up exporting 500 MW to its shivering neighbors in France.[29] On May 25–26, 2012, Germany's solar installatons (mostly distributed rooftop systems) generated a record-setting 50 percent of the country's electricity—22 gigawatts.[30]

The Fall of the Atomic Empire

The entrenched power of the nuclear establishment will not relinquish its economic/political/corporate death grip willingly. Even with a greener leader, France will not face an easy path to a nuclear-free future. In 2012, French citizens were generating more than 75 percent of their electricity from burning atoms.[31] While Germany and Italy are turning toward solar power, the French delegation to a meeting of the European energy ministers in early 2012 insisted that nuclear power remain part of the European Union's energy mix. Former French prime minister and current Socialist Party member Michel Rocard even went so far as to warn that a retreat from nuclear energy would imperil economic growth and lead to civil war.[32]

"[French President] Hollande merely has a popular mandate; of what import is this against the real powers in France?" University of Sydney political economist Evan Jones dared ask in a somber essay in May 2012. "The 'nucléocrates' are committed to third-generation EPR reactors [European pressurized-water reactors]," Jones wrote. They have even planned to construct a buried waste storage facility in Francois Hollande's home base of Corrèze. Jones noted:

> As with Westinghouse and General Electric in the US, EDF pushed a consumerist culture that dramatically expanded electricity usage to cement the profitability of nuclear infrasructure. EDF's motif became 'Toute electrique! Toute nucléaire!'

> As the Americans have their military-industrial-intelligence establishment, immune from democratic and political influence, so too with the French nuclear establishment. It is a state within a state. This establishment is paying not the slightest attention to whatever formal commitments the French government makes to Europe concerning renewable energy targets.[33]

The Demise of the Atom

The meltdowns at Chernobyl, Three Mile Island, and Fukushima illustrate the ultimate insanity of the nuclear option. Of course, the problems of nuclear power extend beyond the most dramatic—and deadly—outcomes. There are many other arguments that can be marshaled against continued reliance on an energy paradigm that was hatched by the Pentagon, promoted by the government, adopted by corporations, and forced on countries around the world under the Atoms for Peace false-flag banner of "clean, safe energy."

Nuclear power was never clean or safe. The Pentagon knew it. The government planners knew it. The nuclear industry knew it. And now the whole world knows it.

Since the first toss of the atomic dice at a desert test site in Alamogordo, New Mexico, incalculable harm has been done to our planet—its air, its water, its land, and its peoples. Tragically, much of this damage will remain as an invisible legacy that will shadow the lives of our children for generations. But if we continue to marshal our outrage, energy, and intelligence in the cause of principled and progressive change, there is still time to start turning our poisoned planet away from the deadly atom and toward a future where the sun shines far brighter than the lethal core of a reactor. We must demand a new paradigm for planetary survival, and a large part of that transformation will require a new conservation ethic and renewable renaissance.

GLOSSARY OF KEY ENERGY TERMS

Power: The rate of doing work, measured in watts (joules per second).

Joule: A unit of electrical energy equal to the work done when a current of one ampere passes through a resistance of one ohm for one second.

Energy: The capacity of a physical system to do work, measured in joules.

Efficiency: The ratio between the useful output of an energy conversion machine and the input, in energy terms.

EROEI: "Energy returned on energy invested," also known as EROI (energy return on investment). The ratio of the amount of usable energy acquired from a particular resource to the amount of energy expended to obtain that resource. Not to be confused with efficiency.

Kilowatt-hour (kWh): The energy used by ten 100-watt bulbs in one hour.

Megawatt (MW): 1 million watts. In 2008, the Energy Information Administration reported that the average US home consumed around 11 MWh per year. A single nuclear reactor may generate 500 to 1350 MW.

Gigawatt (GW): 1 billion watts. This unit is for large power plants or power grids. For example, in 2009 the installed capacity of wind power in Germany was 25 GW.

Terawatt (TW): 1 trillion watts. The total power used by humans worldwide (about 12.5 TW) is commonly measured in this unit.

Boiling water reactors: BWRs operate like fossil-fuel-powered plants, except that a nuclear core is used to heat water to produce steam that drives turbines to generate electricity.

Pressurized water reactors: The most widely used Western reactors, PWRs were originally designed for military use. Water is not allowed to boil but is kept under high pressure in a steam generator, which is used to spin turbines that produce electricity.

Advanced pressurized water reactor: An improved design based on the existing BWR and PWR reactors but incorporating passive safety systems.

GLOSSARY OF KEY RADIATION TERMS

In any given year, the average US citizen can expect to be exposed to around 360 millirems of radiation—from sunshine, rocks, dental exams, and medical X-rays. The NRC advises that such typical exposures are not considered cause for concern. (What may be cause for concern is the fact that the NRC recently nearly doubled the definition of "background" radiation to 620 millirems.)

Here is a brief introduction to some of the new vocabulary we must master if we are to survive in the new evolutionary era known as the Radiocene.

Absorbed dose: Measured in a unit called a gray (Gy). An X-ray involves a small dose, well below 0.1 Gy. Symptoms of radiation poisoning begin with a full-body exposure to 1 Gy. Doses above 6 Gy are untreatable and lethal within two days to two weeks.

Becquerel: A unit of radioactivity, defined as the quantity of a radionuclide that undergoes one decay per second (s–1). One curie equals 3.7×10^{10} becquerels. Abbreviated Bq.

Curie: A measure of radioactivity equal to the amount of a radioactive isotope that decays at the rate of 3.7×10^{10} disintegrations per second. One curie equals the activity of 1 gram of radium.

Picocurie: A picocurie is one trillionth of a curie.

Ionizing radiation: Subatomic particles or electromagnetic waves that are energetic enough to dislodge electrons from atoms or molecules. They include alpha particles, beta particles, and gamma rays.

Rem: A unit measuring radiation exposure, defined as the amount of any ionizing radiation that has the same biological effectiveness as 1 rad of X-rays.

Millirem: One thousandth of a rem. A chest X-ray is equal to about 10 mrem. Most people are exposed to approximately 320 mrem (whole-body exposure) a year—nearly half resulting from routine background exposure. While the EPA recommends limiting additional exposure to less than 100 mrem per year, the NRC has set permissible occupational exposure at 5,000 mrem per year, with public exposure raised to 500 mrem per year.

Sievert: A dosage of ionizing radiation equal to 100 rems (equivalent to 1 gray for X-rays).

Millisievert: One thousandth of a sievert. Abbreviated mSv. Radiation exposure limits for the general public are set at 1 mSv per year (equivalent to 100 mrem/year).

Radioactive Isotopes

Alpha particles: Alpha particles contain two protons and two neutrons. They are produced by the disintegration of uranium-235 and plutonium-239 (with a half-life of 24,000 years). They can be blocked by clothing or a piece of paper and rarely penetrate the skin. While alpha particles are not an external threat, they can prove deadly if inhaled or ingested.

Beta particles: Most of these negatively charged particles travel only a short distance in air. While these particles can be blocked by plastic, wood, or clothing, they can penetrate the skin barrier. Sources of beta radiation include phosphorous-32 and hydrogen-3 (tritium).

Gamma radiation: Unlike alpha and beta particles, gamma radiation takes the form of electromagnetic energy. Gamma rays are the most energetic form of electromagnetic radiation. They are not blocked by plastic, wood, clothing, or skin; you would need 10 cm of lead to block gamma rays.

Neutron radiation: Neutrons are neutrally charged particles created by nuclear power plants by the ionization of plutonium and beryllium and by radionuclides like californium-252. Neutron radiation is highly penetrating and requires shielding with barriers of water and concrete.

Cesium: Naturally occurring in rocks, isotopes of this element are created when uranium fuel is burned in a reactor core. Cesium-137 emits beta rays and strong gamma radiation and has a half-life of 30 years. Exposure to isotopes including cesium-134 and cesium-137 can produce nausea, vomiting, diarrhea, bleeding, and death.

Iodine-131: A radioactive isotope generated by atomic reactors, I-131 has a half-life of eight days. If absorbed by the body, it can cause thyroid cancer and other illness.

Strontium-90: A moderate beta emitter with a half-life of 29.1 years, this isotope is one of the deadliest by-products of nuclear power. Inhaled or ingested, it acts like calcium in the body and consolidates in teeth, bones, and bone marrow, where it triggers bone and blood cancers. The EPA's stated minimum exposure is 4 mrem per year or 8 picocuries per liter.

Tritium: A radioactive form of hydrogen with two neutrons, tritium bonds with oxygen to form radioactive or "tritiated" water. Tritium has a half-life of 12 years and has been linked to increased risk of cancer. The EPA's stated minimum exposure is 4 mrem per year or 20,000 picocuries per liter.

Uranium: All isotopes of uranium are radioactive emitters of alpha and gamma rays. Uranium has a half-life of 4.5 billion years. Because of the long interval as it disintegrates from one element to another until reaching stability as lead, uranium poses a unique health risk that can cause lung cancer, bone cancer, and kidney failure.

Source: "Major Contaminants Glossary: Radioactive Isotopes," State of Washington Department of Ecology, http://www.ecy.wa.gov/programs/nwp/gwcontaminants.htm.

PART ONE

Fourteen Arguments against Nuclear Power

1

The Impossibility of Speedy Deployment

Nuclear power plants cannot be built quickly enough and in a safe and secure manner to be a major global solution for climate change.

—CARNEGIE ENDOWMENT FOR INTERNATIONAL PEACE

IN 2009, the International Energy Agency (IEA) conducted its first detailed assessment of the world's 800-plus oil fields and found that most had already "peaked" while 75 percent of the planet's petroleum reserves were falling at the rate of 6.7 percent a year (twice as fast as estimated just two years earlier). The IEA grimly predicted the world could start running out of oil by 2012—a decade earlier than previous estimates. "We will have to leave oil before oil leaves us," IEA's chief economist Fatih Birol warned, "the earlier we start, the better."[1] Nuclear proponents have argued that "clean atoms" can replace dirty and depleted fossil fuels to provide a hedge against global warming. But facts do not support this pretext for a nuclear renaissance.

More than 200 new reactors have been proposed around the world, but not enough reactors can be built fast enough to replace the world's vanishing fossil fuel resources.[2] In 2004, Princeton researchers Stephen Pacala and Robert Socolow projected that global energy demand would double by mid-century, driving climate-warning carbon output to around 25 billion tons through 2050. Pacala and Socolow estimated it would be necessary to replace 700 gigawatts of fossil-fueled power to remove one billion tons of carbon—a target that would call for building more than 1,000 new reactors.[3] The IEA estimates that renewable energy sources and efficiency measures could produce 10 times these savings by 2050.

The IEA also estimates that cutting CO_2 emissions in half by midcentury would require building 1,400 new 1,000 MW nuclear reactors. That's an average of 32 new reactors every year. But it's taken the industry 10 years to bring just 30 new reactors online, and it usually takes about 10 years from groundbreaking to atom baking. The last US nuclear reactor to be completed, Unit 1 at the TVA's Watts Bar site, took 24 years to build and began operation in 1996.[4] (By contrast, a 1.5 MW wind turbine can be installed in a single day and can be operational in two weeks.) The NRC routinely takes about 42 months to review and approve a reactor application.[5] Given this track record, we can't rely on new nuclear plants as the sole answer to global warming. We just couldn't build reactors fast enough to prevent an irreversible "tipping" of world climate.

What about the inventory of partially built nuclear plants that have yet to be brought online? Of the 35 reactors the IEA listed as "under construction" in mid-2008, a third of them had been under construction for 20 years or longer.[6] Their designs and existing infrastructure are dangerously outdated. For example, in August 2011, the Southern Alliance for Clean Energy (SACE) published a report critical of the Tennessee Valley Authority's plan to complete construction of the long-abandoned Bellefonte Unit 1 power station. Bellefonte was designed in 1968 (so long ago that the engineers who worked on the design were using slide rules). Construction began in 1974 but was discontinued in 1988. The abandoned Unit 1 was subsequently scavenged for parts and scrap metal. SACE, referring to Bellefonte as "America's oldest nuclear power plant that has yet to generate any electricity," was concerned that the Nuclear Regulatory Commission (NRC) gave its blessing for the TVA to complete the plant based on a 44-year-old design that depends on the integrity of an aging foundation and containment building that has been weakened over four decades.[7]

The cost of adding 1,400 new reactors—as envisioned by some advocates of the "nuclear renaissance"—has been estimated at $3 trillion.[8] Even in France (which is frequently cited as a nuclear power success story), the atom's financial record is *un embarras*—between 1970 and 2000, the costs of nuclear construction per kilowatt installed tripled.[9] In the United States, at the start of 2011, the NRC had received applications to build 26 new nuclear reactors (more than half would be modified versions of the Toshiba-Westinghouse AP1000 design) but, given nuclear's high costs and the current availability of cheap natural gas, it is unlikely that more than three of these reactors will actually be built—even with the billions in government loans the Obama administration has offered for the construction of new reactors.

Case in point: one of the top candidates for a federal loan guarantee—a proposed nuclear power plant in San Antonio, Texas, that already had agreements in place to sell power to customers—bailed after skyrocketing cost projections convinced a critical investor to abandon the venture. Financial concerns also prompted the Tennessee Valley Authority to cut plans for four new reactors down to one. In Missouri, the utility provider Ameren pulled the plug on a proposed reactor in Calloway after it lost its bid to start billing utility customers for the plant before construction had even started. And in Florida, pending reactor projects were thrown into question after the state Public Service Commission rejected a request to raise customers' rates to subsidize construction.

Ironically, the twin-reactor Vogtle project in Georgia—the recipient of an $8.3 billion Department of Energy (DOE) construction loan and the first new project to win an NRC construction license in 34 years—had trouble obtaining an NRC permit due to safety shortcomings with the shield design of its new state-of-the-art AP1000 reactors.[10] (Despite the multibillion-dollar loan guarantee, not a single private investor has stepped forward to help match the remaining costs for the Vogtle project.)

"Renaissance" Reactors: Plagued with Problems

The two new reactor designs touted to lead the "nuclear renaissance"—the Toshiba-Westinghouse AP1000 and AREVA's EPR—keep running into roadblocks. While completion of AREVA's showcase plant in Finland remains overdue and overbudget, the AP1000 has spent years mired in the approval process, with the NRC repeatedly expressing dissatisfaction with the plant's design.

In October 2009, the NRC issued a press release noting that the design of the AP1000's shield building could leave the structure unable to withstand buffeting from earthquakes, hurricanes, or tornadoes.[11] The shield building design failed to meet the standards of the American Concrete Institute, and John Ma, the NRC's senior structural engineer, concluded that the building could "shatter like a glass cup" if hit by a plummeting plane or jostled by an earthquake.[12] In November 2009, the NRC informed Toshiba-Westinghouse that the AP1000's steel-and-concrete containment shield did not meet "fundamental engineering standards"[13] after concluding that it might not be able to withstand damage from high winds or handle the weight of the plant's emergency water tank.

In April 2010, alarmed nuclear watchdogs called on the NRC to suspend the operating license for the AP1000 electric pressurized water reactor (EPWR) design after documenting several dangerous defects that NRC regulators apparently failed to detect. The alarms were sounded after 77 cases of corrosion, cracks, and actual holes were found in the containment vessels of existing pressurized water reactors (PWRs). In eight reactors, corrosion had eaten all the way through the steel containment walls. Because visual inspections cannot detect all potential corrosion problems, some of these defects went undetected until the containment vessel was actually breached. In 2009, a finger-size hole was discovered in the containment vessel at the Beaver Valley reactor in Pennsylvania. A nuclear accident at this plant could have exposed residents of Pittsburgh to dangerous levels of radiation.

One analysis concluded that the AP1000—a hybrid design with no operational history—"appears to invite corrosion."[14] Corrosion failures would be critical for the AP1000 because this reactor has no secondary containment protections—in the event of a breach, unfiltered emissions would be vented directly into the sky. Because the Vogtel plant's AP1000 design was new and untested, the NRC was required to convene a public hearing prior to issuing a licencse. During two days of hearings in September 2011, environmental activists and nuclear experts found fault with the AP1000 and issued calls to modify the reactor's design.

The NRC explains that the AP1000 is designed with "passive safety features that would cool down the reactor after an accident without the need for electricity or human intervention." This sounds good, but critics warn that this unique design poses unique risks. Unlike existing PWRs, the AP1000 relies on "a freestanding steel containment and no secondary containment."[15] The "passive safety feature" is, essentially, a huge tank of water installed on top of the reactor. In the event of an accident, 300,000 gallons of water would drop into the containment area to cool the reactor. This could be a one-shot solution, since the gravity-fed tank contains only enough cooling water to last for 72 hours. If the tank cannot be refilled, the AP1000 risks a meltdown.

The AP1000 Gets a Hearing

In June 2010, the NRC's Advisory Committee on Reactor Safeguards listened to an extended presentation by nuclear engineer Arnie Gundersen, who documented seven unreviewed safety concerns. Gundersen began by outlining the AP1000's enhanced potential to experience rust and corrosion. He next drew attention to the containment buildings. While such structures are generally designed to withstand internal pressures of 60 to 200 pounds per square inch (psi), Gundersen noted the AP1000 containment system was engineered to withstand only 59 psi. (The NRC had conveniently calculated that, in the event of an accident, internal pressures would never exceed 58.3 psi. By comparison, pressures inside Fukushima's quake-damaged Unit 1 building topped 120 psi.) Gundersen questioned the NRC's blithe assumption that the containment structures would never leak, that there would be no "inadvertent criticalities" (when reactor fuel reignites on its own following an accident), and that excess heat loads would not raise internal temperatures above safety margins. All of these problems occurred at Fukushima.

Gundersen then focused on a different problem. Unlike most reactors, which fuse a protective metal liner to the inside walls of a concrete containment structure, the AP1000 relies on "a freestanding steel containment and no secondary containment." If corrosion were to develop in this two-inch-thick metal encasement, any resulting leak would be difficult to spot. And if a leak were to occur, the air gap between the metal and the wall could create a "chimney effect" that would suck radioactive gases upward and release them into the atmosphere.[16] In short, Gundersen explained, the AP1000 design would "facilitate the more efficient release of unfiltered, unmonitored radiation" via a "leakage path" that "is not bounded by any existing analysis." Gundersen estimated that a crack in the AP1000's containment system (or a hole the size of the one discovered at the Beaver Valley reactor in 2009) "would create exposure to the public well in excess of the 25 rem limit" permitted under NRC guidelines.[17]

Finally, there is the problem posed by locating multiple reactors at a single site. With the addition of Units 3 and 4, the Vogtle plant would host four reactors. If any Vogtle reactor were to suffer a Fukushima-style explosion, Gundersen testified, it could propel a burst of high-speed shrapnel that could breach the emergency water tank perched on the roof of any neighboring AP1000's containment building. Airborne debris from a nearby reactor accident could choke critical vents, "making it impossible to cool the reactor."[18] (In addition to Vogtle, each of the following US sites hosts two reactors: Amarillo, Clinch River, Comanche Peak, Levy County, Shearon Harris, South Texas Project, Summer, and Turkey Point. There are plans to add additional reactors at Summer and at the abandoned Cherokee nuclear site in South Carolina.)

On November 10, 2011, Friends of the Earth, the AP1000 Oversight Group, and the North Carolina Waste Awareness and Reduction Network (NC WARN) petitioned the NRC to resolve these design problems prior to licensing the Vogtle plant. Nonetheless, in a controversial 4–1 vote in February 2012, the NRC gave the go-ahead for the construction of two AP1000 reactors at the Vogtle site in Georgia. John Runkle, a staff attorney for the AP1000 Oversight Group, accused the NRC of "arbitrarily cutting off its review of unanswered safety issues." Since 1991, Runkle noted, the NRC's policy had been "to resolve safety problems during the certification process and not during construction. Now, the NRC seems poised to certify reactor designs that have not been fully reviewed with many safety issues still unresolved."[19] The consequences can be costly. As Gundersen pointed out, typing in a "ten-dollar modification on a computer now can save a ten-million-dollar modification down the road, if you have to rip out concrete and do it over."[20]

The Vogtle project quickly ran into problems when the construction site was cited for using faulty concrete and steel rebars that were not up to spec. This foreshadowed additional delays and related costs. As nuclear watchdog

Harvey Wasserman noted: "The projected price for Georgia's Vogtle Double Reactor Project has jumped at least $900 million in just three months." Meanwhile, the Southern Company was arguing with the Office of Management and Budget over the terms of the government's multibillion-dollar loan guarantee. Wasserman summed up the squabble as follows: "Vogtle's prime builders want to put up little or no money. They want interest rates lower than what you would pay to buy a new house. They expect you [the taxpayers] to take primary liability for future disasters. They can't say what will happen to the radioactive waste."[21]

The Vogtle project isn't the only one suffering from delays and cost overruns. On April 27, 2012, the NRC approved Progress Energy's proposed twin-reactor complex in Levy County, Florida, ruling that the reactors posed no environmental risks. Still, the plant's prospects remain dicey. In May 2012, Progress again revised its initial estimate for construction costs from $2 billion per reactor to nearly $12 billion per unit—a 600 percent increase. Meanwhile, Progress Energy's under-construction Crystal River reactor in Florida also is in trouble. With more than $2 billion spent on unforeseen repairs, the plant may never open.

Au Revoir, AREVA

AREVA's European Pressurized Reactor (EPR) has been hailed as a new design that would be safer and more reliable, as well as cheaper and quicker to build. But the prototype of this "renaissance reactor"—the 1,600 MW Olkiluoto 3 plant under construction in Finland—remains at least 55 percent overbudget and three years behind schedule (which translates into a $2.8 billion loss). As of March 14, 2012, Greenpeace estimated that the Olkiluoto 3 project had cost Finland's utility customers and taxpayers more than 5.1 billion euros ($3.9 billion).

Olkiluoto's concrete base, reactor vessel, pressurizer, and primary cooling pipes all have failed to meet critical quality and safety standards. By mid-2007, 1,500 safety and quality defects had been recorded.[22] The Olkiluoto reactor is now set to open at the end of 2014, five years behind schedule.

Similarly, in July 2011, Électricité de France (EDF) announced the predicted 2012 opening of its Flamanville 3 EPR would be further delayed for "structural and economic reasons."[23] Flamanville's initial cost estimate of 3.3 billion euros ($4.2 billion) has grown to 6 billion euros ($7.7 billion). After two serious accidents slowed construction in 2011, EDF declared it hoped to finally have the plant online by 2016. Two workers have died during construction, and work has been delayed by flaws in crucial cement work, which will need to be redone.

Despite these setbacks, EDF has continued to push ahead with plans to build four EPR reactors in Britain at an estimated cost of $9.3 billion per unit, as well as reactors in the United States. And AREVA has proposed building new reactors in New York, Pennsylvania, and Missouri, as well as in various other countries. Their plans have not been helped by a November 4, 2010, University of Greenwich study that concluded the EPR reactor should be abandoned because of its excessive costs. To avoid the Atomic Energy Act's explicit ban on direct foreign ownership or control of US plants, these French firms have purchased limited partnership shares in key US energy companies. Foreign energy firms also invest in forging partnerships with Washington. Between 2005 and 2008, AREVA spent $5.2 million lobbying Congress to win support for an atomic rekindling in the United States.[24]

The first US-sited AREVA reactor—a double-size Evolutionary Power Reactor (an experimental design that has never operated anywhere in the world)—was to have been erected as Unit 3 at Calvert Cliffs, Maryland, as part of a US-French push for a "nuclear renaissance." Unistar Nuclear Energy, a joint venture between France's EDF and the US corporation Constellation Energy was formed in 2007 but, by September 2010, the project was near collapse as concerns rose over foreign ownership, insufficient proof of need, and costs. On October 9, 2010, Constellation Energy withdrew from the project, citing unacceptably high costs, and eventually sold its 50 percent share to EDF. EDF and several new partners (including the global construction giant Bechtel) continue to pursue an NRC license to build the Unit 3 EPR.

By early 2012, the World Nuclear Association continued to insist that the hoped-for atomic renaissance was "increasing steadily but not dramatically."[25] There were more than 60 reactors under construction in 14 countries, with most of the new units being built in China (26) and Russia (10).

But it's not enough just to build new power plants. The construction of new nuclear reactors also requires a commitment to mine more uranium ore, to build new uranium enrichment plants, to invest in new reprocessing plants, and to create new waste storage sites. The last element of the fuel loop—waste storage—is a deal breaker, especially in the United States, where the sole proposed waste storage site at Yucca Mountain, Nevada, is apparently off the table. And even if new construction were practical, quick, and affordable, it would be short term—"solving" the global-warming problem for only 40 years, at which point the operators of the plants would need to begin the long and costly process of decommissioning and decontamination.

Meanwhile, clean, low-cost renewable power sources (led by wind turbines and solar technologies) are racing far ahead of nuclear power. This new reality is gaining some powerful advocates. On May 5, 2009, Federal Energy Regulatory Commission chairman Jon Wellinghoff assured reporters that all future US electricity demand can be achieved with a mix of renewable technologies and efficiency measures. "We have the potential in the country," Wellinghoff emphasized. "We just have to go out and get it."[26]

Since 2001, the cost of nuclear power has been increasing at nearly the same rate that solar-electric costs have been falling, on the way to a historic "solar-nuclear crossover" that many observers believe has already occurred. In the first five months of 2011, the cost of solar cells fell 21 percent, leading Mark M. Little, General Electric's director for global research, to predict, "Solar power may be cheaper than electricity generated by fossil fuels and nuclear reactors within three to five years." By 2012, GE hopes to start manufacturing thin-film solar panels with a record 12.8 percent efficiency. GE predicts that it will soon be producing sufficient PV panels to power 80,000 homes a year.[27]

2
Unaffordable Costs

Economies are supposed to serve human ends, not the other way round. We forget at our peril that markets make a good servant, a bad master and a worse religion.

—AMORY LOVINS

WHEN THREE MILE ISLAND'S Unit 2 reactor started operation on December 28, 1978, it was hailed as a state-of-the-art nuclear facility. Three months later, the $900 million power plant turned into a multibillion-dollar liability in a matter of hours. Reflecting on Three Mile Island's ignoble demise, Arnie Gundersen, a former nuclear industry senior vice president with 39 years of experience, shared a catchphrase that is often heard within the nuclear engineering community: "Nuclear power can be cheap and nuclear power can be safe. Just not at the same time."[1] History bears this out. Every spectacular failure of nuclear power has been followed by major increases in the cost of future atomic energy systems.

After the partial meltdown of Three Mile Island in 1979, construction costs for new plants increased 95 percent, while the cost of nuclear electricity jumped 40 percent. After the 1986 explosion and fire at Chernobyl, construction costs soared another 89 percent, while the price for nuclear power rose 42 percent. Ray Rothrock, a former nuclear engineer turned venture capitalist, fears Fukushima could bring investments in "next generation" designs to "a screeching halt."[2]

Between 1950 and 1990, US taxpayers and utilities spent $492 billion on the "direct" costs of nuclear power.[3] Yet by 1980 (30 years after the industry received its first taxpayer handout), nuclear reactors were producing only enough power to provide 11 percent of the country's electricity.[4] Between the early 1970s and mid-1980s, inflation-adjusted capital costs of new plants rose an average of 14 percent each year. Nuclear plants finished in the mid-1980s cost 20 times as much as reactors built in the early 1970s (a sixfold increase when adjusted for inflation).[5] In 2002, the cost of a new reactor was set at $2.3 billion. By 2006, the cost had grown to nearly $4 billion. Electricity from new nuclear plants is at least two to four times more expensive than the cost of simply improving end-use efficiency.[6] Even investment guru Warren Buffett backed out of a nuclear power project after his MidAmerican Energy Holdings Company concluded it did not make "economic sense." Instead, MidAmerican decided to invest in renewable energy systems.[7]

Once again, the Calvert Cliffs case is instructive. In 2004, Constellation Energy (CE) officials believed they could build a third reactor at Calvert Cliffs, Maryland, for between $2 billion and $2.5 billion. The company partnered with France's Électricité de France (EDF) to form UniStar Nuclear Energy. But by the summer of 2008, the estimated cost had ballooned to $9.6 billion. (The actual overall costs were projected to reach between $13 billion and $15 billion.)[8] After spending $600 million between mid-2007 and mid-2010, CE saw a taxpayer bailout as essential. On July 28, 2010, CE's CEO warned investors, "We can't keep going at the rate we're going . . . without [a] loan guarantee."[9]

There wouldn't be any talk of a "nuclear renaissance" were it not for such taxpayer bailouts. The Bush-Cheney administration arranged a massive new handout in 2005. With no new nuclear orders in the United States for three decades, Bush's 2005 Energy Policy Act (EPAct) aimed to jump-start a nuclear revival by providing $10 billion over 23 years. EPAct offered nuclear operators a credit of 1.8 cents per kWh for first eight years of operation, loan guarantees of up to 80 percent of the cost of building new reactors, and $1.8 billion for research and construction of "advanced" reactors. EPAct also allocated $1.1 billion for fusion energy research, and provided another $1.3 billion toward future decommissioning costs. The IRS also offered $1 billion in yearly tax credits to any company that filed a licensing request by the end of 2008, began construction by 2014, and received an NRC certification to put its plant in service by 2021.

Despite billions in direct nuclear subsidies, soaring costs continued to eat away at government's energy bill. In 2007, the $18.5 billion in loans Congress authorized to build six new power plants were only sufficient to cover the construction of two reactors.[10] Undeterred by such reversals, President Obama outdid Bush in February 2010 by asking Congress for $36 billion—double the Bush bailout—to subsidize additional nuclear construction loans.

In December 2011, the nuclear lobby suffered a stunning defeat in the closing days of Congress's lame-duck session. When President Obama tried to tuck an $8 billion nuclear bailout into his omnibus budget plan, angry environmentalists and taxpayers flooded the White House and Congress with 15,000 calls and letters. The nuclear "earmark" was clipped from the budget. "Once again," said Michael Mariotte, executive director of the

Nuclear Information and Resource Service, "taxpayers have been spared the expense of bailing out the wealthy, multinational nuclear power industry."[11]

Although this defeat dumbfounded an industry that had spent $640 million lobbying Washington for handouts, Mariotte correctly predicted, "the nuclear lobbyists will be back [in the new Congress], hat-in-hand, even while distributing campaign checks to their allies." The nuclear bailout battle will continue to simmer as politicians struggle to match pro-growth oratory with spending-cuts rhetoric.

The nuclear lobby has shown the tenacity of a zombie horde. This "living dead" industry has been written off time and time again, yet it continues to clamber back to its feet and lumber blindly onward. One of the historic "blows to the head" that have failed to deter the Mighty Atom's minions was delivered in a withering 1985 cover story in *Forbes* magazine that contained the following assessment:

> The failure of the US nuclear power program ranks as the largest managerial disaster in business history, a disaster on a monumental scale. The utility industry has already invested $125 billion in nuclear power, with an additional $140 billion to come before the decade is out, and only the blind, or the biased, can now think that most of the money was well spent. It is a defeat for the US consumer and for the competitiveness of US industry, for the utilities that undertook the program and for the private enterprise system that made it possible.[12]

The "Invisible Handout"

As things stand today, there is only one precondition to obtaining a government-backed loan: a power company must obtain an NRC license. But as Physicians for Social Responsibility notes, obtaining a license is merely a formality and getting a license is no guarantee that a reactor will actually be built. "On the last round of reactor construction," PSR notes, "over 100 reactors were cancelled after getting a license."[13]

Because nuclear reactors can cost billions to build, they cannot be constructed without massive government subsidies. If Adam Smith's "free market" had been in force, the atom-smiths of the private sector would never have built a single reactor. The nuclear industry is the offspring of federal excess and taxpayer largesse, a classic example of an economic system that might well be called "subsidism." Coal and nuclear plants receive bountiful subsidies, estimated at $250 billion to $300 billion a year worldwide.

Eliminating these subsidies would level the playing field, reveal the true costs of these technologies, and speed the necessary transition to more economically attractive renewables.[14]

Examples of cost overruns abound. In Matagorda County, Texas, after NRG Energy and Toshiba partnered to build two new reactors (Units 3 and 4) at the South Texas Project, costs soared from $5 billion to more than $18 billion. Local activists blasted NRG and Austin's CPS Energy for their lack of "transparency," noting that this was a polite term for "lying to the public and covering up bad news."[15] On April 19, 2011, NRG and Toshiba announced they were writing off their $331 million investment and abandoning plans to complete the reactors. Meanwhile, the estimated costs for building two new reactors at Turkey Point in Florida were topping $24.3 billion,[16] and the fate of Florida's Crystal River facility, which had been shut in 2009, was in doubt—even after owners, investors, and customers had poured $2 billion into questionable attempts to repair the reactor.

In what the *Raleigh News & Observer* called "one of the most exorbitant and bewildering mishaps in the history of the nation's nuclear industry," Crystal River operator Progress Energy attempted to gain access to the plant's steam generator by removing a 23-foot by 24-foot slab from the reactor's four-foot-thick wall. To save money, Progress hired a contractor with no experience. This led to three successive, unplanned, and costly "delaminations" (that is, removals of wall sections, each the size of a basketball court). As the newspaper noted: "The blunder shows that a highly experienced nuclear operator with a sterling reputation in the industry is not immune from a total breakdown in procedures that raise questions about judgment and competence."[17]

The Nuclear Pork Barrel

The nuclear lobby tried to insert $50 billion in loan guarantees into the 2009 federal stimulus package to design new Generation III reactors. That giveaway was stripped from the bill after widespread protests, but nuke proponents quickly responded with another attempt (the fourth in two years) to grab billions in nuclear pork from taxpayers' pockets. In 2010, the Senate proposed a Clean Energy Deployment Administration that would have channeled billions of dollars to coal and nuclear programs. After this bid failed, Lamar Alexander (R-Tenn.), a member of the Senate Environment Committee, tried to include $50 billion in taxpayer loan guarantees in the Clean Energy Development Administration's 2011 budget. The funds were to help kickstart construction of a hundred new reactors over 20 years. CEDA also failed to achieve liftoff.[18] Another failed attempt was made to include a nuclear bailout in a 2011 bill to aid "small business."

Nuclear power has supporters on both sides of the aisle. Under 2011 legislation sponsored by two Democrats, Senate Energy Committee chair Jeff Bingaman (D-N. Mex.) and congressman Jay Inslee (D-Wash.), a "Clean Energy Bank" would be authorized to grant "unlimited" taxpayer loan guarantees to build new reactors.[19] In 2010, Senator Bingaman was among the Democrats pushing for the creation of the Clean Energy Development Administration, despite the knowledge that CEDA was also designed to promote "safe" nuclear power and "clean" coal.[20] And in one especially egregious case of nuclear pandering, in May 2010, the White House (in a shameless, closed-door maneuver) made a failed attempt to tuck $9 billion into the Afghanistan war appropriations bill to bail out the foundering South Texas Project.[21]

Who Benefits from Washington's Nuclear Bailouts?

Taxpayer-backed nuclear loan guarantees are promoted as a means to create US jobs, but the truth is that all of the 18 currently pending US reactors would be designed and built by Toshiba or AREVA, and the high-paid engineering and construction jobs would go to workers in Japan and France.[22] The bailouts planned for Maryland's Calvert Cliffs Unit 3 and South Texas's Units 3 and 4 would mainly benefit Japanese and French nuclear companies. As the Nuclear Information and Resource Service notes: "If American taxpayers were upset about bailing out US banks and car companies, they should be furious about being put at risk in order to fatten the bottom line of overseas nuclear companies." (In another taxpayer giveaway to these foreign corporations, the US Department of Energy announced in May 2010 that a $2 billion loan guarantee would be provided to France's AREVA to build a uranium enrichment facility in Idaho.)[23]

The total cost to taxpayers of these nuclear-renaissance loan guarantees could hit $1.6 trillion.[24] During the 2008 presidential race, senator John McCain called for building 45 new nuclear reactors by 2030. Factoring in the industry's potential for 250 percent cost overruns, that could cost more than $1 trillion, with taxpayers taking a hit for billions of dollars worth of tax breaks, subsidies, loan guarantees, insurance breaks, and bailouts if the builders were to default.

The actual cost of constructing a nuclear reactor can be many times the initial estimate. While the nuclear industry has cited construction costs of about $2,000 per installed kilowatt (kW), out in the real world, Florida Power & Light has tagged the cost for building its two new Turkey Point units at $8,000 per installed kW.[25] The Energy Information Administration found that the average construction cost for 75 US reactors—originally estimated at $45 billion—had ballooned by more than 300 percent to $145 billion.[26] Very few US commercial reactors have proven profitable,[27] and even with federal support, any utility embarking on the nuclear path risks facing a lowered credit rating since cost overruns remain the norm.[28]

No nuclear plant has ever been completed on budget.[29] One of the most embarrassing cost overruns occurred when essential elements of California's Diablo Canyon reactor were installed backward and upside down. Even worse, New York's Shoreham plant fell 10 years behind schedule, went 10 times overbudget, and never even opened.[30]

The Price-Anderson Bailout

One of Washington's biggest-ever handouts to private industry was the 1957 Price-Anderson Act, which provides "limited liability protection" to power plant operators. This subsidy is worth several billion dollars annually.[31] Without Price-Anderson, the utilities would have been forced to purchase expensive liability insurance that would have raised the cost of nuclear electricity to prohibitive levels.[32] Federal estimates have placed the potential cost of a "non-worst-case" accident at $500 billion.[33] Bush's EPAct extended Price-Anderson liability exemptions through 2025. EPAct also provided up to $2 billion to reimburse owners for licensing delays caused by legal challenges over plant safety.[34]

Following the September 11th attacks, US nuclear insurers raised premiums on US reactors by 30 percent, bringing the Price-Anderson subsidy up to around $32 million per reactor.[35] Although US taxpayers would bear the major costs in the aftermath of a nuclear accident, they have no say in the reactor permitting process.

When Price-Anderson was first introduced, electricity was provided by regulated monopoly utilities subject to state and federal oversight. The proposed new reactors, however, would be owned by independent power producers (a.k.a. "exempt wholesale generators") that have no obligation to serve the public or submit to oversight by state utility commissions.

As author and anti-nuclear activist Harvey Wasserman has observed, "If there is a warning light [for nuclear insurance bailouts], . . . it is the Deepwater Horizon disaster, which much of the oil industry said (like Three Mile Island and Chernobyl) was 'impossible.' Then it happened. The $75 million liability limit protecting BP should be ample warning that any technology with a legal liability limit (like nuclear power) cannot be tolerated."[36]

Stranded Nuclear Assets

Under President George W. Bush, the 2009 US budget included $1.419 billion for programs to support the nuclear industry, including $301.5

million for an Advanced Fuel Cycle Initiative reprocessing plant and another $487 million for building a mixed-oxide fuel fabrication facility. The National Energy Policy Act of 2005 called for the United States to build additional facilities to extract uranium from used fuel rods and to produce new fuels needed "for the next generation of nuclear energy systems."[37] Bush's proposed budget also contained more money for promoting Generation IV nuclear reactors and the Nuclear Hydrogen Initiative, which would use the heat from hotter-running fourth-generation reactors to produce hydrogen for fuel. Under President Barack Obama, the 2011 budget request for the Department of Energy's nuclear power programs hit $912 million—more than the DOE's request for solar, wind, geothermal, biomass, and hydrogen technology combined. In addition to these direct investments of federal tax dollars, the proposed budget also called for $56 billion in loan guarantees to build nuclear plants—but only $3 billion to $5 billion to support clean energy projects.

Meanwhile, the ongoing global economic meltdown promises to turn nuclear reactors into (quite literally) "toxic assets"—for utilities, investors, and taxpayers. In June 2008, Moody's Investors Service Global Credit Research warned that new "nuclear plant construction poses risks" to the credit rating of power utilities, which could see a 30 percent deterioration in "cash-flow-related credit metrics." Even without the substantial costs of delivering electric power to customers, the expense of fueling, operating, and maintaining a new reactor continues to hover around 25 to 30 cents per kWh.[38]

This crushing disadvantage is worsened by comparison with burgeoning renewable energy sources such as solar and wind, which are inherently decentralized and thus can have lower distribution costs. Even if a new nuclear plant could produce power at 15 cents per kWh, this would still be 50 percent higher than the national average utilities currently charge their residential customers. To pay for such costs, customer rates would have to rise drastically—as they already have for some nuclear-powered utilities. (One reason Georgia may have been selected for a federal subsidy to build the first new nuclear plants is that Georgia law puts customers on the hook to pay reactor construction costs—even if no power is ever actually generated.)

High costs are not a US monopoly. In India, the average cost for building the country's last 10 reactors soared three times overbudget.[39] Some reactor construction costs come from the fact that the huge containment vessels must be imported from Japan, the only country with a steel industry robust enough to tackle the chore.[40] In order to finance construction of these pricy power plants, many countries have become indebted to multinational firms like British Nuclear Fuels, AREVA, General Electric, Mitsubishi, Siemens, and Westinghouse. The Philippines wound up paying the Export-Import

Bank of the United States $155,000 a day in interest for the Bataan nuclear facility, a $2.3 billion atomic power plant that took eight years to build, racked up a list of more than 4,000 defects, and never went operational.[41] In Finland, AREVA's supposedly vanguard Olkiluoto project was at least $3 billion over its projected $4.2 billion budget as of 2009.[42] Meanwhile, Atomic Energy of Canada, once seen as the nuclear industry's flagship organization, burned through $1.64 billion Canadian tax dollars in 2008, and in 2009, the conservative government announced plans to sell its nuclear white elephant.

The British Renaissance Reels from Soaring Costs

On May 15, 2006, British prime minister Tony Blair likened global climate change to the threat of terrorism and vowed to address the problem by building six new nuclear power plants. Six years on, Britain's would-be fleet of reactors appeared to be on the rocks. First, NuGen, a French-Spanish nuclear power consortium with plans for UK expansion, announced it was suffering setbacks. In March 2012, two German power companies bolted from a £15 billion ($23.7 billion) joint venture to build two new British reactors—one at Wylfa in North Wales and another at Oldbury-on-Severn in Gloucestershire. (The German companies pulled back after chancellor Angela Merkel called for the closure of Germany's nuclear power sector—a decision that meant the German firms would need to start plowing billions of dollars into the costs of decommissioning existing plants.)

At the same time, EDF, France's national power company, was reconsidering its joint venture with a British partner to build two new reactors at Hinkley Point in Somerset and Sizewell in Suffolk. In May 2012, with work on the Hinkley reactor more than two years behind schedule, EDF stunned its British partners with the announcement that it was increasing the projected cost for the two reactors by 40 percent—to £7 billion ($11 billion) per reactor. EDF's shift was magnified by a seismic event that rocked the French political landscape—a national election that knocked nuclear-friendly Nicolas Sarkozy from power and ushered in president Francois Hollande, a Socialist candidate who campaigned on a promise to close 24 of the country's 58 reactors and drastically reduce France's reliance on atomic power.

The triple punch of nuclear setbacks stemming from Fukushima, Berlin, and Paris had the UK's energy planners on the ropes. With North Sea oil reserves set to run dry and nuclear joint ventures suddenly in disarray, the future looked uncertain. "With all of our current nuclear power plants due to stop working within 10 years," London *Independent* editor Oliver Wright asked in May 2012, "what will it cost to keep on the lights?" Desperately clinging to the waking dream of a nuclear solution, Wright reported,

Britain's leaders had indicated that they had "no objection to allowing the contracts for new nuclear plants to pass to nuclear operators such as China, Russia, and Japan to keep the program alive."[43]

By the summer of 2012, however, the cost of electricity for the average British consumer hovered around £45 ($71) per MWh, while the cost of new nuclear was pegged at anywhere from £60 to £90 ($142) per MWh. At this rate, financing the completion of just four of the ten reactors envisioned by Britain's nuclear coalition would require what London's *Guardian* newspaper called "eye-watering levels of financial support"—i.e., a government subsidy of £60 billion ($95 billion) per MWh over 30 years. "No reactor has ever been built anywhere in the world without substantial government subsidy," the *Guardian* noted, adding that, with average reactor costs now rising 15 percent per annum, it is likely that "no reactor ever will be built without substantial government funding in [the] future."[44]

A lot of tea must have been spilled on the morning of April 20, 2012, when British readers opened *The Guardian* and read the headline: "Ministers Planning 'Hidden Subsidies' for Nuclear Power." According to a previously secret document obtained by the paper, government ministers were "planning to subsidize nuclear power through electricity bills—despite their promise not to." *The Guardian* reported, "This is the clearest evidence yet of government plans to subsidize nuclear power through the back door, by classifying it with renewables as 'low-carbon power.'" The leaked document also revealed a plan to permit "nuclear operators to reap higher prices for their energy than fossil fuel power stations."[45]

Decline and Decommissioning

If nuclear reactors are costly to build and run, that's nothing compared to the financial impact when they fail. Dealing with the lasting damage of the Chernobyl disaster will have cost Ukraine an estimated $170 billion by 2015.[46] Similarly, it is impossible to know the ultimate costs from Fukushima—the loss of homes, land, and national treasure drained by continued monitoring of the reactors, unprecedented cleanup costs, and long-term care for people who will suffer long-term health effects. One fiscal repercussion of Fukushima is already traceable, however: in November 2011, the International Atomic Agency reported that the quake-and-tsunami-induced failure had caused global nuclear construction costs to shoot up another 5 to 10 percent.

Even when nuclear plants are operating properly, they generate radioactive wastes that need to be securely stored for thousands of years. Because of this, nuclear power's economic deficiencies will plague the world's economies long after individual plants are finally shut down. The cost of decommissioning ranges between $655 million and $1 billion per reactor, so the total obligation for the United States' 104 reactors alone could top $100 billion.[47] Unfortunately, the Wall Street meltdown devastated many of the trust funds created to pay for eventual decommissioning in the United States,[48] making proper dismantling and decontamination less likely.[49] (See chapter 11, "The Decommissioning Dilemma.")

The most devastating costs, of course, result when a reactor fails, melts down, or explodes. The multiple meltdowns at Fukushima bear witness to the environmental, human, and financial burdens that can mark a nuclear accident site for generations. In March 2012, the Japanese electric utility TEPCO announced losses of $9.8 billion for fiscal year 2012 (ending on March 31, 2012). TEPCO president Toshio Nishizawa asked Tokyo for permission to hike electricity rates for residential users an average 10.28 percent. Some homeowners were warned that their summertime rates might rise by as much as 30 percent. On the positive side of TEPCO's ledger, the company accepted $12 billion in government bailout funds (in addition to $13 billion in loans it had previously received). Tokyo also has set aside $30.1 billion in taxpayer funds to help TEPCO cover compensation claims from its Fukushima plant's many victims. By May 2012, TEPCO had been handed a total of $45 billion in government money. Looking ahead, the Japan Center for Economic Research has estimated that the cost for decommissioning Japan's nuclear plants and compensating all the Fukushima victims will exceed $250 billion.

Meanwhile, nuclear offers little promise for the planet's two billion people who remain without access to electricity. In 1988, the United Nations estimated that it would cost as much as $46,000 per kilometer to build transmission lines to all the earth's two million remote villages (that's $83,674 in 2010 dollars).[50] In such situations, renewable technologies offer a cheap, quick, and readily available solution.[51] The cost of a 1,000 MW nuclear plant is now on par with retail costs of 1,000 MW worth of solar panels.[52]

The latest estimates for generating nuclear power run as high as 30 cents per kWh. This is far higher than the costs of more readily available renewable technologies and 10 times the cost of energy efficiency measures,[53] which can be realized for as little as 1 to 3 cents per kWh.[54] So-called negawatts (the energy saved through efficiency upgrades) can easily accomplish greater reductions in greenhouse gas emissions—and at a much quicker pace.[55]

On June 28, 2011, the US Energy Information Administration announced that—for the first time—renewable energy had trumped nuclear as a major source of power. For the first quarter of 2011, renewables provided

5 percent more electricity than nuclear reactors and matched three-quarters of the power provided by domestic crude oil. Taken together, renewables provided nearly 12 percent of US energy output. The leading contributors were biomass/biofuels (48 percent), hydropower (35 percent), and wind (12 percent), followed by geothermal and solar. But when it comes to growth, the big winner was solar. Compared to the first quarter of 2010, solar-electric production soared 104 percent, outpacing even wind (40 percent) and hydropower (28 percent). While renewables surged, coal fell 5 percent, and nuclear flatlined.

In summary, nuclear power's huge costs for construction, decommissioning, and waste storage make it a doomed technology. Far from offering a technological breakthrough, the supposed "nuclear renaissance" is merely a renaissance in government subsidies—or more precisely, a renewed expansion of the same subsidies that have maintained the industry on life support since its inception. There are much better ways to spend that money.

HANFORD'S RADIOACTIVE HEADACHE: A COSTLY CASE STUDY IN WASTED EFFORT

Washington State's Hanford Nuclear Reservation is both a monument to the Atomic Age and an orphan of the Cold War. Originally built to abet the top-secret Manhattan Project, Hanford is the plant whose ring of nuclear furnaces cooked the plutonium cakes that fueled the world's first nuclear test (the "Trinity" blast in Alamogordo, New Mexico) and the Pentagon's first atomic bomb (dropped on the people of Hiroshima in World War II).

By 1989, nine of Hanford's military-grade reactors had been shuttered, leaving behind the 1,150-MW Columbia Generating Station, the state's sole surviving atomic power plant. Columbia Station has one of the worst performance records of any US reactor. The Institute of Nuclear Power Operations has fingered the plant as one of the two worst when it comes to safety improvements and "human performance." Columbia's single, longest stretch of uninterrupted operation lasted just 505 days. The reactor produces only 10 percent of the electricity generated by its operator, Bonneville Power Administration; the other 90 percent is clean and renewable, thanks to the 31 hydroelectric dams installed on the Columbia River. Yet despite the Columbia Station's dismal record and capacity, the NRC has agreed to extend the reactor's life another 20 years when its 40-year operating license expires in 2024.

During Hanford's Cold War years, nearly 1.7 trillion gallons of contaminated wastes—a simmering stew of radioactive sludge generated by a near half-century of plutonium production—were heedlessly dumped into unlined trenches. Today, an additional legacy of radioactive residues lurks out of sight in 177 massive, underground tanks and, under a 2009 DOE plan, Hanford is set to receive an *additional* three million cubic feet of radioactive waste from the country's other nuclear facilities.[56]

While liquid wastes are secured inside 28 new, double-shelled tanks, the bulk of the sludge and "hardened" wastes is stored in 149 single-shelled tanks. These tanks are now nearly 70 years old and more than 60 of them are leaking. Although cleanup costs have been running around $2 billion per year, millions of gallons of poisonous wastes already have spilled into the soil beneath the 586-square-mile facility. The great fear is that the deadly underground plume will eventually reach the nearby Columbia River, which marks the border between Oregon and Washington State.

The stockpile of contaminated sludge festering on the Hanford Reservation is so toxic that only a few hours of exposure could lead to death. In 1989, faced with an obvious and growing threat to Portland (home to more than half of Oregon's citizens), state officials and the EPA reached an agreement with the DOE to contain the site's storage problems by building a $12.3 billion state-of-the-art waste treatment facility.

"Sealed in Amber"

In 1992, the DOE embarked on a decade-long project to build the treatment facility—a 65-acre complex including four massive buildings, each with a footprint larger than a football field and towering as high as a 12-story building. Using a process known as vitrification, the plant was supposed to separate low- and high-level radioactive wastes and permanently isolate them in a composite of unbreakable glass—"sealed in amber," as it were. The process would require vast amounts of energy to superheat the mix to 2,100°F (1,149°C). The boiling, radioactive soup would then be poured into huge 4- to 7-ton steel canisters and left to congeal until safe for long-term storage.

The cleanup of Hanford's stored wastes was to take 30 years. However, it turns out that building nuclear waste containment plants is just as unpredictable as building nuclear power plants. Originally budgeted at around $4 billion and targeted for completion in 2011, the project had cost taxpayers $12.3 billion by 2011 and its opening date had been set back to 2019. In November 2011, the DOE announced that it would need to add another $800 million to $900 million to the plant's cost.

In early 2012, the project's senior government scientist, Donald Alexander, spoke to *USA Today*. Because of the pressure "to make progress," Alexander complained, "the design processes are cut short, the safety analyses are cut short, and the oversight is cut short." Even worse, Alexander confessed, "we're continuing with a failed design."[57]

The problems begin in the storage area, where robotic tractors and powerful jets of high-pressure chemicals are supposed to turn hardened wastes into tractable sludge. The crushed and partially dissolved extracts then would be piped four miles to a "pretreatment" complex, before proceeding to other buildings for processing.

Driven by the desperate need to address a paramount danger in the shortest possible time-frame, the DOE relied on the same "design-build" approach (a top-down, authoritarian "master builder" style of decisionmaking) that has caused engineering headaches at nuclear power plants. In the name of expediency, *USA Today* reported, DOE was "moving ahead with construction of the vessels before tests are complete. That means the equipment might have to be re-engineered and modified, or replaced entirely, if it fails."

Plutonium Pudding

Case in point: In the process of designing the plant, engineers were surprised to discover that the plutonium particles they were supposed to move through the plant's pipes were actually 10 times larger than they had anticipated. This meant the pipes were susceptible to clogging. To prevent hydrogen gas explosions and uncontrolled nuclear reactions inside the pipes, engineers decided to install "pulse-jet mixers" to keep the radioactive, fudge-like paste moving. But as Walter Tamosaitis, the project's supervising engineer, pointed out, there was one hitch, "No one can stand up and say with any certainty that [the mixers] will work."[58]

Pushing plutonium slush through metal pipes is expected to cause erosion that could cause containment systems to fail well before their planned 40-year lifetimes. In addition, the use of mixers to push the piped wastes into metal processing vessels adds another level of abrasiveness that could cause the vessels to begin leaking after only 10 years' use.

USA Today concluded that the "design-build" imperative was part of an overall "progress-at-all-costs climate that discourages scientists from raising design concerns." This fear was echoed by a January 2012 DOE report that pointed to "significant concerns . . . for managing safety issues" at the Hanford project, citing the specific risk of a hydrogen gas explosion.[59]

The Hanford project has been the target of at least three federal investigations, including one that cited Bechtel National, the primary contractor, for "potential nuclear safety non-compliance" in the facility's design and construction. The DOE's Inspector General and the Government Accountability Office have launched separate probes of construction delays, cost overruns, and the installation of substandard equipment.

Because the piping already installed may not be able to handle the "plump plutonium particles," engineers are contemplating the construction of another building—a *pre*-pretreatment facility. As it now stands, the current plant design will only be able to treat about *half* of Hanford's stock of liquid, low-level radioactive waste. As *USA Today* discovered: "No plan has been finalized—or budgeted—for how to process the rest."[60]

In the meantime, DOE officials asked for another $840 million to fund the project through 2012—up 22 percent over initial estimates—but Congressional budget-cutters trimmed the request to $740 million.

Donna Busche, the nuclear safety manager for URS Corp. (a Bechtel National subcontractor), offered a summation of the Hanford Headache. When safety questions are raised, Busche told *USA Today*, "the first question that gets asked is not 'how are we going to solve it?' but 'how much is it going to cost?' . . . I've never seen this sort of flagrant disregard for technical issues."

3

Inefficient and Unreliable

What exactly is nuclear power? It is a very expensive, sophisticated, and dangerous way to boil water.

—HELEN CALDICOTT

ONE OF THE NUCLEAR INDUSTRY'S main complaints about renewable energy is that it is not dependable—the sun doesn't always shine, the wind doesn't always blow. But this argument ignores the nuclear industry's own operating history. Of the 253 nuclear power plants originally proposed for the United States, only 132 were ever built, and 21 percent of those were later permanently closed because of cost or safety problems. More than a quarter of US reactors have at some point completely failed for a year or more—some more than once. With renewable energy, you can be sure the sun will rise the next day and the wind will resume blowing. But when a nuclear power plant is shut down for repairs, there is no certainty about when—or if—it will resume operation.

Industry performance records reveal that, between 2003 and 2007, US nuclear plants were shut down 10.6 percent of the time while the failure rate for solar stations and wind farms was typically around 1-2 percent.[1] Meanwhile, the nuclear power industry suffers from the opposite problem: once started, a chain reaction is difficult to stop. Forced shutdowns and costly restarts put added stress on the integrity of the physical plant.

Even when a reactor operates perfectly, it still needs to be shut down every 18 to 24 months for maintenance and refueling—which requires replacing one-third of the reactor core's fuel rods.[2] These shutdowns average 39 days, and once a reactor is restarted, it takes several weeks to achieve full electric production. The LaSalle 1 reactor in Illinois holds the record for the longest refueling operation—just a tad over two years. And the cost of refueling isn't cheap. Replacing one-third of the fuel rods in a typical 1,000 MW boiling water reactor (BWR) or pressurized water reactor (PWR) can cost about $40 million. Fuel represents about 30 percent of a reactor's operating costs.[3]

The Union of Concerned Scientists cites a number of examples of the industry's blemished performance record. In 1983, the Pilgrim reactor in Massachusetts was taken off-line for ten months to repair a cracked recirculation system. In 1984, the Peach Bottom 2 reactor in Pennsylvania was shut down for more than 14 months. In 1993, one of the two reactors at the Sequoyah power station in Tennessee was closed for repairs for more than a year, while the second unit suffered an outage that lasted seven months. In 1997, seven reactors were shut down for an entire year. And, bear in mind, there is a financial disincentive to shutting down nuclear reactors to address safety problems or repairs: whenever shutdowns occur, costly replacement energy must be purchased on the spot market.[4]

And reactors don't always shut down only when planned. In August 2003, more than a dozen reactors in the northeastern United States and Canada were forced to shut down when a power blackout cut all outside electrical power. Only two of the reactors delinked from the grid and eased into shutdown mode as designed. Ten of the reactors unexpectedly collapsed into full shutdowns. The blackout revealed that most of the reactors lacked properly maintained emergency diesel generators.

When outside power is interrupted and backup power fails, reactor operators can lose the ability to monitor and control the reactor core, which can quickly lead to a catastrophic meltdown. The problem isn't restricted to faulty generators in North America. In Belgium, for example, two out of three backup power systems failed during a test at the Tihange 2 reactor in 2005. And reactors are also at risk from natural disasters such as floods, hurricanes, and tsunamis, which can disrupt off-site power, endangering plant operation and safety. Not to mention the fact that an earthquake, as we learned from the Fukushima disaster, can cause displacement of storage-pool cooling water. A quake can even cause a shifting of spent-fuel storage racks, bringing fuel rods into contact and risking a nuclear fission chain reaction.

Operating Extensions and "Uprates"

While there are plans to add another 16.6 GW of unproven fourth-generation nuclear capacity in the United States, 4.5 GW are expected to be *lost* as the old plants are retired. Under pressure from the reactor operators, the NRC has begun extending the 40-year operating life of 52 aging US reactors to 60 years and, in June 2012, met to consider extending some operating permits for up to 80 years—twice the reactors' intended operating life.[5] Former DOE policy advisor Robert Alvarez greeted the news with alarm. "The idea of keeping these reactors going for 80 years is crazy!" Alvarez declared.

Greenpeace nuclear policy analyst Jim Riccio claimed the meeting "shows the extent to which the NRC is captured [by the nuclear industry]. Nuclear regulators know that embrittlement of the reactor vessels limits nuclear plant life but are willing to expose the public to greater risks from decrepit, old and leaking reactors. As we learned from Fukushima, the nuclear industry is willing to expose the public to catastrophic risks."[6]

In the meantime, the stopgap plan is to partially fill the lost nuclear output with 2.7 GW coming from "uprates" of existing plants.[7] These "uprates" have allowed half of US reactors to exceed their original power generation maximums by replacing aging turbines and pipes to boost performance. Permitting these plants to operate at 20 percent over their intended design capacity now accounts for 25 percent of the total generating capacity of US reactors.[8] Instead of running these old reactors with heightened caution, the NRC is allowing owners to try to extract extra profit by running their aging nuclear cores "hotter and harder."

With nuclear construction halted or stalled, other nuclear nations also are extending the lives of plants that were designed to run for only 30 to 40 years. In most cases, aging reactors are being relicensed for another 10 years of use, but two Finnish reactors have been granted waivers to keep them running for as long as 60 years.[9]

The practice of extending a reactor's service beyond its intended life (while simultaneously allowing it to operate at more intense levels of generation) has been referred to as a game of "Chernobyl roulette." In terms of road wear, many of these reactors would qualify as "jalopies." You can't keep changing the tires and replacing parts forever.

While the nuclear industry likes to boast that the United States has a fleet of 104 working reactors, it prefers to draw attention away from its nuclear scrap heap. While the industry aggressively pursues uprates and operating-life extensions for the reactors it has managed to get working, over the past 40 years nuclear operators have been forced to shut down 44 reactors either during construction or before reaching the end of their permitted operating life—an industry-wide "success rate" of only 70 percent.[10]

A Cumbersome Technology

Some industries (like the electronics and information technology sector) evolve rapidly, with impressive rates of innovation and remarkable profitability. Some industries (like the Detroit auto industry) go through long periods of technological stagnation. The nuclear industry, with its cumbersome scale, its daunting hazards, its inability to control construction costs, and its profound unprofitability (sans massive taxpayer subsidies), has evolved only haltingly. In December 2010, the US Energy Information Administration increased the estimated cost for constructing a nuclear plant by 37 percent. By contrast, the agile renewables industries—wind power, solar, and biomass—continued to leap ahead of their cash-intensive competition, surpassing not only nuclear but also natural-gas turbines and other old technologies in speed of deployment, cost, efficiency, and safety.

At present, the fuel rods that power nuclear reactors don't even last as long as a compact fluorescent bulb—they burn out after two to three years, becoming intensely radioactive waste that needs to be safely stored for tens of thousands of years. But the nuclear industry has promised a new generation of "greener and safer" reactors. Instead of relying on mechanical pumps and manual safety controls, some of these new designs would trust "natural forces" like gravity and convection to safeguard reactors without human intervention. Engineers claim that massive reactors like the 1,150 MW Toshiba-Westinghouse AP1000 would produce 10 percent less radioactive wastes than current reactor designs[11] (not an impressive reduction). But the AP1000 and smaller reactors like the 25 MW Hyperion exist only on paper. (The first AP1000, under constructed in Georgia, will not be completed for several years.) And some of these "new-generation" reactors come with an added burden: they produce plutonium as a by-product, which would make it easier for countries with state-of-the-art power plants to also start making state-of-the-art atomic bombs.[12]

Dependent on Imported Fuel

While nuclear power advocates stress the importance of gaining independence from costly foreign oil, the nuclear industry itself is critically dependent on foreign resources. The US nuclear industry draws most of its fuel from the same 10 mines in six countries that supply 85 percent of all the world's mined uranium. The United States imports around 60 percent of its uranium (mostly from Canada and Kazakhstan).[13] The world annually mines between 36,000 and 60,000 tons of uranium,[14] but this is barely enough to supply existing commercial and military reactors. Uranium, like oil, is finite. As supplies diminish, costs will rise. Since 2005, the price of mined uranium has soared from $12 to $45 a pound.[15] With fewer than 5.5 million tons of recoverable uranium worldwide [16] (and 2009 consumption running around 67,000 tons), uranium reserves are not expected to last much beyond 70 years.[17] Building more reactors would only accelerate the depletion of the world's limited uranium supplies.

Even now, the only way the United States is able to power existing reactors is by relying on "secondary sources"—reprocessing spent fuel, recycling plutonium, and re-enriching depleted uranium. Surprisingly, nuclear disarmament treaties turn out to be another part of the bailout infrastructure, since they were intentionally written to allow the United States and Russia to use old nuclear weapons as a fuel source. Many warheads contain highly enriched uranium-235, and some contain plutonium-239, which can be used in mixed-oxide (MOX) reactors. Since 2000, 30 tons of decommissioned warheads have been "diluted" and used to produce around 10,600 tons of uranium oxide per year, accounting for about 15 percent of the world's reactor fuel.[18]

As uranium stocks are depleted, uranium-fueled light-water reactors will need to be replaced by plutonium-fueled systems supported by breeder reactors and reprocessing.[19] (Breeder reactors create—or "breed"—more radioactive fuel than they consume. Reprocessing involves recovering usable fissile material from radioactive waste.) But breeder technology and fuel reprocessing produce such massive volumes of toxic by-products that the United States and Germany have abandoned their fast-breeder plans. Even the celebrated French Superphénix, hailed as the world's largest fast-breeder reactor, was shut down in 1998 after numerous technical problems and an investment of $10.5 billion. (The Superphénix is now known as "The Grand Failure.")[20]

All three US reprocessing plants have been shut down. Britain's $2.6 billion Thorp reprocessing site at Sellafield was set to close in 2010, but as of March 2011, Sellafield was reportedly still dumping eight million liters of nuclear waste into the Irish Sea—every day. (Sellafield, which was shut for repairs in 2005 following the disappearance of 20 tons of high-level radioactive wastes, has been called a "white elephant" that has never operated properly. Some 100 tons of plutonium will remain in storage at Sellafield when the site does finally close. A December 2010 report by the Norwegian Radiation Protection Authority warned that an accident at the Sellafield storage site could expose Norway to seven times more radioactive fallout than rained down after the Chernobyl explosion.[21]) This leaves only one commercial plant capable of reprocessing spent fuel—the facility at La Hague, France.[22]

Inefficiencies and CO_2 Burdens

The claim that nuclear energy can help address global warming fails to pass the efficiency test. Atomic energy looks attractive in light of the fact that producing onc kWh of electricity from coal emits nearly a kilogram of carbon dioxide. However, while it is true that a nuclear reactor can produce electricity without generating a coal plant's CO_2, it is also true that wind power and cogeneration are 1.5 times more cost-effective at displacing a coal plant's CO_2. Efficiency is an even better bet, beating nuclear by a 10-to-1 margin, coming in at one to three cents per kWh. Energy expert Amory Lovins calculates that, from an efficiency standpoint, investing in nuclear power would make global warming worse, since pouring money into reactors instead of renewables produces "two to ten times less climate-solution per dollar."[23]

When the full life cycle of a nuclear plant is factored in (construction, mining, shipping), the industry's true CO_2 burden becomes overwhelmingly clear. Meanwhile, day in and day out, nuclear power plants routinely discharge clouds of 1,000°F (538°C) steam directly into the atmosphere. According to one estimate from the University of Chicago, the waste heat from all the world's electric power plants (including coal, oil, and nuclear) amounts to more than 27,640 trillion BTUs per year.[24] The added heat from an expanding "nuclear renaissance" would only further fuel global warming.

Ironically, nuclear plants themselves are sensitive to global warming. Because reactors rely on oceans or rivers to provide water for cooling, summer heat waves have forced many reactors in the United States and Europe to shut down to avoid overheating. French reactors have been forced to power-down during summer months when the heated water discharged by their coolant systems raised river temperatures far beyond legal limits. In other parts of Europe and the United States, the loss of coolant water from reduced river flows (due to climate change) have forced reactor shutdowns during hot summer months when electricity demand peaks. (See chapter 9, "Climate Change: Droughts, Floods, and Solar Flares.") Forced shutdowns and the resulting restarts not only are costly but also put added stress on the integrity of the physical plant.

Efficiency issues continue to dog the industry. On April 19, 2011, the nuclear renaissance suffered another blow when NRG, a New Jersey–based energy producer, along with its Japanese partner, Toshiba, abandoned plans to build two massive South Texas reactors. The *New York Times* reported that the collapse of "the largest nuclear project in the United States" was due to "uncertainties created by the accident in Japan."[25] That was only part of the story. The project also was ditched because of growing concerns that the reactors would never prove competitive given the fact that Texas enjoys a large inventory of low-cost natural gas and a surplus of electricity. In addition, nearly 8 percent of the state's electricity (620 MW) is generated by wind power, while solar installations (which grew by more than 100 percent in 2010) now provide Texans with more than 944 MW of clean, safe, efficient, and reliable power.

WERE FUKUSHIMA'S REACTORS FATALLY FLAWED?

There may have been ample evidence to anticipate and prevent the explosions of Fukushima's Mark 1 reactors—if only the nuclear industry had taken heed of an event that happened in North Carolina in 1972.

Following the Fukushima explosions, nuclear engineer Arnie Gundersen spent hours poring over TEPCO's performance records for the period following the earthquake but before the tsunami. While studying one eye-boggling page filled with hundreds of numbers densely arrayed in a matrix of tight columns and rows, he spotted something remarkable. A close look at the numerical web for the data charting the pressure inside one of the overheating reactors revealed the pressure rising implacably upward. This was no surprise: the overheating reactor was turning the surrounding water into hydrogen gas. An explosion looked inevitable. But when the readings reached 100 pounds per square inch (psi), the pressure suddenly stopped rising and fell back to safer levels. What could have caused this?

Gundersen thought back to an event that happened at a US reactor decades ago that suggested a solution to the puzzle.[26] In 1972, technicians tested the containment vessel of a Mark 1 reactor at the North Carolina's Brunswick plant using high-pressure air. Internal pressure began to climb until, at around 100 psi, the readings suddenly dropped. An investigation revealed that the intense pressure had forced the restraining bolts on the containment vessel's huge metal cap to stretch, allowing the lid to rise. Eventually, this allowed the pressurized air to escape. Fortunately for the technicians at Brunswick, the escaping gas was not hydrogen. TEPCO's employees at Fukushima were not so lucky.

In 1985, an NRC analysis concluded there was a likely chance that the Mark I design would fail within hours of a core melt. In 1986, Harold Denton, the NRC's top safety officer, warned that the "Mark I containment, especially being smaller with lower design pressure, in spite of the suppression pool, . . . [has] something like a 90 percent probability of that containment failing."[27]

Gundersen believes the flaw that triggered the Fukushima explosions is the same one detected in the Brunswick reactor 40 years earlier. If so, this danger remains inherent and unresolved—and haunts every Mark 1 reactor still in operation today.[28]

The sequential triple explosions of Fukushima's reactors provide an important lesson that needs to be emphasized. This. Was. Not. A. Coincidence. This was a design characteristic. The evidence now suggests that General Electric's Mark 1 reactors were not only designed to produce electricity, they were also inadvertently designed to explode. If the Mark 1 reactor were a commercial product like a Toyota, a toy, or a toaster, it would already have been subject to a recall by the US Consumer Product Safety Commission.

4

Catastrophic Dangers: From Three Mile Island to Fukushima

> *Given the present level of safety being achieved by the operating nuclear power plants in this country, we can expect to see a core meltdown accident within the next 20 years, and it is possible that such an accident could result in off-site releases of radiation which are as large as, or larger than, the releases estimated to have occurred at Chernobyl.*
>
> —NRC COMMISSIONER JAMES ASSELSTINE, TESTIFYING BEFORE CONGRESS IN 1986

ON MARCH 29, 2009, the thirtieth anniversary of the Three Mile Island (TMI) meltdown, former Nuclear Regulatory Commission member Peter A. Bradford told a subcommittee of the Senate Committee on Environment and Public Works: "Among the lessons of TMI is that nuclear power is least safe when complacency and pressure to expedite are highest." Bradford attributed the TMI catastrophe to "a technology that had rushed far ahead of its operating experience." The result was "a landscape dotted with nine-figure cost overruns, a nine-figure accident, eight-figure cancellations, and eight-figure mishaps in such areas as steam generator tubes, pressure vessels, seismic design, and quality assurance."[1] The industry's current rush to extend the operating lives of aging reactors—and to call for a taxpayer-financed "nuclear renaissance"—gives new urgency to Bradford's warning.

Bradford offered another sobering lesson from TMI: granting an operator a license is no guarantee that the operator won't inadvertently "turn a billion-dollar asset into a two-billion-dollar clean-up in ninety minutes." Bradford noted that the "near-rupture" of a reactor vessel at Ohio's Davis-Besse reactor in 2002 was partly due to NRC's "undue solicitude for the profits of the licensee." The NRC had agreed to delay inspections of a vital safety component to accommodate the plant's owner. This allowed corrosive liquids to eat through six inches of the reactor vessel's carbon-steel top, leaving only a half inch of metal behind to prevent a massive explosion.[2]

When Three Mile Island's Unit 2 reactor started operation on December 28, 1978, it was hailed as a state-of-the-art facility. Three months later, it was a multibillion-dollar liability. As Bradford told the Senate subcommittee, "The melting of the core during the early hours of the accident was far more severe than was known at the time."[3] The truth was that half of the core melted.

The story Bradford told the Senate was quite different from the story that was given to the public at the time. For days, public officials issued one false statement after another, including: there were no radiation releases; there were radiation releases but they were "controlled"; radiation releases were "insignificant"; there was no melting of the reactor fuel; there was never any danger of an explosion; there was no need to evacuate the surrounding communities. It was nine years before TMI's owners finally admitted there had been a partial meltdown at the reactor. It wasn't until 1983 that a sonar probe revealed extensive damage caused by a partial meltdown that destroyed nearly half the reactor core. TMI's containment vessel had failed, releasing a cloud of radiation that was about 100 times more significant than the initial estimates offered by the industry and the NRC. Thirty years after the accident, the NRC still does not know how much radiation was released or where it went.

The TMI accident was followed by a die-off of birds and bees. Diseases killed neighboring livestock. Dogs and cats were born dead or deformed.[4] Local, state, and federal officials promised to conduct thorough health studies of the residents exposed to TMI's radiation. Instead, the state hid the health impacts by deleting cancer reports from the public record, abolishing the state's tumor registry, and ignoring an apparent tripling of the infant death rate around the damaged reactor. Independent surveys by local residents found high incidences of cancer, leukemia, and birth defects. Cancer rates within a 10-mile radius of TMI jumped 64 percent between 1975 and 1985. Some 2,400 residents filed a class-action lawsuit demanding compensation, but the federal courts still refuse to hear the case. TMI's owners have paid at least $15 million in out-of-court settlements to parents of children born with birth defects—on the condition that the families remain silent about the settlements.[5] In the mid-1980s, the people of Harrisburg voted three-to-one to shut down TMI's Unit 1. The Reagan administration ignored the vote and ordered Unit 1 back online. The reactor is still operating.

A Chernobyl in the United States?

The Chernobyl reactor, which detonated in a rural region of Ukraine, has caused more than $500 billion in cleanup costs to date. Initial government reports claimed that only 31 "liquidators" (members of the cleanup crew) died. By 2005, however, between 112,000 and 125,000 of these "liquidators" had succumbed to radiation exposure. Over the past 23 years, according to the most recent estimates, the Chernobyl blast may have caused the cancer deaths of nearly a million people worldwide.[6]

While Chernobyl was built in a remote location, most US reactors are located near major population centers, including Baltimore, Baton Rouge, Boston, Chicago, Los Angeles, Manhattan, Miami, Minneapolis, and Philadelphia. US government reports estimate that the meltdown of a single nuclear reactor could irradiate an area as large as Pennsylvania.[7]

When the accident at Three Mile Island happened, nuclear reactors were providing 12 percent of US electricity. Thirty years later, with reactors providing 20 percent of the country's electricity, nuclear power was often taken to be a safe, reliable technology. Fukushima changed that. Still, most Americans remain unaware of an ominous spate of recent near-disasters that have occurred at reactor sites around the world. (This ignorance could be linked to the fact that two powerful defense contractors with interests in nuclear technology have also controlled two of the country's three major TV news networks: until 2000, Westinghouse owned CBS, while General Electric (which designed the Fukushima reactors) is part-owner of NBC Universal, which co-owns MSNBC through a joint venture with Microsoft.)[8] The public also remains largely unaware of the many operational "incident reports" at US reactors that are filed with the NRC every year.

Most US reactors are rated "green" in the NRC's four-stage color-coded safety-rating system. The next level is "white," followed by "yellow" and, finally, "red"—the last warning issued before the NRC can order a plant to cease operations. The NRC has rarely shut down a US reactor, no matter the severity of the problems. The only US reactor that is "red-tagged" as of this writing is Alabama's Browns Ferry plant. Browns Ferry was tagged following an incident on October 23, 2010, when plant operators were unable to shut down the Unit 1 reactor after a valve failed. It was subsequently discovered that the valve had been inoperable for 18 months. Disaster was averted when plant operators managed to cool the reactor using a fire hose installed after a large fire damaged the facility in 1975.

In the past, the NRC has red-tagged only four other reactors. Indian Point in New York was tagged after operators neglected to correct cracking in the Unit 2 reactor's steam generator (July 1, 2001); Point Beach in Wisconsin was flagged after operator error caused a failure of cooling pumps (February 28, 2002); Davis-Besse in Ohio was cited for failing to control leaks from the coolant system (August 9, 2002); and Point Beach was again sanctioned for design inadequacies and safety shortcomings at the Unit 2 reactor (March 24, 2003).[9]

Redefining the Risks

The NRC rates nuclear mishaps on the seven-point International Nuclear and Radiological Event Scale. "Incidents" (meaning events or failures that result in no off-site radiation releases or severe equipment damage) are rated 1 through 3. "Accidents" (meaning radiation is released off-site or plant equipment is severely damaged) are rated 4 through 6. The worst-case scenario is a "level 7 major accident." But the NRC's categories are deceptive.

National University of Singapore public policy professor Benjamin Sovacool investigated the implications of revising the NRC's definition of an "accident" to include incidents involving a loss of life or more than $50,000 in damages. With these parameters, Sovacool concluded, "a very different picture emerges." He has identified at least 99 such accidents that occurred worldwide between 1952 and 2009 and caused total of $20.5 billion in financial losses. This would work out, on average, to more than one nuclear accident and at least $330 million in damage every year for the past 30 years and, as Sovacool notes, "this average does not include the Fukushima catastrophe."[10]

Sovacool's research also reveals that "nuclear power ranks higher than oil, coal, and natural gas systems in terms of fatalities, second only to hydroelectric dams. There have been 57 accidents since the Chernobyl disaster in 1986. While only a few involved fatalities, those that did collectively killed more people than have died in commercial US airline accidents since 1982."

Sovacool cites a more comprehensive scale—one expanded to include injuries, worker radiation exposures, and malfunctions that fell short of causing leaks or shutdowns. Using these parameters, the nuclear industry's history has recorded 956 serious reactor incidents between 1942 and 2007. Another index referenced by Sovacool documented more than 30,000 mishaps at US nuclear power plants between 1979 and 2009—many of which had the potential to cause serious damage and meltdowns.

According to Sovacool's estimates, "The meltdown of a 500-megawatt reactor located 30 miles from a city would cause the immediate death of an estimated 45,000 people, injure roughly another 70,000, and cause $17 billion in property damage."

Meltdowns and Mishaps

When it comes to US nuclear meltdowns, Three Mile Island is not the only site where plant operators have watched helplessly as their reactor core turned to hot, radioactive porridge. In 1966, Michigan's Fermi 1 plutonium breeder reactor suffered a partial core meltdown. In 1975, Alabama's Browns Ferry reactor caught fire, resulting in a 20 percent core damage probability (CDP, an estimate of the likely damage to a reactor core in the event of an accident). In 1978, a steam generator failure at California's Rancho Seco station resulted in a 10 percent CDP.

In 2006, the emergency power failed at Sweden's Forsmark plant, affecting 4 of the country's 10 nuclear power reactors. Plant officials later admitted, "It was pure luck there wasn't a meltdown."[11] In another close call, a bolt of lightning disabled the Yankee Rowe reactor in Massachusetts in 1991. The owners refused to follow up with safety tests until ordered to do so by Congress.[12]

New Jersey's Oyster Creek reactor—a General Electric Mark 1 Fukushima-style reactor and the oldest US plant to win an operating extension—was not even designed to survive a direct hit from a plane. Its containment vessel is essentially identical to the ones that ruptured in Japan. Its reactor shell (which is supposed to control a radioactive steam leak in an emergency) is too corroded to guarantee safety. And despite visible rust around the reactor core and a history of steam and radiation leaks, the NRC did not require that the plant be tested for metal fatigue before relicensing it.[13]

Official Risk Assessments

The Atomic Energy Commission (precursor to the Nuclear Regulatory Commission) first asked the Brookhaven National Laboratory to assess the consequences of a domestic US nuclear reactor accident back in the 1957. (In an attempt to allay the public's fears about radiation exposures, the AEC also promoted the idea that there was a "threshold" level for radiation exposure, below which there was no cause for alarm.) In its 1965 WASH-740 update to that assessment, Brookhaven concluded that a reactor accident could cause 45,000 deaths and $17 billion in property damage over an area equal to that of the state of Pennsylvania. (WASH-740 was followed by WASH-1400 in 1977 and NUREG-1150 in 1991.)

In 1982, the NRC invited scientists at the DOE's Sandia National Laboratories to update the government's official risk assessment report by estimating the financial and human costs of a meltdown and containment breach for each of the country's atomic reactors. The projections in Sandia's Calculation of Reactor Accident Consequences (CRAC-2) report were near apocalyptic. In the case of the Indian Point twin-reactor complex north of New York City, Sandia projected the damage from an accident at the Unit 3 reactor alone would cause 50,000 early fatalities, 167,000 early injuries, 14,000 cancer deaths, and $314 billion in property damages. An accident at both reactors would result in an estimated 96,000 early fatalities, 308,000 early injuries, and 27,000 cancer deaths.[14] (Note: Those were 1980 dollars. Adjusted for inflation, an Indian Point meltdown would cost $864 billion in 2012 dollars.

Another critical factor is population growth. Around 8.2 million people now live in New York City, an increase of more than a million aince 1980.)

Damage from a catastrophic radioactivity release from just one of Pennsylvania's Peach Bottom reactors was set at 72,000 early fatalities, 45,000 early injuries, 37,000 cancer deaths, and $119 billion in property damages ($732 billion in 2012 dollars). But CRAC-2 was based on 1970 census data. Populations have grown significantly over the intervening years, so casualty figures would now be much worse.

Such shocking figures may explain why the NRC attempted to conceal these findings from the public. Fortunately, US Rep. Ed Markey (D-Mass.) made the information public in congressional hearings. Given the real-life case studies provided by Three Mile Island, the Chernobyl disaster, and Fukushima, these estimates are now considered unrealistically low.

Washington Rewrites the Risks

In 2012, under the Obama administration, the NRC produced an update of CRAC-2. The results left nuclear watchdogs slack-jawed with disbelief. Instead of improving on the previous accident assessments, NRC's new State-of-the-Art Reactor Consequences Analyses (SOARCA) came to the stunning conclusion that, even with a fleet of aging reactors, the "risks of public health consequences from severe accidents" were likely to be "very small." Moreover, the "long-term risk" of anyone dying from cancer from a nuclear accident was downgraded to less than one in a billon. How could the NRC account for such a remarkable reversal of earlier warnings? Because, the SOARCA explained, "successful implementation of existing mitigation measures can prevent reactor core damage or delay or reduce offsite releases of radioactive material."[15] It was as if TMI, Chernobyl, and Fukushima had never happened.

"NRC should immediately withdraw its absurd SOARCA report," Beyond Nuclear's Cindy Folkers scoffed, suggesting that nuclear safety would be better served if the NRC devoted its time to "protecting the public health, safety, and the environment—its mandate—rather than doing the nuclear power industry's bidding."[16]

No reactor—not even the latest state-of-the-art reactor—will ever be inherently safe, because most nuclear accidents (such as those at Chernobyl, Three Mile Island, the Fermi 1 reactor in Michigan, the Tokaimura plant in Japan) are caused by human error rather than purely technological failure. Until you have foolproof humans, the notion of an inherently safe and foolproof reactor remains absurd. As nuclear physicist Edward Teller once observed: "There's no system foolproof enough to defeat a sufficiently great fool."[17]

FUKUSHIMA'S DEVASTATING LEGACY

On March 11, 2011, a 9.0-magnitude earthquake off the coast of Japan triggered the automatic shutdown of 11 reactors at four coastal sites (including Kashiwazaki-Kariwa, the world's largest nuclear plant, which had been severely damaged by a 6.8-magnitude earthquake in 2007). A churning wall of seawater followed, rushing ashore and erasing homes, buildings, farms, and lives. The coastal village of Sendai ceased to exist. As the floodtides retreated, an invisible tide of radiation from Fukushima Daiichi's melting reactors was only beginning.

On March 12, a hydrogen explosion ripped the roof off Fukushima's Unit 1 containment building. On March 14, a second explosion destroyed the Unit 3 building. On March 15, a third hydrogen detonation demolished the containment structures around the Unit 2 reactor and a fire erupted at the adjacent Unit 4 reactor, followed by two large airborne releases of cesium-137. Out at sea, a radioactive cloud washed over the USS *Ronald Reagan,* on patrol in the Pacific. Helicopters flying within 60 miles of the coast returned to the aircraft carrier coated with radiation 400 times above normal. The on-deck crew received "a month's worth of radiation in about an hour."[18] The US Geological Survey estimated that it took only 18 days for the initial radiation to circle the earth.

Japanese officials initially downplayed the danger (while secretly mulling the evacuation of Tokyo). At first they reported radiation levels were only one-tenth those registered at Chernobyl. A few weeks later, officials reluctantly doubled their estimate. (In October 2011, the journal *Nature* would report that the initial radiation was actually *more than double* the levels at Chernobyl.)

While the world was initially led to believe that Fukushima's Unit 5 and Unit 6 reactors were undamaged, stable, and secure, this turned out not to be the case. On May 28, 2011, the cooling pump feeding seawater into the Unit 5 reactor failed, causing temperatures to rise dangerously in the reactor core and spent fuel pool. It took workers 15 hours to install a backup pump. By the time it kicked in, the core was within less than seven degrees Celsius of a boil-off that could have led to a fourth meltdown.[19]

TEPCO's Other Reactors

With all attention focused on Fukushima Daiichi, few people were aware that another near-disaster was playing out at Fukushima Daini, a TEPCO plant located seven miles away. Although the site's four reactors shut down as designed, the tsunami's waves damaged the cooling systems in three of the reactors. TEPCO workers managed to stabilize one reactor in a few hours but had to work for nearly five days to save the remaining three units. Fortunately, the Fukushima Daini plant did not suffer the loss of its outside 500 kV power line.[20] It wasn't until January 2012 that Japan's Nuclear and Industrial Safety Agency revealed that all four of the diesel-powered emergency power generators and the spent fuel pool cooling system at Fukushima Daini had been damaged.[21]

In Tokyo, chief cabinet secretary Yukio Edano was reassuring the public that there was no health risk from the multiple nuclear disasters, but inside the confines of government, Edano was warning colleagues about a possible "demonic chain reaction." If radiation continued to spread from the Daiichi site, Edano feared, TEPCO's workers might choose to abandon reactors at at the Daini and Tokai sites—each site progressively closer to Tokyo. "We would lose Fukushima Daini, then we would lose Tokai," Edano fretted. "If that happened, it was only logical to conclude that we would also lose Tokyo itself."[22]

The reason this "demonic" event did not happen is attributable to the actions of prime minister Naoto Kan. TEPCO president Masataka Shimizu had called the prime minister with news that the company had decided to abandon the out-of-control reactors. Shimizu argued that the government had no power to order TEPCO employees to risk their lives. Kan refused to accept this as an option. In the early hours of March 15, Kan burst into TEPCO's headquarters and angrily demanded that the company take responsibility for containing the damage—no matter the risk, no matter the cost. Kan threatened to "demolish" TEPCO if it fled from the fight. TEPCO backed down and cancelled its planned evacuation. A brave cadre of volunteers—including the "Fukushima 50," the first TEPCO employees to reenter the stricken site—went back to work, dousing the shattered reactors with seawater (using the same hoses and water cannon that had been used in previous months to battle anti-nuclear protesters).

Japan's "Nuclear Samurai"

Within the first three weeks, more than 500 "nuclear samurai" fought to prevent further damage within a fog of radiation that was 100,000 times normal background levels. They waded through water that was so contaminated it burned their boots off. In those first weeks, 19 employees received radiation doses exceeding 100 millisieverts. (In the United States, worker exposure is limited to 50 millisieverts per year.)[23]

Faced with growing public anger over the handling of the Fukushima disaster, Kan was forced to resign on August 29, 2011. It wasn't until the Rebuild Japan Initiative Foundation (RJIF) published its 400-page investigation of the disaster that Kan's "failure is not an option" moment was revealed. "Prime Minister Kan had his minuses and he had his lapses," RJIF founder Yoichi Funabashi declared, "but his decision to storm into TEPCO and demand that it not give up saved Japan."[24] For its part, TEPCO (which refused to participate in the RJIF investigation) now stands to receive an 11-trillion-yen ($137 billion) government bailout, after which the company will be nationalized.

On June 8, 2011, Japanese authorities admitted for the first time the possibility that the molten fuel in all three reactors had escaped the steel containment vessels, moving from a "meltdown" to a more serious "melt-through." With seeming indifference to the suffering of the Japanese people, TEPCO officials apologized for the fact that "this accident has raised concerns around the world about *the safety of nuclear power generation*" (emphasis added).[25] TEPCO did *not* issue an apology when—just five days later—six TEPCO cleanup workers received radiation doses exceeding TEPCO's 250-millisieverts-per-year limit.

The battle to contain the nuclear dragon raged through the summer as thousands of TEPCO employees, conscripted workers, and volunteers knowingly risked the their health and their lives in a desperate struggle against unknowable dangers.

On August 15, 2011, steam was reported to be spewing from cracks in the earth surrounding the plant. In mid-October, hydrogen gas was reported to be accumulating in the pipes at Unit 1, and TEPCO admitted that severe damage to the spent fuel pools at all three units had caused the release of incredibly high levels of cesium-137 and cesium-134. (A month before, the NRC had assured Congress and the American people that the spent fuel had not been damaged.)

The Unit 4 Threat

On September 29, 2011, one week after Japan's newspapers reported that melted fuel rods from all three reactors could be "sinking into the ground" at the rate of 17 meters per year, Akio Matsumura (the first secretary-general of the Green Cross International) published a chilling report, "The Fourth Reactor and the Destiny of Japan." Matsumura warned that the 1,535 used nuclear fuel rods at Fukushima's Unit 4 reactor posed a grave hazard. Owing to the design of all General Electric Mark 1 reactors, the dangerously radioactive rods were stored on the outside roof of a building that was structurally damaged and tilting. "If the structure collapses," Matsumura wrote, "we will be in a situation well beyond where science has ever gone. The destiny of Japan will be changed and the disaster will certainly compromise the security of neighboring countries and the rest of the world."[26]

If Unit 4's 460 tons of spent fuel were to crash to earth, it could damage the 6,375 fuel rods in a nearby "common pool." Together, Fukushima's 11,138 spent fuel assemblies contain 134 million curies of cesium-137, which the US National Council on Radiation Protection estimates is roughly 85 times the amount of cesium-137 released at the Chernobyl accident. In a March 25, 2012 letter to the United Nations secretary-general Ban Ki-moon, Matsumura warned that a collapse at Unit 4 could "destroy the world environment and our civilization. . . . This is an issue of human survival. It is no exaggeration to say that the fate of Japan and the whole world depends on No. 4 reactor."[27]

After visiting the wreckage of the Fukushima reactors in 2012, Oregon senator Ron Wyden wrote, "The scope of damage to the plants and to the surrounding area was far beyond what I expected. . . . The precarious status of the Fukushima Daiichi nuclear units and the risk presented by the enormous inventory of radioactive materials and spent fuel in the event of further earthquake threats should be . . . a focus of greater international support and assistance."[28]

The fate of the Unit 4 reactor should be a matter of urgent, international concern. The world community, the United Nations, and the United States (which lies downwind of the teetering Unit 4 reactor building) should be working side by side with TEPCO and Tokyo to stabilize the facility and devise an effective means to remove the precarious stores of spent fuel. The lessons learned could prove invaluable should an earthquake shatter a similar Mark 1 reactor in Alabama, Georgia, Illinois, Iowa, Maine, Minnesota, Nebraska, New Jersey, North Carolina, Pennsylvania or Vermont.

A China Syndrome in Japan?

By December 2011, TEPCO was admitting that about 60 percent of the fuel in Units 2 and 3 had melted down, while the molten fuel in Unit 1 had breached the pressure vessel and poured onto the floor of the outer containment vessel. According to TEPCO's best estimates, the hot fuel had burned through 65 cm (25.6 inches) of concrete floor and was hovering only 37 cm (14.6 inches) above the reactor's final outer steel wall—the last line of defense against a "China syndrome" that would release the fuel into the ground—and groundwater—beneath the plant.

In December 2011, the Nuclear Information and Resource Service (NIRS) assessed the potential that the molten "corium" might contribute to a China syndrome and concluded, "This is not the case." NIRS pointed out that concrete doesn't "melt" but it does "crumble"—at temperatures above 1,000 degrees centigrade. This could certainly happen, since nuclear fuel begins to melt at 2,200 degrees centigrade. Since September 2011, however, temperatures inside the Fukushima reactors have been hovering at around 100° centigrade. NIRS suggested it would be more likely to see the corium penetrate the containment walls, which are much thinner than the floor. NIRS did offer one critical caveat, however: "A new disruption to the cooling water system now in place, such as another major earthquake/tsunami. . . , certainly can't be discounted."[29]

A Misleading Claim of "Cold Shutdown"

On December 19, 2011, TEPCO and the government announced that their workers and volunteers had achieved a "cold shutdown" at the three damaged reactors. Japan's Nuclear and Industrial Safety Agency (NISA) was a bit more circumspect. NISA announced that the reactors had attained "a condition equivalent to a 'cold shutdown.'" TEPCO's claim was quickly challenged by critics who pointed out that the term "cold shutdown" refers to *undamaged* reactors operating under *normal* conditions where cooling water has been brought below the boiling point with *no continuing core damage or radiation release.*

Nature brought a quick reminder of just how precarious the situation remained. On January 1, 2012, a 6.8 quake jolted Fukushima. This was followed by a significant increase in radiation around the plant.

Despite December's "cold shutdown" claim, workers continued the battle to keep the broken reactors under control. In January, the water temperature inside the Unit 2 reactor soared to 164°F (73.3°C). With temperatures still rising, TEPCO increased the flow of cooling water from the usual 1 ton per hour to 9.5 tons per hour. On February 8, the temperature inside Unit 2 was finally brought down to 68.5°C after workers increased the flow of cooling water to 13.5 tons per hour—the highest level since December 2011. They also added boric acid to the cooling water to prevent a "recriticality." This precaution was seen as another troubling sign that the announcement of a "cold shutdown" may have been premature.

Another indication that the situation remained dangerously out of control came with the announcement that TEPCO planned to drop 3.5 tons of ice from a helicopter in an attempt to cool the spent fuel stockpiles outside the Unit 1 reactor. This ill-advised plan was dropped because of extreme radiation over the Unit 1 site.[30]

Radiation levels inside the damaged reactors remained dangerously high, with the deadliest radiation streaming from Unit 3. A remote inspection of the Unit 2 containment chamber on March 28 revealed radiation levels 10 times the fatal dose. A remote robotic probe discovered that levels of cooling water had fallen from 33 feet to barely 2 feet. More than half of the fuel had spilled from the broken reactor vessel and was producing lethal waves of radiation hitting 73 sieverts per hour. (Exposure to 7 sieverts over an hour can cause death within a month; 20 sieverts can kill within days.)[31]

TEPCO admitted that the radiation inside Unit 2 had become so lethal that not even mechanized robots could survive the bombardment for more than two hours. This means that in order to clean up the still largely unknown and unmapped mess of melted fuel inside the damaged reactors, TEPCO needs to invent a whole new class of robots built to operate while exposed to previously unimagined levels of radiation.[32]

The Nuclear Dragon Is Still Not Dead

By May 2012, work was well under way to create a new nuclear watchdog agency under the aegis of Japan's Environment Agency, but the opposition Liberal Democratic and New Komeito parties were insisting on an independent commission with full autonomy. At the same time, prime minister Yoshihiko Noda let it be known that, despite widespread opposition, he intended to seek the restart two reactors at the Oi facility in Fukui Prefecture. This announcement came as a rebuff to 66 municipal leaders who had formally petitioned the government to permanently close and decommission all of Japan's reactors. (Noda's stance is completely at odds with that of his industry minister, Yukio Edano, who insists, "I would like to break away from the reliance on nuclear plants and reduce dependence to zero as promptly as possible.")

In a statement marking the one-year anniversary of the Fukushima disaster, NIRS executive director Michael Mariotte issued the following statement:

> One year after the Fukushima accident began—an accident that still has not ended—at least 80,000 Japanese people have lost their homes and livelihoods. Hundreds of thousands more are living in contaminated zones and afraid (often for good reason) of the food they eat and water they drink. The nuclear industry and radiation deniers claim that no one has died because of Fukushima. That's only because the cancers that are coming take more than one year to appear. While the nuclear industry appears determined not to learn any lessons from Fukushima, the public understands the lessons very well. And the most important lesson is that we must end the use of nuclear power and move as quickly as possible to clean and sustainable energy sources.[33]

As if to underscore Mariotte's message, 29 new earthquakes rattled Japan within the week, including two measuring 6.1 and 6.8, and a 7.4 jolt that struck on the first anniversary of the Tohoku quake. Scientists at the University of Tokyo warned that the 9.0 quake that struck on March 2011 had realigned Japan's offshore tectonics, raising the chances of a devastating 7-plus-magnitude quake even more likely in the near future. While government officials predicted a 70 percent chance a severe quake could strike within 30 years, the university scientists concluded there was a 75 percent chance of another monster quake striking within four years.[34]

MAJOR NUCLEAR ACCIDENTS (PARTIAL LIST)

Canada (1952)—Loss of coolant at the Chalk River reactor causes a partial meltdown and a hydrogen gas explosion that destroys the reactor-vessel seal. Some 100,000 curies of radiation are released into the air and 4,500 tons of radioactive water spill into the facility's basement and ditches along the Ottawa River.

USA (1953)—An explosion at the Savannah River reprocessing plant in South Carolina releases plutonium sludge and 10 times as much radioactive iodine as the TMI meltdown.

Soviet Union (1957)—An unknown number of people are killed in an explosion at the secret Mayak nuclear reprocessing site in Chelyabinsk; 270,000 are evacuated from 217 nearby cities and villages. Hundreds of survivors died from cancers. The land remains uninhabitable.

United Kingdom (1957)—On October 10, a radiation release during a reactor fire at the Windscale facility contaminates 35 workers and sends a radioactive cloud over northern Europe.

USA (1959)—A partial meltdown at a sodium reactor contaminates the Simi Valley, north of Los Angeles, with the third-largest release of radioactive iodine in history. The accident was covered up for more than 45 years. (See the box, "Simi Valley," on page 62.)

USA (1961)—Three US Army technicians are killed in an SL-1 reactor explosion in Idaho. Probable cause of death: a "love-triangle" suicide. All three bodies are buried in lead-lined coffins in three different states and remain intensely radioactive.

USA (1966)—The reactor core melts when the cooling system fails at Detroit's Enrico Fermi 1 breeder reactor.

Scotland (1967)—Fuel element melts and catches fire at the Chapelcross reactor.

Switzerland (1969)—A coolant leak in an underground reactor in Lucens causes an explosion, a partial core meltdown, and severe radioactive contamination. The reactor is closed and the facility is permanently sealed.

USA (1972)— Two workers at the Surry reactor in Virginia are killed due to failure of faulty welds. The next year, two more workers die in an explosion while inspecting defective valves.

Russia (1974)—Three people are killed during an explosion and radiation leak at a nuclear power station in Leningrad.

USA (1975)—A worker using a candle to check for air leaks sets fire to Alabama's Browns Ferry plant, which burns for 7.5 hours with two General Electric reactors operating at full power. One reactor goes "dangerously out of control," resulting in a 20 percent core damage probability (CDP)—a near-meltdown. The plant remains shut for two decades.

Slovakia (1976)—An explosion at a Bohunice nuclear plant kills two workers.

USA (1978)—A steam generator failure at California's Rancho Seco plant results in a 10 percent CDP.

USA (1979)—Three Mile Island: thousands of Pennsylvanians are evacuated in the worst US commercial reactor accident.

France (1980)—Fuel bundles rupture at a Saint Laurent reactor, causing a radioactive leak.

United Kingdom (1981)—Radiation leaking from the Sellafield reprocessing plant (née Windscale) contaminates local dairy pastures. Leukemia rates soar to triple the national average.

USA (1981)—California's San Onofre plant is closed for 14 months for repairs to 6,000 leaking steam tubes. During the restart, the plant catches fire, knocking out one of two backup generators.

Argentina (1983)—An engineer is killed by radiation exposure following a criticality accident at a research reactor in Buenos Aires; 17 others are injured.

United Kingdom (1983)—Sellafield radioactive fallout contaminates the coastline and ocean.

United Kingdom (1984)—Sellafield radioactive fallout triggers a nine-month closure of local beaches. The plant operators face criminal charges.

Russia (1985)—A steam explosion at the Balakovo nuclear plant kills 14 workers.

Japan (1985)—A reactor at Fukushima complex catches fire during a routine shutdown.

Russia (1986)—Chernobyl: the world's worst nuclear accident to date. The number of dead remains unknown; estimates run from 300,000 to nearly 1 million.

Germany (1986)—A fuel accident releases fallout up to two kilometers from the Hamm-Ueutrop nuclear plant.

USA (1986)—Four workers are killed in an explosion at the Surry nuclear power plant.

USA (1989)—Eight employees are irradiated at the Savannah River reprocessing plant.

Germany (1989)—The Greifswald reactor narrowly avoids meltdown after "technical failure."

France (1990)—Two workers are exposed to radioactive contamination during refueling operations at Blayais plant.

France (1992)—An accident at the Dampier power plant irradiates two workers.

Japan (1993)—Two workers are injured and one killed in high-pressure steam accident at Fukushima reactor.

Russia (1993)—An explosion occurs at the Tomsk-7 nuclear complex.

Japan (1997)—An explosion at the Tokaimura plant leaves 35 workers exposed to high levels of radiation.

Japan (1999)—An uncontrolled nuclear reaction takes place at the Shika reactor. The incident is covered up by plant operators.

Japan (1999)—Two workers are killed at the Tokaimura plant during an unplanned chain reaction; 116 workers are exposed to radiation.

USA (1999)—An explosion rocks Oregon's Trojan nuclear plant as the zinc-carbon coating on spent fuel cylinders begins to release hydrogen gas while the cylinders are being moved into dry storage.

Germany (2001)—The emergency cooling system fails in the Philippsburg reactor.

Germany (2001)—A hydrogen explosion occurs in a Brunsbüttel boiling water reactor.

USA (2002)—Undetected corrosion on the pressurized chamber of the Davis-Besse reactor comes close to causing a massive radiation accident.

Hungary (2003)—Ruptured fuel elements at the Paks reactor cause a radioactive leak.

Japan (2004)—A steam explosion kills four at a Mihama reactor.

Bulgaria (2005)—A control rod jam in the Kosloduy-5 reactor prevents safe shutdown.

Sweden (2006)—A catastrophic core meltdown of the Forsmark reactor is barely averted after an external short circuit causes a failure of the emergency power system.

Germany (2007)—A transformer fire results in failure of the emergency power supply at the Krümmel reactor near Hamburg. Operators cover up failure of emergency systems and damage to reactor.

Japan (2007)—A transformer fire following an earthquake triggers a radioactive leak at the Kashiwazaki-Kariwa plant. The quake shuts down the 8,000 MW reactor in 90 seconds. Seven reactors are damaged by quake.

France (2008)—A major leak at the Tricastin nuclear facility in southeastern France spills 30 cubic meters of uranium-rich water onto the plant's grounds.

Russia (2009)—The Leningrad nuclear power plant is shut down after a crack is discovered in a pump, threatening a "potentially catastrophic technical malfunction."

France (2009)—A "significant" incident causes the evacuation of the reactor unit in the Gravelines nuclear plant in northeastern France—the fifth "level-1" emergency in three years.

Japan (2011)—Six coastal reactors at Fukushima are disabled by an earthquake and resulting tsunami. A cooling system failure leads to explosions, partial meltdowns, and massive releases of radiation.

France (2011)—A furnace explodes at the state-owned Centraco nuclear waste treatment facility, killing one worker and injuring four others.

Sweden (2011)—A severe fire erupts at the Ringhals Unit 2 reactor during a test of the containment vessel, causing $267 million in damages. The incident is triggered when a vacuum cleaner, left in the containment area, catches fire.

Additional significant accidents have occurred at the following US plants: Byron, Braidwood, Comanche Peak, Diablo Canyon, Grand

Gulf, Marble Hill, Midland, Nine Mile Point, Seabrook, Shoreham, South Texas, WPPSS, and Zimmer.

Sources: 1996 Greenpeace report, "Calendar of Nuclear Accidents and Events" (http://archive.greenpeace.org/comms/nukes/chernob/rep02.html), Greenpeace's 2006 report, "An American Chernobyl: 'Near Misses' at US Reactors Since 1986" (http://www.greenpeace.org/usa/Global/usa/report/2007/9/an-american-chernobyl-nuclear.pdf), and "List of Civilian Nuclear Accidents" (Wikipedia). On March 14, 2011, London's *Guardian* newspaper published a review of "Nuclear Power Plant Accidents: Listed and Ranked Since 1952." *The Guardian* noted that the International Atomic Energy Agency, "astonishingly, fails to keep a complete historical database."

5
Aging Reactors

Nuclear reactors, which only ran half-marathons during their youth, are being asked to run full marathons in their old age.

–AILEEN MIOKO SMITH, GREEN ACTION ACTIVIST

AS OF MARCH 2012, the owners of 71 tottering US reactors had applied to the NRC for 20-year operating extensions. All 71 applications were approved. If allowing reactors to operate 50 percent beyond their intended lifetime seems a risky proposition, consider this: The average lifetime of a modern automobile (for which improved designs are introduced on a yearly basis) is 12 years. In comparison, 79 percent of US reactors are more than 26 years old, and they are running on designs that haven't changed since the 1960s.

When it comes to age, the Oyster Creek BWR (50 miles due east of Philadelphia) and the two Indian Point PWRs (25 miles north of New York City) are nuclear Methuselahs. Philly's BWR (dubbed "Oyster Creak" by its critics) is the country's oldest operating reactor. Boiling water since 1969, the Oyster Creek reactor was due to be retired in 2009. Instead, the NRC extended its 40-year license by another 20 years. (Oyster Creek may actually be shut down in 2019 because its operator, the Exelon Corporation, has refused to pay the $750 million cost of building cooling towers ordered by state officials to protect sea life in nearby Barnegat Bay.)

The NRC appears to have a soft spot for the special needs of elderly reactors. In 1995, the nuclear industry successfully lobbied the NRC to change the rules for granting license renewals for aging reactors by narrowing the scope of the required safety and environmental review. (Think of it as a medical exam that shrugs off preexisting conditions.) Relying on an "aging facility plan" provided by the industry, the NRC lowered requirements for full inspections, seismic oversight, and review of emergency plans. As a member of the Union of Concerned Scientists put it, "If you don't look for problems, you can't find problems."

A 1995 Oak Ridge National Laboratory report concluded that, over the course of seven years, aging reactors were responsible for 19 percent of the "scenarios" that could have resulted in serious accidents. In 2001, the Union of Concerned Scientists noted that age-related incidents had forced eight reactor shutdowns in just 13 months. An Associated Press review of NRC reactor safety notifications from 2005 to 2011 found 26 alerts attributed to rust and clogged or leaking pipes, while "aging was a probable factor in 113 additional alerts." In 2008, one NRC survey reported "degraded conditions" were responsible for 70 percent of the "potentially serious safety problems" threatening US reactor safety.[1]

The Most Alarming Maladies of Aging Reactors

It doesn't take a major failure to damage a nuclear reactor. Even the slow, ongoing degradation that inevitably afflicts aging equipment poses unacceptable risks. Something as small as peeling paint—if sucked toward the pumps that move the coolant—could cause a serious accident. Here are seven major areas of concern identified by the AP investigation.

Brittle Vessels

Under the constant bombardment of radiation, steel reactor vessels can experience "embrittlement." In PWRs, this can cause cracking that could spill radioactive debris from the core into the outside environment.

Leaking Valves

BWR operators have a hard time abiding by the "leakage standard" intended to contain radioactive steam in the event of an accident. In 1999, because so many main steam isolation valves were leaking more than the allowed 11.5 cubic feet per hour, the NRC changed the rules to permit plants to leak up to 200 cubic feet per hour. But even this wasn't enough to keep Georgia's Hatch Unit 2 in compliance. In its review of NRC reactor safety notifications, the AP discovered a 2007 memo in which Hatch operators reported a leakage of 574 cubic feet per hour from Unit 2.

Cracked Tubes

The steel alloy used in the tubes of PWR steam generators has proven susceptible to cracking. The AP noted that this "rampant" cracking was alarming because the tubes contain radioactive coolant. A burst coolant pipe could discharge radioactive gases outside the containment building, while a corresponding loss of coolant could precipitate a core meltdown. In 1993, the AP found evidence of seven cracking incidents that had caused "outright ruptures," and workers at the Catawba plant near Charlotte, North Carolina, found that more than half of its steel alloy tubes—more than 8,000 pipes—had developed cracks. "There is no end in sight to the steam generator tube degeneration problems," an internal NRC memo confessed. "Crack depth is difficult to measure reliably and the crack growth rate is difficult to determine."

Corroded Pipes

Corrosion of pipes installed underground or in damp settings continues to be a growing problem for the industry. The explosion of a corroded pipe at Virginia's Surry 2 reactor in 1986 killed four workers who were engulfed in a blast of superheated steam. An "internal industry document" uncovered by AP reporters further revealed that the number of leaks from underground pipes had increased fivefold between the years 2000 and 2009. And in September 2010, the Union of Concerned Scientists reported evidence of 400 radioactive leaks linked to corroded pipes at US reactors.

Corroded Containment

In 2001, the NRC was confronted with 66 cases of unexpected damage to containment buildings surrounding reactor vessels. Severe corrosion was a factor in a quarter of the reports. The NRC found the critical steel containment liners at Virginia's North Anna and North Carolina's Brunswick sites had been penetrated by rust. In 2009, inspectors discovered a corrosion hole that had completely penetrated the liner at Pennsylvania's Beaver Valley Unit 1.

Compromised Cables

Faulty electrical cables have been a problem as far back as 1993. The AP discovered an "official use only" report from that year in which one NRC employee had expressed alarm that electrical systems were experiencing unexpected age-related breakdowns. Cables buried underground or strung through wet or hard-to-access areas were particularly at risk. In 1993, the NRC reported that nearly one-fifth of the electrical cables had failed during tests designed to simulate 40 years of use. In 2008, the NRC reported the problem had only grown worse.

As of 2008, the NRC had set the total number of "known" incidents linked to cable failure at 269. Failure of the electrical system could trigger additional problems, some of which could risk potentially catastrophic damage to a reactor core. On January 22, 2011, for example, an electrical cable failed at the Palisades reactor, a 39-year-old plant in Michigan. This, in turn, caused a blown fuse and a valve failure that allowed radioactive steam to spill into the air outside the plant.

Cracking Nozzles

In 2001, the NRC discovered caustic chemicals eating away at the metal nozzles on reactors. When the NRC got around to inspecting plants a year later, they discovered a football-size hole burned into the vessel head of Ohio's Davis-Besse unit. There was less than an inch of protective liner left. NRC commissioner Peter Bradford estimated that Ohio came within two months of a "near rupture" that would have coated the state in radioactive fallout. According to NRC reports, another 10 plants developed similar cracks during 2001–2003. An NRC spokesperson assured the public that the agency had "learned from the incident and improved resident inspector training . . . to ensure that such a situation is never repeated."[2]

Incredibly, on the same day that the NRC released that statement, workers at the Davis-Besse plant discovered that 24 of the 69 nozzles on a replacement vessel head also had been damaged—in this case, by corrosion. Despite the additional discovery of extensive cracking in the concrete shield surrounding the reactor, the NRC gave Davis-Besse the green light to restart of the reactor on December 6, 2011.

PALISADES: ONE OF THE OLDEST, ONE OF THE WORST

Although it was completed in 1967, the Palisades nuclear power plant didn't begin generating power until 1971. In 2011, the plant, which sits on the shores of Lake Michigan, marked its fortieth birthday as one of the country's 10 oldest reactors. Despite a long list of accidents, emergency shutdowns, and safety citations at the plant, the NRC has issued Palisades a license to operate for another 20 years, through 2031.

On February 29, 2012, the NRC convened a meeting in South Haven, Michigan, to reassure residents after a flurry of accidents had struck the plant in 2011. Instead of quieting local concerns, the NRC representatives were forced to confess that the plant contained the country's most brittle reactor vessel. The watchdog group Beyond Nuclear has accused the NRC of repeatedly lowering its "pressurized thermal shock" safety regulations "to accommodate Palisades [and] to enable this dangerously degraded reactor to operate."[3]

While most US reactors are rated "green" in the NRC's color-coded rating system, Palisades was downgraded to "white" in 2008, and as 2012 began, the NRC was considering a further downgrade to "yellow." (A "red" rating could require the plant to cease operations if improvements are not made.) The NRC has vowed to close the plant if its safety record is not improved.

The NRC's concerns were heightened on September 25, 2011, when an accidental loss of power caused an automatic shutdown at the plant. The shutdown unleashed a chain of events that, the plant manager confessed, "could have killed somebody." It started when a worker dropped a tool while trying to repair a burned-out bulb on an indicator button. This caused an electrical arc that disabled half the control room gauges and prompted the emergency shutdown.

"Fortunately, the Emergency Core Cooling System did not inject cooling water into the hot core," Beyond Nuclear observed. This could have "risked a fracture of the RPV [reactor pressure vessel] like a hot glass under cold water." This, in turn, could have caused a loss-of-coolant accident and a possible meltdown. Beyond Nuclear added that the accident "came precariously close to completely filling the pressurizer and one of the steam generators with water, which would have meant loss of control over core temperature and pressure, and could have broken pipes."[4]

The Palisades plant went on to suffer four other emergency shutdowns in 2011—due to leaks, pump failures, and malfunctioning valves—forcing NRC staff to put in a thousand hours of overtime to run extra inspections. Nuclear watchdogs blamed the owner, Entergy Nuclear, for the plant's malfunctions. When Entergy bought the plant from Consumers Energy in 2007, it promised to replace a host of major components the original owner identified as seriously degraded. These included a corroded reactor lid and a reactor vessel that the *Detroit Free Press* described as "possibly the most brittle in the country."[5] Entergy failed to make the recommended replacements. This came as no surprise to industry critics at Beyond Nuclear, who warned in February 2012 that Entergy was infamous for "buying reactors cheap, then running them into the ground."[6]

The previous month, NRC regional administrator Cynthia Pederson had also scolded Entergy officials. "Quite frankly, we find your performance troubling," Pederson said. "What we want to see is a change in performance." Pederson spelled out her criticism of plant management for a story in the *Detroit Free Press* citing "organizational failures, a plan for change that came only after performance had declined steeply, poor instructions for work that needed to be done, failing to follow procedures, poor supervision and oversight, poor maintenance and multiple events caused by human errors or equipment failures."[7]

Adding to Palisades' uncertain future, in January 2012, the NRC cited Entergy for a "legal violation" after a supervisor lost his temper, stormed out of the plant, and disappeared—leaving no one in charge of the main control room.

"If all these failings and accidents line up in just the right way," warned Kevin Kamps, a local resident and a Beyond Nuclear activist, "we could have a very bad day at Palisades. It's an accident waiting to happen."[8]

FUKUSHIMA-STYLE REACTORS IN THE UNITED STATES

(Names and Locations of General Electric Mark 1 Reactors)

Reactor	Location
Browns Ferry 1*	Decatur, Alabama
Browns Ferry 2*	Decatur, Alabama
Browns Ferry 3*	Decatur, Alabama
Brunswick 1*	Southport, North Carolina
Brunswick 2*	Southport, North Carolina
Cooper*	Nebraska City, Nebraska
Dresden 2*	Morris, Illinois
Dresden 3*	Morris, Illinois
Duane Arnold*	Cedar Rapids, Iowa
Fermi 2	Monroe, Michigan
Fitzpatrick*	Oswego, New York
Hatch 1*	Baxley, Georgia
Hatch 2*	Baxley, Georgia
Hope Creek**	Hancocks Bridge, New Jersey
Monticello*	Monticello, Maine
Nine Mile Point 1*	Oswego, New York
Oyster Creek*	Toms River, New Jersey
Peach Bottom 2*	Lancaster, Pennsylvania
Peach Bottom 3*	Lancaster, Pennsylvania
Pilgrim**	Plymouth, Massachusetts
Quad Cities 1*	Cordova, Illinois
Quad Cities 2*	Cordova, Illinois
Vermont Yankee 1*	Vernon, Vermont

*Reactor has received a 20-year license extension from the NRC.
**A 20-year license renewal extension for the reactor is under review by the NRC.

Source: Nuclear Information and Resource Service, June 2012. For more information, go to http://www.nirs.org/reactorwatch/accidents/gemk1reactorsinus.pdf.

6

Environmental Pollution: Water, Air, and Land

If it is true that fusion will put unlimited amounts of energy into our hands, then I'm worried. Our record on this score is extremely poor. It seems that every time mankind is given a lot of energy, we go out and wreck something with it.

–DAVID R. BROWER

NUCLEAR ENERGY is not the "clean" energy its backers proclaim. For more than 50 years, nuclear energy has been quietly polluting our air, land, water, and bodies. Every aspect of the nuclear fuel cycle—mining, milling, shipping, processing, power generation, waste disposal, and storage—releases greenhouse gases, radioactive particles, and toxic materials that poison the air, water, and land. Nuclear power plants routinely expel low-level radionuclides into the air in the course of daily operations. While exposure to high levels of radiation can kill within a matter of days or weeks, exposure to low levels on a prolonged basis can damage bones and tissue and result in genetic damage, crippling long-term injuries, disease, and death.

Water Pollution

The 104 US reactors operating in 40 of the 50 states routinely discharge used coolant water into the nation's major streams, the Great Lakes, the Gulf of Mexico, and the Atlantic and Pacific oceans. While much of a reactor's coolant water is released as steam (which heats the atmosphere), the remainder—heated up to 25°F over ambient water temperatures and sometimes tainted with radioactive isotopes—is discharged back into local waters, where it wreaks damage on river and ocean life. Thermal pollution of the Hudson River from the Indian Point plant kills more than two billion fish a year. Thermal discharges from the Salem Nuclear Generating Station (which swallows three billion gallons of Delaware Bay water every day), have been linked to a 31 percent decline in bay anchovy. California's two coastal plants at San Onofre and Diablo Canyon suck in nearly a million gallons of seawater every minute to use as free coolant. San Onofre's two reactors alone pour 2,400 million gallons of water (heated to 19 degrees Fahrenheit above ambient ocean temperatures) back into the Pacific every day.

When it comes to producing electricity, nuclear is an extravagantly water-wasting technology. A nuclear power station requires 20 to 83 percent more water than any other kind of power plant. Even Toshiba-Westinghouse's state-of-the-art, supposedly efficient AP1000 needs to consume 750,000 gallons per minute to operate safely.

Air Pollution

Whenever a nuclear accident spews an unusual amount of radiation into the local air, industry spokespeople invariably reassure the public that the release "poses no immediate harm." The critical word is "immediate." The fact is there is no "safe level" of nuclear exposure: all exposure to ionizing radiation is potentially harmful.[1] And the industry knows this.

An average 1,000 MW reactor contains approximately 16 billion curies of radioactive material—the equivalent of 10,000 Hiroshima bombs.[2] Tritium, krypton, xenon-135, iodine-131, and iodine-129 (with a radioactive half-life of 16 million years) are routinely vented into the air. Tritium, a radioactive form of hydrogen, is dangerous if inhaled or ingested. It can combine with oxygen to form tritiated water molecules that can be absorbed through pores in the skin, leading to cell damage and an increased chance of cancer. The government considers these tritium releases "permissible," so no attempt is made to monitor or regulate them, even though tritium exposure is known to damage human cells and cause cancer, leukemia, birth defects, immune system damage, and genetic mutations. And genetic damage imposed by radiation exposure is passed on from one generation to the next.

We will never know how many people have died as a result of fallout from atmospheric nuclear bomb tests, the Chernobyl explosion, the Fukushima meltdowns, and fallout from the Windscale reprocessing facility on the British coast. One Indian researcher has offered the astounding estimate that infant deaths in India attributable to fallout from the Nuclear Age—from 1945 to 1999—may have topped 22.6 million.[3]

Assessing the health risks from routine nuclear power operations and from major accidents is made more difficult by the fact that agencies assigned to monitor and assign radiation risks often have a bias toward the industry. The watchdog group Beyond Nuclear has charged that estimated doses of Chernobyl radiation published by the UN Scientific Committee on the Effects of Atomic Radiation (UNSCEAR) "grossly underestimated the true exposure."[4]

"Permissible Doses"

In 1934, the International Commission on Radioligical Protection (ICRP) established idea that there was a "permissible" dose of acceptable radiation. This standard has since been revised four times, prompting the British medical journal *The Lancet* to ponder whether the ICRP's "reluctance to recommend more protective standards was due to concerns about potential financial consequences to the nuclear industry." The ICRP has even lowered the perceived risks by discounting some radiation-linked tumors as "partial cancers" that need not be counted.[5]

In testimony before the National Academy of Sciences' Nuclear and Radiation Studies Board on February 10, 2010, Beyond Nuclear cited a scientific paper that concluded that the "NRC allows public exposures to radiation at risk levels one hundred to ten thousand times higher than the federal government permits for any other carcinogen. EPA has previously opposed NRC's radiation standards for this reason, asking why radiation should be treated as a 'privileged pollutant,' permitted to expose people to cancer risks at levels far above that allowed for any other pollutant."[6]

Beyond Nuclear has asked the NRC to "justify to members of the public why ionizing radiation from nuclear facilities is conferred this special status."[7] Meanwhile, the NRC has found another way to downplay radiation risks—by increasing the estimate of "background radiation." The NRC has upped its assessment of "normal" radiation several times, most recently from 360 millirem to 620 millirem.

After the Fermi 2 nuclear plant began operating in Michigan's Monroe County in 1988, the cancer rate for people under the age of 25 living near the plant rose to more than triple the state average, and radioactive iodine has been found in the milk of cows grazing downwind from the plant. Nevertheless, the plant's operator, DTE Energy, has applied to build a new reactor. While neither the plant operator nor the NRC is required to monitor cancer rates around nuclear reactors,[8] Michigan residents living within 10 miles of a nuclear power plant are supposed to draw some comfort from that fact that they are now eligible to receive government-issued potassium iodide (KI) anti-radiation pills to be taken "in the event of an accident."[9] (The NRC requires only that plant operators "consider" providing KI to nearby residents in the event of an accident. Only 21 of the 34 states with large populations living within 10 miles of a reactor currently have KI distribution plans.)

Chernobyl's Global Toll

The Chernobyl explosion and fire released a globe-girdling cloud of radiation that the US Lawrence Livermore National Laboratory estimates to have exceeded 4.5 billion curies. Other estimates range as high as 9 billion curies. While the International Atomic Energy Agency claims that only 56 people perished as a direct result of Chernobyl's fallout, a 2010 report by the Belarus National Academy of Sciences attributed an estimated 93,000 deaths and 270,000 cancers to fallout from Chernobyl. Estimates from the Ukrainian National Commission for Radiation Protection set the death toll at 500,000. Chernobyl survivors—and their children—continue to suffer from high rates of leukemia and thyroid cancer.[10] Studies published in *Radiation and Environmental Biophysics* (1997)[11] and the *European Journal of Cancer Care* (2007)[12] found a significant correlation between cesium exposure and perinatal mortality in "downwind" countries seven months after the Chernobyl disaster.

Some 24 years after the disaster, the journal *Ecological Indicators* reported on the largest study of wildlife in the Chernobyl "exclusion zone," which revealed that the damage of low-level radiation was not limited to vegetation and insects. Biologists found "overwhelming" damage to reptiles, amphibians, mammals, and birds, including barn swallows showing tumors on feet, on necks, and around the eyes.[13]

In 2009, Germany's Environment Ministry was forced to pay $550,000 to compensate hunters for wild boars that were deemed too radioactive to consume. The boars had dined on mushrooms contaminated with cesium-137 from Chernobyl. In Germany, the "safe" level for consumption is 600 becquerel per kilogram, but these boars averaged 7,000 becquerels per kilogram.[14]

Nuclear Plants, Cancer, and Leukemia

Around the world, nuclear workers, their families, and people living near nuclear facilities suffer elevated risks of cancer and leukemia. Since Entergy's Vermont Yankee nuclear plant opened in 1972, the rate of cancer deaths in Windham County has risen 5.7 percent above the national average.[15] A 1990 study found that residents of Harrisburg, Pennsylvania, the site of the Three Mile Island accident, suffered from an increased incidence of cancers.[16]

In a detailed 1996 investigation, radiation researcher Jay M. Gould, PhD, illustrated the cancer-reactor link by poring over 50 years' worth of studies by the National Cancer Institute and the Centers for Disease Control. Gould analyzed health statistics for of the country's 3,000-odd counties, nearly half of which were identified as "nuclear counties" because they were located within 100 miles of a reactor site. Gould's research concluded that these counties had "more than two-thirds of all breast cancer deaths in 1985–89 and a combined age-adjusted breast cancer mortality rate . . . that is significantly higher than that of all remaining counties."[17] Gould demonstrated how government studies minimized the apparent impact of nuclear exposures by assuming radiation emissions never moved beyond the county boundaries. Gould also pointed out how averaging cancer rates within a 50- or 100-mile circumference could be deceiving since major cancer increases tend to be isolated in the smaller, concentrated areas found downwind from the reactors. Though there is as yet no proven causal link, that could change once the US National Academy of Sciences releases an update of the National Cancer Institute's 1990 survey of diseases and deaths that have occurred near nuclear plants. An initial scoping study was begun in September 2010 and five "Phase 1" meetings were set for 2012. At this rate, publication of a new update is, most likely, several years away.

Children are especially vulnerable to harm from radiation, and around the world, children living near nuclear power plants are suffering from leukemia. In the United States, epidemiologists found that childhood leukemia deaths near newer nuclear power plants were 9.4 percent above the US average; they were even higher near older plants, coming in at 13.9 percent above the US average.[18] European studies have found similar increases of leukemia in children living near nuclear power plants. The GeoCap Study, a six-year study published 2012, found that French children living within five kilometers of nuclear plants were twice as likely to suffer from childhood leukemia.[19] A 2008 study of children living within five kilometers of German reactors reported that they were 2.19 times more likely to develop childhood leukemia.[20]

George W's Farewell Gift

In addition to offering billions in taxpayer bailouts to the nuclear industry during his presidency, George W. Bush handed one last gift to Big Nuke. On January 15, 2009, in one of the last acts of the Bush administration, exiting EPA administrator Marcus Peacock radically rewrote the 1992 Protective Action Guides (PAGs). The PAGs are the federal guidelines that regulate US radiation exposure safeguards. Ignoring the criticisms of more than 60 environmental and public health organizations, the Bush administration's PAGs rewrite dramatically increased the permissible airborne radiation exposure for Americans.

Under the new rules, permissible levels of strontium-90 fallout in drinking water were raised 1,000-fold, while allowable exposures to iodine-131 were raised 100,000-fold. Some new standards were "seven million times more lax than permitted under the Safe Drinking Water Act."[21]

Ignoring the National Academy of Sciences' warnings that cancer risks had increased significantly since publication of the 1992 PAGs, Bush's EPA relaxed radiation exposure limits across the board. The new PAGs also reduced standards for cleaning contaminated sites by offering token "benchmark" goals that (according to the government's own estimates) would expose one in four people to cancer risk. Bush's PAGs also permitted cleanup goals to be curtailed to accommodate "economic considerations."

The new rules were intentionally submitted late to delay publication in the *Federal Register* until after President Obama's inauguration. Even then, two-thirds of the new standards remained unpublished, assuring that they would not be subject to public review. It was a remarkable parting gift from a Republican president to the nuclear industry, the DOE, and the NRC.

In fairness, Bush's move to raise the "safe" level of radiation exposure was not without precedent. In 1964, Democratic president Lyndon Johnson ordered a 20-fold increase in the permitted exposure to iodine-131 and strontium-90 in milk.

Land Pollution

By 1978, the US "uranium rush" had left 140 million tons of crushed-rock tailings at 16 operating mills and 22 abandoned sites, with additional wastes piling up at an average of 6 to 10 tons a year. The 1.7-million-ton tailings pile at Shiprock, New Mexico, covers 72 acres. All tailings piles release radon gas and long-lived radioactive isotopes into the air, rivers, arroyos, and aquifers. Radon gas (believed responsible for a fivefold increase in lung cancer among uranium miners) continues to poison the winds blowing over abandoned piles of mining wastes that lie scattered around the world.

In 1979, 94 million gallons of contaminated liquid tailings burst from a containment dam in New Mexico, sweeping 1,100 tons of radioactive wastes into the Rio Puerco, which flows into the Little Colorado River and on to Lake Mead, a major source of drinking water for Las Vegas and Los Angeles.[22] In 1984, a flash flood flushed four tons of tailings into a tributary of the Colorado River, which provides irrigation for farms and drinking water for cities in Nevada and Southern California. Less dramatic but also deadly is the imperceptibly slow, toxic seepage from tailing ponds that has steadily poisoned critical subsurface aquifers across the Colorado Plateau.[23]

The devastation to portions of America's landscape has been so vast and long-lasting that the government has no hope of ever repairing the damage. Instead, it has created a term to describe these irreparably damaged, nuclear no-man's-lands: they are called "National Sacrifice Zones." (The term was reportedly coined by Energy Department engineers to describe abandoned nuclear facilities and Superfund sites.[24])

Despite the environmental and health damages wrought by uranium mining, there have never been any binding standards requiring mine operators to minimize harm to the local land or people. The World Nuclear Association (a trade body representing 90 percent of the industry) is considering a "charter of ethics," but it would be voluntary and self-policed. At best, some local activist communities have been able to demand a higher price for the ore extracted from their damaged lands. For example, in 2008, in a rare victory, the people of Niger forced the French firm AREVA to increase the price of a kilogram of uranium.

Back in the United States, environmentalists and preservationists had cause to celebrate when the Interior Department secretary Ken Salazar announced a 20-year ban on hard-rock mining and uranium extraction on nearly a million federal acres surrounding Arizona's Grand Canyon National Park. But shortly after Salazar's January 9, 2012, announcement, the Nuclear Energy Institute and the National Mining Association filed suit against the government, demanding a suspension of the ban in order to permit industry to open as many as 30 new uranium mines near a popular tourist destination visited by 4.5 million people a year. Actions like this can give rise to the conundrum: "What is the half-life of corporate greed?"

SIMI VALLEY: THE COVER-UP OF AMERICA'S "WORST NUCLEAR ACCIDENT"

In the days of the Cold War, parents in Southern California would sometimes roust their children in the dark hours before dawn to scan the eastern mountains for the brief, unearthly flash that accompanied aboveground tests of atomic bombs at the Nevada Test Site. In the San Fernando Valley, north of Los Angeles, Cold War kids also gathered in the evening to wait for flickers of flame that jetted into the night skies above the Simi Valley hills. The red flares accompanied test-firings of the prototypes of rocket engines that would eventually carry John Glenn into orbit and send men to the moon.

It never occurred to those awestruck youngsters that the flares in the north were linked to the flashes in the east. Residents of the San Fernando Valley knew that the Rocketdyne testing site (a.k.a. the Santa Susana Field Laboratory) also housed a 20 MW sodium reactor, but they had no cause for concern: the reactor was just another exotic manifestation of Progress—and the lab's operators assured everyone that the reactor was perfectly safe.

It took many decades of disease and suffering before a lawsuit finally revealed the truth: in 1959, the sodium reactor suffered a partial meltdown, showering the downwind hills and meadows of the 2,850-acre site with a fog of chromium and radioactive isotopes, including iodine-131. Fallout from the meltdown contained an estimated 1,300 curies of iodine-131—80 to 100 times the amount of iodine-131 released at Three Mile Island. "That would make it the third largest release of iodine-131 from a reactor accident in the history of nuclear power," Dr. Arjun Makhijani (an expert witness in a 2005 class-action lawsuit against Boeing) explained during a 2006 radio interview. "First there was Chernobyl, then Windscale in England in 1957, and the third worst would be this sodium reactor experiment in Simi Valley."[25]

The cover-up went all the way to the top. Five weeks after the accident, the Atomic Energy Commission published a report that insisted no radiation had been released. Other accidents at the reactor were ignored or minimized, including a second serious accident, in 1964, in which 80 percent of the cladding on the reactor's fuel rods evaporated in a meltdown. The plant was then permanently shut and decommissioned in 1965.

It was local doctors who managed to unravel the mystery. Four decades later, a doctor began to wonder why many older residents were starting to drop in with complaints of thyroid problems and unusual cancers. A check with other doctors revealed a similar pattern—a pandemic of atypical cancers was ravaging the valley. A UCLA investigation determined that the death rate in the area exposed to fallout from the 1959 meltdown was six to eight times higher than normal.

In September 2005, Boeing (Rocketdyne's parent company) paid $30 million to compensate local residents for early mortalities and a range of rare diseases. But Boeing refused to offer a settlement to any of its former workers. "They're still denying it," Bonnie Klea told an interviewer with Public Radio International's *Living on Earth*. Klea, who worked at the Rocketdyne site for 11 years, now suffers from a rare form of bladder cancer. "They're still lying to employees, telling them that their jobs didn't give them cancer when we're all sick and some of us are dead."[26]

Boeing conducted its own studies to show the workers had not been harmed, but the company's credibility was undercut by its refusal to release details of urine tests conducted on the workforce. Boeing continues to claim that "there is no evidence that working conditions caused increased mortality in the Rocketdyne workforce." Boeing also insists that the 1959 incident was not a "meltdown," contending that "releases were contained and controlled in accordance with regulatory guidelines."[27] Of course, if the problem had actually been "contained and controlled," there would have been no "releases."

In the 1980s, the US Department of Energy (DOE) claimed that the long-hidden fallout zone in the Simi Valley had been successfully decontaminated and was now available for development. Because there was money to be made housing a growing suburban population, bungalows and schoolyards soon started springing up at the scene of the country's worst radiological accident. But in February 2012, the EPA returned with some bad news for the valley's new residents: more than 52 years after the long-forgotten meltdown, soil tests had revealed that the ground was still contaminated with cesium that was radiating at levels 9,328 times greater than background levels and nearly 1,000 times the EPA's threshold for safety.[28]

Dan Hirsch, a local resident and president of the Committee to Bridge the Gap (the group that exposed the meltdown cover-up), was shocked to discover the valley was "still contaminated, half a century after the event and after two prior cleanups." While the surface contamination was bad, digging into the soil turned up even higher levels of radiation. Strontium-90 radiating at levels 71 times the background level was found two feet below the surface. In nearby Runkle Canyon (an area that is scheduled to become the site of a major housing development), strontium-90 levels were recorded at 284 times the background radiation levels. Soil samples also detected dangerous levels of carbon-14, cobalt-60, and neptunium-239, a signature decay product of plutonium-239.[29]

In December 2010, the DOE signed an agreement to clean up and decontaminate the site. Millions of dollars later, it is clear that effort has failed. It is beginning to look as though the Simi Valley may never be decontaminated. Instead, it could become a "National Sacrifice Zone," sealed off from human habitation. Meanwhile, the radiating remains of the country's worst nuclear accident continue to pose a risk to the millions of Californians who live along the headwaters of the Los Angeles River.

7

Damage to Indigenous Peoples

Nuclear power is unacceptable because it unavoidably inflicts cancer and genetic injury on people. It is mass premeditated murder.

–JOHN GOFMAN, ATOMIC ENERGY COMMISSION DIRECTOR OF RADIATION AND HEALTH RESEARCH AND FOUNDER OF THE COMMITTEE FOR NUCLEAR RESPONSIBILITY

BRINGING POWER to the cities of the industrialized world has often brought destruction to the lands of indigenous cultures. As of 2012, about 64 percent of the world's uranium came from mines in Kazakhstan and Canada, followed by Australia, Namibia, Niger, and 20 other countries, and 85 percent of that uranium was marketed by just eight companies.[1] Seventy percent of this uranium is found on lands held sacred by indigenous peoples.[2]

Navajo, Acoma, and Laguna miners harvested the yellow dirt of the Colorado Plateau to extract the uranium that powered the Cold War, often working in unventilated shafts 2,700 feet underground and exposed to radiation levels 750 times above even the lax safety standards of the 1950s. These native miners, mill workers, truckers, and their families suffered the multiple consequences of "red lung" disease—lung cancer, pulmonary fibrosis, kidney damage, and birth defects. From the early 1970s to the late 1990s, cancer rates among the Navajo doubled. As Judy Pasternak recalls in her poignant book, *Yellow Dirt: An American Story of a Poisoned Land and a People Betrayed*, "The symptoms became familiar. First, a lump on the neck. Then shortness of breath. Then, spitting up blood." Between 1969 and 1993, Navajo miners who had toiled to pull ore from sacred lands in America's Southwest suffered from lung cancer rates 28 times greater than Navajo men who never worked in the mines.[3] In a 1987 report prepared for the National Institute for Occupational Safety and Health, Richard Hornung reported that Navajo miners were five times more likely to develop lung cancer than the general population.

Dumping on Native Lands

Native lands have provided both the raw uranium ore and, more often than not, the final resting place of hazardous radioactive wastes. Producing a ton of uranium fuel can leave behind 20,000 tons of waste rock and more than 4,000 tons of toxic tailings salted with elements that remain deadly for hundreds of thousands of years. Tailings can contain as much as 85 percent of the ore's original radioactivity, and the rubble routinely releases alpha particles from thorium-230, lead-210, polonium-210, and radium-226. In a strong breeze, the radon gas released by tailing piles can travel 1,000 miles in a day.[4]

Now, having exploited native lands and poisoned native lives, the nuclear waste-makers want to turn native lands into dumps for nuclear waste. Because tribal lands are considered "sovereign" territory, federal environmental and health standards need not be strictly enforced. In addition, the promise of million-dollar contracts can be attractive to Native American families that are twice as likely as the average American to be living beneath the national poverty level. Originally, 17 of 20 potential sites identified for federal interim storage of high-level waste were located on Native American tribal holdings. Among those sites initially targeted for "repositories" were lands of the Mescalero Apaches in New Mexico and the Skull Valley Band of Goshutes in Utah. Both were proposed as "National Sacrifice Zones" suitable for fulfilling the government's need for monitored retrievable storage facilities for radioactive waste. With the Obama Administrations decision to close the Yucca Mountain storage facility (which occupies land held sacred by the Western Shoshone), there is increased pressure to store wastes on other native lands. Native activists argue that these dump sites violate promises contained in the 1863 Treaty of Ruby Valley.[5] Groups like Honor the Earth (founded by author and activist Winona LaDuke) are working to convince tribal leaders that no amount of money is worth the lingering health risks of harboring nuclear waste. In October 2011, LaDuke's organization succeeded in convincing the Goshutes to reject a lucrative waste-site contract. Meanwhile, Canada continues with plans to store its high-level waste on Cree and Dene lands in Saskatchewan.

The EPA has established radiation emission limits for waste sites in the United States and monitors those sites, but in most less-developed countries, tailings are simply dumped and abandoned.[6] Uranium tailings dumped in the deserts of Namibia and Niger, for example, poison the food, water, and grazing lands used by nomadic and desert tribes, such as the Tuareg of Mali and Niger. When the desert-dwelling Tuaregs' complaints became too militant, the Niger and Mali governments responded by labeling them "terrorists."[7]

Canada dominates Africa's uranium trade, with operations in 35 countries, including Namibia, Africa's top uranium producer. In Namibia, open-cast mining processes have irrevocably destroyed vast stretches of the desert, and future mining operations threaten to devastate 80 percent of the desert homeland of the Nama and Topnaar peoples. Canada's Xemplar Energy Corporation holds an exclusive license to explore for uranium in Namibia's ecologically fragile Namib Naukluft Park, home to rare plant and animal life. So while the Topnaar have been told they are no longer permitted to hunt for food or harvest melons in the park, Canadian geologists are free to scavenge the park for uranium ore.

Broken Promises in Australia

In Australia, the Commonwealth Radioactive Waste Management Act of 2005 allowed the government to dump nuclear waste on Aboriginal land without consultation or consent. In 2012, Australia's Labor government finally made good on its overdue campaign promise to repeal this act. The victory was diminished, however, by the discovery that the government still plans to expand the Muckaty nuclear dump located on traditional Aboriginal Northern Territory lands—and to enlarge uranium mines in South Australia and the Northern Territory. At the same time, the West Australian and Queensland governments continue to call for a ban on uranium mining.

The Olympic Dam mine at Roxby Downs is largely exempt from Australian law, Aboriginal land rights legislation, environmental protection, and freedom-of-information requirements. The owner, BHP Billiton, pays nothing for the 33-35 million liters of water it extracts from the Great Artesian Basin every day, though it has destroyed the ancient mound springs on native Arabunna land.[8] Pollution from the Ranger mine has poisoned the wetlands in Kakadu, home to the Bininj/Mungguy people. Australia's environment minister Peter Garrett has approved expansion of the Beverly mine, despite the fact that the facility has polluted groundwater on traditional Adnyamathanha land.[9]

Significantly, when the United Nations voted 144 to 4 in favor of the UN Declaration on the Rights of Indigenous Peoples, Australia was one of only four countries to cast a negative vote. The others were Canada, New Zealand, and the United States.

Remining the Southwest

In the deserts of the southwestern United States, the uranium miners' legacy lingers. One study prepared by the Navajo AML Reclamation Program[10] and presented to the May 1, 2003 annual meeting of the Navajo Abandoned Mines Collaboration, estimated that 1,200 untended uranium mines remained on Navajo lands, with nearly half requiring millions of dollars' worth of environmental restoration work.

While abandoned mines continue to haunt the Southwest, talk of a supposed "nuclear renaissance" now poses another threat—the specter of renewed mineral extraction triggered by a modern land rush to open new uranium mines. On December 5, 2008, in one of its final acts, the Bush Administration overturned a Congressional resolution banning uranium mining on one million uranium-rich acres of federal land surrounding the Grand Canyon. In response to the ruling, foreign mining firms immediately besieged the Bureau of Land Management and filed 1,100 claims to mine for uranium within five miles of the Grand Canyon National Park. The foreign interests that stand to benefit from Bush's untaxed and essentially free "job creating" mining program include firms from Canada, France, South Korea, and Russia's state-owned atomic energy corporation, Rosatom.

In addition to undermining national sovereignty, Bush's controversial mining permits pose a direct threat to the park's land, water, and air. The Colorado River also would be at risk from radioactive pollutants and toxic releases of mercury, arsenic, cyanide, and selenium—for decades, if not longer. You don't have to look far for an example of the lasting damage such mines can cause. One Grand Canyon uranium mine that was closed in 1969 is still producing radiation levels 450 times normal background levels, creating a radioactive "no-go zone" inside the park.[11]

In 2009, incoming Interior Secretary Ken Salazar ordered a two-year moratorium on new mining claims near the park and, on January 9, 2012, President Obama announced a 20-year ban on any new claims. Unfortunately, this action came too late to have any effect on the 3,000 mining claims that were ultimately filed as a result of Bush's parting gift to the uramium mining industry.

It is somewhat ironic that the United States, whose energy companies have plundered so many foreign lands in pursuit of nuclear ore, is now set to find itself on the receiving end of a phalanx of foreign-owned mining operations. But, sadly, some old realities will remain unchanged. The new mines will continue to rip open the lands of native communities, exposing new generations to the deadly dust that radiates from the extraction of atomic ore. The region's long-suffering native peoples, along with their sacred lands, deserve better—not just in the United States, but around the globe. It is time for the United States, along with Australia, Canada, New Zealand, and other uranium-addicted nations to join the rest of the world community in ratifying—and respecting—the UN Declaration on the Rights of Indigenous Peoples.

8

Earthquake Risks

Nuclear reactors in the central and eastern US face previously unrecognized threats from big earthquakes, the Nuclear Regulatory Commission said Tuesday. Experts said upgrading the plants to withstand more substantial earth movements would be costly and could force some to close.

–*THE WALL STREET JOURNAL*, FEBRUARY 1, 2012

MODERN ENGINEERING sometimes commits a critical error: assuming a fundamental, unchanging stability that planet Earth does not, in fact, offer. Without the blinders of hubris, cities would never have been built in Kobe, Mexico City, Christchurch, or San Francisco. It's bad enough that people build cities in historic floodplains (only to suffer monumental losses when a "200-year flood" turns homes to flotsam), but what kind of denial explains building great cities atop primordial slabs of continental rock that have a historical record of snapping with the seismic force of a buried nuclear blast? Around the world, in one location after another, the human community has bet its future on the unknowable whims of fault lines.

There are basic rules involved in mastering long-term survival on a living planet, and one of those ground rules is: "The ground rules." You can't take the dirt beneath your feet for granted—especially if you live in earthquake country. The International Atomic Energy Agency (IAEA) estimates that 20 percent of the world's reactors are currently operating in regions of known seismic activity. A growing concern with "beyond design basis" accidents prompted the IAEA to create the International Seismic Safety Centre in 2008. (A "beyond design basis event" refers to any incident that genereates greater stress than a nuclear plant was designed to withstand.)

Japan's Fukushima facility was supposedly designed to withstand the greatest credible seismic threat. Suddenly, in the aftermath of the 2011 quake and tsunami, TEPCO officials faced problems "beyond the design capacity" of the plant. Japan's suffering has caused understandable alarm in the United States, where 23 Fukushima-style General Electric Mark 1 reactors sit at 16 sites in Alabama, Georgia, Illinois, Iowa, Massachusetts, Minnesota, Missouri, New Jersey, New York, North Carolina, Pennsylvania, and Vermont. (See "Fukushima-Style Reactors in the United States" on page 55.)

In the United States, 11 reactors are located at five different sites on the Pacific and Gulf coasts, situated near active earthquake faults. These include California's Diablo Canyon and San Onofre, Louisiana's Waterford station, North Carolina's Brunswick plant, and the South Texas Project. These are among the first reactor sites that should be targeted for decommissioning, given that seismic activity on our planet is chronic and increasing. For example, a study of more than 386,000 earthquakes between 1973 and 2007 shows seismic activity increasing fivefold over a 20-year span. According to Tom Chalko, the Australian scientist who conducted the survey, "The most serious environmental problem we face . . . [is] rapidly and systematically increasing seismic, tectonic, and volcanic activity."[1]

The NRC Rethinks Earthquake Hazards

In the aftermath of the Fukushima quake and meltdowns, the NRC promised to take a harder look at seismic risks in the United States. Some industry critics dismissed this promise as little more than a public relations exercise. The skepticism was born out by the release of internal NRC emails obtained under a Freedom of Information Act request. The documents revealed that, at the same time NRC officials were assuring the public that a Fukushima-style disaster was unlikely in the United States, the commission's own staff had privately expressed concern that scores of US reactors were vulnerable to a potentially severe seismic calamity. "[We] need to get a handle on [these] external events' hazards," one NRC staffer emailed. In another internal memo, Brian Sheron, head of the NRC's Office of Nuclear Regulatory Research, offered a chilling rhetorical question: "Isn't there a prediction that . . . the West Coast is likely to get hit with some huge earthquake in the next 30 years or so? Yet we re-license their plants."[2]

NRC staff first suggested updating seismic risks in 2005. In September 2010, five years later, the NRC finally filed its Safety/Risk Assessment report. The conclusions were unsettling. With only four exceptions, every US reactor was found to be at a greater risk of seismic damage than had been assumed, with the median risk found to be *triple* previous estimates. Nonetheless, the NRC concluded that, despite the new findings, "operating nuclear power plants remain safe with no need for immediate action."

After the Fukushima quake left a collection of Japanese reactors in ruins, the NRC decided it was time to shift from studies to regulation and to identify those US reactors in need of "backfitting" (that is, in need of critical retrofits to safeguard against quake damage). The chore was complicated by the fact that the NRC previously had required "plant fragility" data for only one-third of the country's reactors—those falsely believed to be at greatest risk. But instead of taking an aggressive lead in the project, the NRC simply sent a request to plant operators inviting them to submit details on soil conditions and seismic safety. NRC staff scientists would then estimate earthquake risks based on the information supplied by the reactor operators. With uncanny timing, the NRC announced the release of its new post-Fukushima seismic model on January 31, 2012, the same day a 3.2-magnitude quake struck the twin reactors at Virginia's North Anna site. (This was not the first quake to hit the complex: a 5.8 quake had damaged the facility on August 23, 2011.)

The NRC's $7 million Central and Eastern United States Seismic Source Characterization for Nuclear Facilities (CEUS) model[3] took an overdue look at an "expanded data set" that studied major ground-motion events as far back as 1568. This CEUS review revealed that not enough attention had been focused on the risk of earthquakes striking in the central and eastern regions of the United States. The examination of historic tremors revealed that the most vulnerable US site is neither Diablo Canyon nor San Onofre —both famously perched near faults along the windswept California coast. Instead, America's most vulnerable reactor turns out to be Indian Point Unit 3, which sits on the East Coast, just 24 miles north of New York City. The second most at-risk reactor is Pilgrim Unit 1 in Plymouth, Pennsylvania. In third place are the twin reactors in Limerick, Pennsylvania.

Despite its expressed concerns to implement "lessons learned from events at the Fukushima Daiichi nuclear power plant," the NRC's engagement remained strictly advisory. As the commission went on to explain in a press release: "The NRC is *requesting* US nuclear power plants to re-evaluate seismic hazards using this information as well as other guidance." [Emphasis added.][4]

Quake Risks—From the Mississippi to Dixie

While the reactors sitting on the seismically unstable West Coast were all built with earthquakes in mind, the rest of the country was not considered seismically active (at least by nuclear power advocates; geologists knew otherwise). A quick glance at the historical record shows that the 16 most powerful temblors to hit the continental United States struck in the Midwest—15 in Missouri and 1 in South Carolina. A mere 200 years ago, the Midwest was ground zero for two of the strongest jolts ever to the hammer the North American landmass. In the winter of 1811, and again in 1812, temblors from the New Madrid Fault erupted beneath what are now Arkansas, Kentucky, Illinois, Missouri, and South Carolina. The 7.2- to 8.0-magnitude quakes were so powerful that they caused the Mississippi River to flow backward and set church bells ringing in the steeples of Boston. Another huge quake struck South Carolina in 1886, crumbling more than 14,000 brick chimneys in Charleston. The Great Charleston Quake, as it came to be known, was the deadliest earthquake ever to hit the East Coast. It jostled 30 states and was felt all the way to the New York–Canadian border.

Today, the largest concentration of US reactors near a major earthquake zone can be found in the Southeast, where nine nuclear facilities (containing a total of 17 reactors) lie within a 300-mile radius of Charleston, a city rebuilt atop the debris of the Great Charleston Quake of 1886. Three of these plants (Oconee, Robinson, and Summer, with a total of five reactors among them) can be found on the NRC's list of power plants at the greatest risk of suffering core damage during an earthquake.

The extensive geological evidence of midcontinental megaquakes (which is known to go back at least 4,500 years) was overlooked during the headlong rush into the Atomic Age. As a result, 96 aging reactors that now populate the United States east of the Rockies were built, quite literally, on a "faulty premise." Far from being "quake-free," the rock beneath the eastern United States is notably dense and expansive, providing a geologic rigidity that amplifies temblors and can broadcast shock waves across hundreds—even thousands—of miles.

The 14 reactors that appear to be most at risk from a reprise of the New Madrid quake include those located at Arkansas Nuclear, Callaway, Browns Ferry, Sequoia, Watts Bar, Oconee, Virgil C. Summer, Catawba, and H. B. Robinson.

(See "The 27 US Reactors Most at Risk from Earthquakes," page 79.)

The Specter of Inland Quakes

On March 19, 2011, in the aftermath of the Fukushima calamity, the NRC issued a special public report based on its seismic reassessments of US reactor risks. The report "NRC Frequently Asked Questions Related to the March 11, 2011 Japanese Earthquake and Tsunami," identified 22 Midwest plants that posed a "potential hazard," while another 27 plants were deemed worthy of further study.[5] Again, the NRC concluded there was "no need for immediate action" and merely recommended additional studies to determine whether some plants should receive safety upgrades—at some point in the future. In an attempt to quiet public's concern over the images being beamed in from Japan's coast, the report concluded: "It is important not to extrapolate earthquake and tsunami data from one location of the world to another when evaluating these natural hazards."

We now know that the 92 percent of US reactors built inland—far from the coasts and the threat of hurricanes and tsunamis—are increasingly vulnerable to natural disaster. Thanks to modern satellite mapping and advances in the science of geology, the US Geological Survey (USGS) now warns that inland quakes are expected to be more common and stronger than existing nuclear reactors were built to handle. A 2005 NRC investigation warned that older US plants might not survive a major quake. Following the Fukushima disaster, the NRC recommended that all US reactors be reexamined for quake risks and revamped to meet heightened safety standards.[6]

In September 2011, an Associated Press review of government data concluded that the risk of an earthquake triggering a serious nuclear accident was "greater than previously thought—24 times as high in one case." (That would be the Perry 1 reactor in Ohio. Also at a much higher risk than originally thought are River Bend 1, Dresden 2 and 3, Farley 1 and 2, and Wolf Creek 1.) In light of this new understanding of quake risks (and in the aftermath of the August 23, 2011 temblor that struck the North Anna facility), the NRC called for safety modifications at a quarter of the country's nuclear facilities.[7]

US Reactor Damaged by Massive Quake

The problem was underscored on August 23, 2011, when the two 980 MW reactors at Virginia's North Anna were tripped off-line by the largest quake to hit the region in 117 years. The plant's operator, Dominion Power, assured the public that "no release of radioactive material has occurred *beyond those associated with normal station operations*" (emphasis added).[8] This marked the first time a US reactor was shut down because of an earthquake. (The NRC had identified North Anna as the seventh likeliest reactor to be at risk from a quake.)

The epicenter of the 5.8-magnitude quake, only 12 miles from the plant, released powerful shock waves that reverberated from North Carolina to Toronto. In Washington, D.C., the quake damaged the Capitol Building, closed Union Station, and left cracks in the Washington Monument. Although it lasted only 3.1 seconds, the quake caused vibrations that were double the force the reactors were designed to handle. The shaking was so intense that 25 of the plant's 27 steel spent fuel storage containers—each standing 16 feet tall and weighing 115 tons—were knocked as much as 4.5 inches out of position. The jolt broke the concrete face on several of the plant's 50-ton, 16-foot-long horizontal storage bunkers—the first time in history that nuclear storage casks had been damaged by ground movement.[9]

Some geologists believe the quake may have been triggered by hydraulic fracturing ("fracking") operations.[10] As far back as 1966, both the US Army and the US Geological Survey concluded that deep-injection wells could trigger quakes. The USGS explicitly linked quake activity to the "secondary recovery of oil" (that is, "fracking").[11] And sure enough, when US oil companies started fracturing America's underground rock formations in earnest around 2004, soon thereafter "freak earthquakes" began occurring across Arkansas, New York, Oklahoma, Texas, and West Virginia—states that are home to most of the country's nuclear plants.

Ground Acceleration: A Core Concern

In planning for North Anna, Dominion Power had estimated that only six major quakes would be expected to strike the region over the course of 10,000 years. Dominion was so unconcerned about the threat that it removed the plant's seismographs in the 1990s as a cost-cutting measure. In response to the August 2011 quake, the Virginia reactor's first seismic reevaluation in 20 years concluded that North Anna was 38 percent more likely to sustain quake-related damage to its reactor cores than had been believed. And while the plant was designed to withstand a "peak ground acceleration" (PGA) of 0.12g, new estimates revealed that the reactors were at risk of being hit with a PGA of 0.535g—more than four times greater. (Surprisingly, while seismic risk assessments for city buildings are undertaken every five to six years, there is no similar requirement for reactors. Only Diablo Canyon—whose two reactors were built near a known earthquake fault—is subject to regular risk reassessments.)

North Anna is only *operating* plant to have experienced an event in excess of the NRC's operating basis earthquake (OBE) standard. (OBE refers to the strongest earthquake a reactor can withstand while still continuing to operate reliably and safely.) South Carolina's Summer plant experienced an OBE in 1979, and Ohio's Perry reactor experienced one in 1989, but both occurred while those reactors were awaiting a license to operate. Despite evidence of their unexpected vulnerability, both reactors received operating licenses.

The two existing reactors at Georgia's Vogtle plant were engineered to shrug off a PGA of 0.2g. But seismic projections for the two new Toshiba-Westinghouse AP1000 reactors that the NRC licensed for construction in February 2012 predict a potential PGA of 0.266g at the Vogtle site. An official at Georgia Power (a subsidiary of plant operator Southern Company) informed the *Wall Street Journal* that the new calculation "doesn't necessarily mean the current plants aren't safe." Asked about the troubling discrepancies, the official replied, "Industry and the NRC are working together to decide what to do about the differences."[12]

Industry Resists Earthquake Protections

It is now accepted that earthquakes pose the single greatest danger to nuclear reactors and spent fuel storage facilities. Nonetheless, the NRC and the nuclear industry continue to debate the proper response to these "low likelihood/high consequence" events. On July 19, 2011, NRC chairman Gregory Jaczko announced a new set of safety standards that included improved ventilation to prevent the accumulation of explosive hydrogen gas and more robust electric cooling systems to protect reactors and spent fuel ponds. The NRC also required plant operators to provide emergency backup power for at least eight hours.

However, nuclear industry leaders objected that the NRC's proposed 90-day deadline for action was unrealistic, contending that it could take years to implement the changes.[13] Alexander Marion, vice president of the Nuclear Energy Institute, asked why the industry should spend billions to protect a reactor from a major quake that "would only happen every 100,000 years."[14]

While the US nuclear operators dithered, the rest of the world was thrown into action by the Fukushima meltdowns. The European Union called for mandatory "stress tests" of its reactors; Soviet-era plants in eastern Europe were reinforced after they failed to pass seismic inspections; Japan began a major reassessment of the ability of its 50 reactors to survive quakes and tsunamis; France moved to reinforce its 58 nuclear plants to handle a quake twice as violent as the largest foreseeable "1,000-year event."

In the United States, it should be imperative that the NRC and state officials take action to address the unique hazard posed by the five Fukushima-style General Electric Mark 1 reactors at immediate risk of catastrophic damage during an earthquake. These "twice-damned" reactors (burdened by both a dangerous design flaw and perilous placement) are the Duane Arnold unit in Iowa, Dresden units 2 and 3 in Illinois, and Peach Bottom units 2 and 3 in Pennsylvania.

The United States Faces a Future of Super Quakes

Chris Goldfinger, director of Oregon State University's Active Tectonics and Seafloor Mapping Lab, is one of a growing community of scientists who believe the world is facing a shaky future. "Make no doubt about it,"

Goldfinger warned in August 2011, "we're in the middle of a global cluster of megaquakes."[15] Six of the 16 fiercest megaquakes on record occurred between 1952 and 1964—all along the "Ring of Fire" that roughly marks the continental boundaries of the Pacific and South Pacific oceans. Five of the 10 largest megaquakes registered since 2004 have struck along the Ring of Fire. Japan, Chile, Indonesia, and New Zealand have all recently experienced historic megaquakes.

"There is little doubt now that earthquakes do tend to occur in clusters," *Newsweek* reported in a cover story the week after the Tohoku quake. "A significant event on one side of a major tectonic plate is often—not invariably, but often enough to be noticeable—followed some weeks or months later by another on the planet's far side." In the case of the Pacific Plate, "that leaves just one corner unaffected—the northeast." *Newsweek* concluded that a megaquake could shake loose "the San Andreas Fault, underpinning the city of San Francisco."[16]

The Pacific Plate travels about 3.5 inches a year in a west-northwest direction toward the North American Plate at the Cascadia Subduction Zone (CSZ). Caught between these two major plates is the Juan de Fuca Plate, a small but active region that parallels about 300 miles of the Northwest coast. A recent study of the sediment left by submarine landslides in the ocean off California and Oregon has revealed disturbing evidence that 19 "monster quakes"—measuring 9.0 and higher—have hammered the West Coast over the past 10,000 years. The most violent US quake—a 9.0-magnitude monster triggered by the CSZ—hit off the Oregon coast on January 26, 1700, rocking the unsettled wilderness of the Pacific Northwest. Another 22 quakes measuring between 8 and 8.3 (both larger than the 7.9 quake that hit San Francisco in 1906) have visited the region, hitting, on average, once every 240 years. Scientists who have studied the region believe there is a one-in-three chance that a monster 9.0 quake could strike the West Coast within the next 50 years.[17]

A Lesson from Fukushima

The Japanese tsunami of March 2011 was neither "unprecedented" nor unpredictable. Japan is one of the planet's most earthquake-prone countries. It is located atop a convergence of four tectonic plates. The resulting geological instability causes around 1,000 tremors a year. In 869 AD, the Jogan tsunami—which was at least as powerful as the 46-foot-tall wave that swept away the town of Sendai—devastated the same section of Japan's coast.

As Greenpeace International noted in a February 2012 report, back in 1971 (the year Fukushima Daiichi commenced operation) the US government had warned that the plant's design was susceptible to a "lethal nuclear explosion and widely scattered radioactive fallout" in the event of a cooling system failure.[18] In 1991, the NRC's NUREG-1150 report again warned the Mark 1 design was likely to fail in an earthquake.[19] And in October 2006, nuclear physicist Hidekatsu Yoshii had advised the Japanese government, "There is a risk of meltdown due to failure of the cooling systems in 43 nuclear power plants (including Fukushima 1), because they are so designed that power transmission lines would be damaged by earthquakes, thereby causing a complete power failure; or the supply of cooling water would be disrupted in the event of large tsunami waves."[20] In 2007, after the powerful 6.8 Chuetsu quake severely damaged and shut down Japan's massive Kashiwazaki-Kariwa nuclear power plant, the government drew up plans for a combined earthquake-and-nuclear-disaster drill, but after the Nuclear and Industrial Safety Agency (NISA) objected that such an exercise might cause "unnecessary anxiety and misunderstanding" among local residents, the drill was reframed as a response to a "heavy snow."[21]

The dangers were not unknown; they were merely ignored—and covered up.

A SHORT HISTORY OF NUCLEAR REACTORS AND QUAKES

The restless earth has provided a number of recent warnings that nuclear power is no match for unpredictable tectonic forces:

- In 1986, a 5.0-magnitude quake shook Ohio, damaging the Perry nuclear plant, cracking concrete and causing pipes to leak. (The plant was awaiting a new load of radioactive fuel—the very next day.)
- In November 1993, a 5.8-magnitude quake in Japan's northeast Honshu region caused the shutdown of the 497 MWe Onagawa 1 reactor.
- In August 1999, a 6.5-magnitude quake in Japan's Tokai region shut down two of the five reactors at the Hamaoka nuclear plant. One reactor remained shut for 12 years; two others were scrapped.
- On September 21, 1999, a devastating 7.6 quake in northern Taiwan killed thousands and triggered the shutdown of three reactors.

- In 2000, a 7.3-magnitude quake struck a part of Japan with "no known geological faults," prompting a national review of reactor safety.
- In May 2003, a 7.1-magnitude quake tripped an automatic safety shutdown at the Onagawa 3 reactor in Japan.
- In October 2004, Nigata Prefecture in Japan was rolled by a 5.2-magnitude quake that caused a shutdown of the Kashiwazaki-Kariwa 7 reactor.
- In December 2004, a 9.0-magnitude quake in the waters off Sumatra kicked up a massive tsunami that raced across the Bay of Bengal and caused a shutdown of the Kalpakkam nuclear power station on India's east coast.
- In 2004, 2005, 2007, and 2009, severe quakes caused the automatic shutdown of dozens of reactors in Japan.
- On August 16, 2005, three reactors at Japan's Onagawa facility automatically shut down when a 7.2-magnitude earthquake hit northeast Honshu. The reactors did not return to full operation until May 2007.
- On July 16, 2007, a 6.8-magnitude quake rocked Japan's 7,965 MW Kashiwazaki-Kariwa plant causing a shutdown of four TEPCO reactors. The reactors went into automatic shutdown—but not in time to prevent a minor release of radioactivity.
- On March 11, 2011, the Tohoku quake and tsunami shut down 11 of Japan's nuclear reactors—including four of the six reactors at Fukushima Daiichi, four at Fukushima Daini, three at Onagawa, and one at Tokai.

Source: "Nuclear Power Plants and Earthquakes," World Nuclear Association, November 2011, http://www.world-nuclear.org/info/inf18.html.

THE 27 US REACTORS MOST AT RISK FROM EARTHQUAKES

In 2011, the NRC identified the following 27 US nuclear reactors as being in need of upgrades to better withstand earthquakes:

Crystal River 3 (Florida)
Dresden 2 (Illinois)*
Dresden 3 (Illinois)*
Duane Arnold (Iowa)*
Farley 1 (Alabama)
Farley 2 (Alabama)
Indian Point 2 (New York)
Indian Point 3 (New York)
Limerick 1 (Pennsylvania)
Limerick 2 (Pennsylvania)
North Anna 1 (Virginia)
North Anna 2 (Virginia)
Oconee 1 (South Carolina)
Oconee 2 (South Carolina)
Oconee 3 (South Carolina)
Peach Bottom 2 (Pennsylvania)*
Peach Bottom 3 (Pennsylvania)*
Perry 1 (Ohio)
River Bend 1 (Louisiana)
Saint Lucie 1 (Florida)
Saint Lucie 2 (Florida)
Seabrook 1 (New Hampshire)
Sequoyah 1 (New Hampshire)
Sequoyah 2 (New Hampshire)
Summer (South Carolina)
Watts Bar 1 (Tennessee)
Wolf Creek 1 (Kansas)

*Indicates Fukushima-style Mark 1 reactors.

Source: Amanda Peterson Beadle, "Report: 27 US Nuclear Reactors Need Upgrades to Avoid Severe Damage from Earthquakes," Climate Progress (blog), September 2, 2011, http://thinkprogress.org/climate/2011/09/02/310816/report-27-u-s-nuclear-reactors-need-upgrades-to-avoid-severe-damage-from-earthquakes/?mobile=nc.

9

Climate Change: Droughts, Floods, and Solar Flares

Nuclear power is not a solution for climate change. It is a cynical gambit on the part of the global nuclear power industry to save itself from being phased out.

–IRENE KOCK, NUCLEAR AWARENESS PROJECT

THE ARGUMENT that "clean" nuclear power can help fight global warming has been given the cold shoulder by climate activists who point out that nuclear reactors are only marginally useful in countering climate change. An examination of the complete life-cycle costs reveals that nuclear energy actually helps stoke global warming. While it is true that a fully functioning reactor releases little CO_2, an honest greenhouse-gas assessment must include the significant volumes of CO_2 generated by the overall operations of the nuclear industry.

Vast amounts of CO_2 are generated by all the fossil-fuel-powered drills, trucks, locomotives, and cargo ships involved in mining the ore and delivering it to refineries, enrichment facilities, power plants, and, ultimately, to radioactive waste storage sites. Fossil fuels also are consumed (and CO_2 released) in the fabrication of the thick concrete housings and assembly of the huge metal parts that go into making a nuclear power plant. It takes many years for a fully operational nuclear plant to generate sufficient energy to offset the energy consumed in the plant's construction.

When the entire fuel cycle is considered, a nuclear reactor burning high-grade uranium can produce a third as much CO_2 as a gas-fired power plant. But there's another problem: the world's supply of high-grade ore is running out. If reactors are forced to burn enriched, low-grade ore (containing only one-tenth the amount of uranium), the nuclear fuel cycle will begin to pump out more CO_2 than would be produced by burning fossil fuels directly.

Nuclear reactors would be a lot safer if they were constructed on a dead moon—or on a planet without a molten core, extreme climate changes, polar reversals, and tectonic plates. There is growing evidence that nuclear reactors will become increasingly vulnerable to onslaughts of extreme weather triggered by climate change. As a 2008 study by the watchdog group Beyond Nuclear concluded, "Nuclear power reactors will not be able to operate safely or reliably under the increasingly unstable weather conditions caused by climate change."[1] Not even the crippled Fukushima reactors escaped the weather's wrath. In late January 2012, a week of unusually cold winter weather caused pipes at the damaged Unit 4 reactor to freeze and burst in at least 30 locations. One broken pipe spilled 8.5 tons of radioactive water.[2]

The nuclear industry has spent heavily to push the message that nuclear power is a "solution" to global warming. In fact, climate change appears likely to hamper any nuclear revival. In the United States, the rising temperatures associated with climate change already have forced the closure of overheating reactors in Alabama, Illinois, Michigan, New York, and Pennsylvania. Soaring external heat and increased temperatures of river water used for cooling reactors has caused similar shutdowns or power-downs at plants in France, Germany, Romania, and Spain.[3]

In the summer of 2003, French firefighters had to be dispatched to hose down overheating reactors at the Fessenheim nuclear plant.[4] In July 2010, the three nuclear reactors at Browns Ferry in Alabama were forced to cut power production 50 percent when the temperature of the river water upstream from the plant's discharge pipes hit 89 degrees. (When the river's upstream temperature exceeds 90°F, state regulations require that the reactors shut down.)[5] The owners of the three Palo Verde reactors in bone-dry Arizona have become so desperate to ensure a steady supply of coolant that they now contract with Phoenix to cool the plant's cores with city sewage, which it taps via a 30-mile-long pipeline.

The planet's rising temperatures also increase the likelihood of fires. In the summer of 2011, a monster wildfire swept across New Mexico and veered frighteningly close to the Los Alamos Nuclear Weapons Lab and its collection of tens of thousands of barrels filled with plutonium-contaminated wastes. The Union of Concerned Scientists has summed up the climate problem in a single, scintillating sentence: "Mercury rising means nuclear power and electrical output and safety margins falling."[6]

Those coastal reactors that draw their cooling water from the oceans are at risk from oil spills. After oil from BP's Deepwater Horizon disaster spilled into the Gulf of Mexico, every coastal reactor from Texas to south Florida faced a possible shutdown. As the DOE's Office of Electricity Delivery and Energy Reliability warned, "If water supply for these facilities becomes contaminated with oil, cooling water systems could be damaged."[7]

Hurricanes, Tornadoes, and Floods

In addition to reactor shutdowns occasioned by rising temperatures and warming rivers, a host of new threats are (literally) appearing on the horizon. Hurricanes, tornadoes, and epic floods have measurably increased in frequency and intensity since the 1990s.

Residents of the East Coast and the Gulf of Mexico have seen homes, roads, farms, and cities ravaged by more than one "once-in-a-lifetime" flood or hurricane. The United States has 24 reactors running at 14 sites located in proven hurricane corridors along the Atlantic coast and the Gulf. Incredibly, the majority of the new reactors proposed for construction under the banner of a "nuclear renaissance" would be located smack in the middle of the Atlantic and Gulf storm tracks.

In 1992, Hurricane Andrew slammed into Florida's Turkey Point reactor near Miami, forcing the plant to rely on its emergency power generators. When the plant came dangerously close to running out of diesel fuel, Turkey Point's operators had to commandeer fuel from nearby hospitals to keep the generators running and prevent a meltdown.

In June 1998, Ohio's Davis-Besse reactor was hit by a tornado that cut the plant off from the electric grid. Emergency generators kicked in, but the power outage lasted so long that the generators nearly failed. Without emergency cooling, the spent fuel in on-site storage pools could have overheated. Fortunately, outside power was restored shortly before the emergency generators would have run out of fuel.

In 2005, Hurricane Katrina forced the shutdown of the Waterford nuclear plant in Louisiana, and storm-related flooding along the Mississippi scuttled the state's River Bend reactor.

In 2010, a tornado took down the power grid near Monroe, Michigan, damaging the 1,140-megawatt Fermi 2, and leaving the largest Fukushima-style Mark 1 BWR reactor on earth without access to outside power needed to continue running emergency generators. Fermi's storage pool contains more high-level waste than the four damaged Fukushima reactors combined. Fortunately the tornado hit four years after the plant's owners had repaired Fermi's emergency systems. The repairs were begun in 2006 after workers discovered the reactor's emergency power generators had been out of commission—for 20 years. (A 1966 core meltdown at the Fermi 1 reactor became the subject of John G. Fuller's 1975 bestseller, *We Almost Lost Detroit.*)

In August 2011, Hurricane Irene put 14 nuclear power plants from Maryland to New Hampshire (and a nuclear fuel production facility in North Carolina) under an NRC storm watch. The 2,1110 MW Millstone plant in Connecticut and the Brunswick plant on the North Carolina coast were both forced to reduce power as the storm blew past. In New Jersey, the Exelon Corporation took its Oyster Creek reactor off-line. The worst damage was recorded at the Calvert Cliff plant in Maryland, where the Unit 1 reactor was forced into automatic shutdown when a transformer was damaged by flying aluminum siding ripped loose by the storm.

Another long-term climate threat will come with the slowly rising sea levels that accompany the melting of the planet's polar ice packs. As the oceans gain a higher profile, the world's coasts (and the reactors built on them) will become more vulnerable to storm surges, hurricanes, and tsunamis. As nuclear engineer Arne Gundersen has wryly noted, "Sandbags and nuclear power plants should not be in the same sentence."[8]

If a storm were to cause a blackout and damage a reactor so severely as to cause a release of radioactivity, the surrounding communities might not realize the danger. In a blackout, emergency sirens designed to warn nearby residents of danger would not work, since they also draw their power from the electrical grid.

Tsunamis in the Heartland

The United States has recently seen how Midwest reactors can be threatened by "inland tsunamis" released by extreme storms that cause rivers to overflow. There are 64 nuclear reactors sitting near rivers or reservoirs prone to flooding. (A total of 88 reactors—85 percent of the country's atomic power plants—are at risk of flooding from swollen rivers or hurricanes.) The storm-tossed Great Lakes are surrounded by 23 nuclear reactors (13 in the United States and 20 in Canada). A radioactive accident at any of these Great Lakes reactors could poison the drinking water used by 40 million people—a body of water that represents 20 percent of all the fresh water on the surface of the earth.

In the 1990s, flooding on the Missouri River caused the loss of emergency backup systems at Nebraska's Cooper plant. Similar flooding along the Mississippi damaged emergency systems at the Prairie Island reactor in Minnesota. In both cases, the rising waters flushed radioactive contamination into the environment and cut off designated "evacuation" routes. An accident at the Prairie Island site could potentially contaminate the 1,245,000-square-mile Mississippi watershed, which includes 40 percent of the US landmass and 40 percent of the country's drinking water.

In April 2011, the Fort Calhoun reactor, 20 miles north of Omaha, was engulfed by floodwaters as the rain-swollen Missouri River overflowed its banks. Although the plant had been shut down for maintenance, the flooding was followed by an electrical fire that erupted in a critical safety system.

The cause of the fire was not determined, but NRC investigators believed a badly designed breaker system was a contributing factor. (Nearly a year after the fire, NRC inspectors were still unsure of the cause, attributing it to either a badly designed part or improper installation.)

With the reactor still closed in February 2012, the owners were working their way through a list of 600 improvements the NRC had ordered completed before the plant could reopen. The extended closure forced the Omaha Public Power District (OPPD) to spend $32 million for replacement power through the end of the year. The flood alone was estimated to have cost OPPD $75.9 million as of December 2011. The utility predictably passed the costs along to its customers and proposed hiking electricity rates by 5.9 percent.

Climate-related shutdowns are expected to increase as the planet's weather grows more unruly. "And when reactors partially power down, or shut down entirely . . . it often happens at a time when reliable electricity supplies are needed most," Beyond Nuclear notes. The irony in this, for the nuclear renaissance crowd, is that it is renewable energy—not nuclear power—that is proving itself to be "more reliable in a global-warming world."[9]

SPACE WEATHER: REACTORS AND SOLAR FLARES

Solar energy may soon eclipse nuclear power—only not in the way we hoped. According to NASA, the planet will soon face an outbreak of powerful solar flares capable of collapsing global power grids. Were this to happen, the world's reactors could be left to run wild, overheat, melt down, and explode.

More than 869,600 miles wide and large enough to contain more than a million Earths, the sun accounts for 99.86 percent of the mass of the entire solar system and radiates about 383 billion trillion kilowatts of energy. Some of that immense energy is discharged in solar flares, also known as coronal mass ejections (CMEs). And according to researchers, CMEs are expected to reach peak levels in 2013.

A CME can bathe the planet in magnetic tides of high-energy particles, immense tsunamis of electrified gas that can easily exceed a billion tons. (To visualize the magnitude of such a high-energy CME impact, imagine Mount Everest turned into a boiling blast of energy and hurled directly into Earth's path.) These high-energy particles can trigger geomagnetic disturbances that can disrupt power lines. Since the 1970s, the array of high-voltage transmission lines spanning the United States has grown tenfold. NASA warns that these interconnected networks can be energized by a solar flare, causing "an avalanche of blackouts carried across continents [that] . . . could last for weeks to months."[10]

There have been two massive CMEs over the past two centuries. The 1859 Carrington Event irradiated Earth for nine days, causing the Northern Lights to dance in the skies as far south as Hawaii.[11] On May 14, 1921, a GMD lit up northern skies as far south as Puerto Rico. Both flares disrupted telegraph communication around the world, causing some equipment to burst into flames.

But twentieth-century telegraph lines were more resilient than today's electronics. Solar flares can bake the fragile circuitry that controls aircraft, banking, GPS, radio, TV broadcasts, iPods, and the Internet. As NASA solar physicist Lika Guhathakurta put it, "A similar storm today might knock us for a loop."[12] A National Academy of Sciences report estimates that a "century-class" solar storm could affect more than 100 million Americans and cause 20 times the damage as Hurricane Katrina ($2 trillion) while "full recovery could take 4 to 10 years."[13] Physicist Michio Kaku has characterized the threat of a massive solar flare as "a potential Katrina from space. . . . We'd be thrown 100 years into the past."[14]

On March 13, 1989, a 90-second solar blast slapped Hydro-Québec's transmission system and left six million Canadians without electricity for nine hours. The storm cooked transformers in Great Britain and triggered 200 "anomalies" at oil-, coal-, and nuclear-fueled facilities across the United States, even melting a transformer at New Jersey's Salem 1 reactor.

A Carrington-size geomagnetic disturbance could damage thousands of extra-high-voltage (EHV) transformers around the world. These transformers can weigh up to 300 tons and cost more than $1 million. Power grids cannot operate without them. Because each is custom-built to regional specifications, procuring a new EHV can take up to three years and rebuilding a damaged grid could take decades.

That could be the best-case scenario. More worrisome is imagining what would happen to nuclear power plants that are reliant on a functioning electrical grid.

A 2011 Oak Ridge National Laboratory report warned of a 33 percent likelihood that a solar flare could lead to long-term power loss over a nuclear reactor's 40-year life. With 440 nuclear power plants in 30 countries—and 250 research reactors—there are nearly 700 potential Fukushimas waiting to be unleashed by a super flare.[15]

Faced with a grid collapse, nuclear plants must rely on backup power to cool reactor cores and spent fuel ponds. But the NRC requires reactors to have only eight hours of battery power and enough fuel to run emergency generators for a week. Restoring outside power to Fukushima's damaged reactors was a daunting task even when Japan had a functioning grid to fall back on. If the sun sends a geomagnetic tsunami sweeping across Earth, it could become impossible to provide *any* form of traditional power.

NASA has proposed a "solar shield" to detect incoming CMEs and warn operators to shut down power grids until the danger passes. Unfortunately, the plan is experimental and has never been field-tested. The United States could protect its grid by spending around $1 billion to "harden" 350 key EHV transformers and stock blast-proof warehouses with replacement parts. Transformers also could be protected with ground resistors. Costing about $40,000 each, they could be installed on 5,000 critical transformers for less than $200 million—about one-tenth the cost of a B-2 stealth bomber.

In August 2010, the US House of Representatives unanimously approved a bill that would have allocated funds to protect the grid against geomagnetic storms. But the Senate failed to act and the bill foundered.[16] In June 2011, the United States and United Kingdom announced plans to mandate "controlled power cuts"—shutting down portions of the electric infrastructure to protect the larger grid—but that's only two countries. Until *every* nuclear nation is prepared for grid collapse, the inevitable cascade of core meltdowns resulting from solar flares could mark the sudden end of the industrialized world.

10

Radioactive Wastes

No currently available or reasonably foreseeable reactor and fuel cycle technology development—including advances in reprocess and recycle technologies—have the potential to fundamentally alter the waste management challenge this nation confronts over at least the next several decades, if not longer.

–BLUE RIBBON COMMISSION ON AMERICA'S NUCLEAR FUTURE, JANUARY 2012

ON JUNE 11, 2012, a sunny and breezy day in San Francsico, Marvin Fertel, the CEO of the Nuclear Energy Institute, settled into a chair at the Commonwealth Club. The occasion was a panel discussion on the future of nuclear power and the audience was salted with nuclear critics ready to pepper the panelists with questions. Fertel was joined by two nuclear critics—Joe Rubin, an investigator with the Center for Investigative Reporting and former California Energy Commissioner Jim Boyd.

"With nuclear," Boyd said, citing the disaster in Japan, "it has very high rewards alleged, but incredibly high risks associated with anything that goes wrong." Fertel responded by insisting nuclear "is not a troubled industry." Nucelar reactors, he assured the audience, were a safe and proven means of generating electricity.

During the Q and A session, one audience member walked to a microphone and took issue with Fertel's claim that nuclear reactors generate electricity. "That is technically incorrect," the speaker observed. "Reactors produce heat, not electricity. There are many sources of heat. It is the steam passing through the turbines that generates the electricity." Fertel smiled and nodded his head to concede the point. The speaker wasn't finished. "There is only one thing that nuclear reactors *do* produce," he continued, and that is something that nobody needs, nobody wants, and nobody can figure out what to do with. "That product is atomic waste." Fertel's smile faded.[1]

The average nuclear power plant creates 20–30 tons of high-level waste per year, plus another 70 tons of low-level wastes (workers' clothing, tools, and cleaning materials),[2] but no proven long-term waste storage sites have yet opened.

The Demise of Yucca Mountain

The US government's showcase deep-storage site at Nevada's Yucca Mountain was originally projected to cost $58 billion and to be open for business in 1998. In 2010, after 30 years and an investment of $10 billion, President Obama cut funding for Yucca Mountain, due to the project's irresolvable design and environmental problems. Had Yucca Mountain opened when planned, the total operational costs (including its planned decommissioning in 2133) would have topped $96 billion.[3] With a planned working life of 125 years, Yucca Mountain would not have solved the long-term waste disposal problem. It might not even have solved the short-term waste disposal problem, since, by law,[4] the repository could not have stored more than 70,000 metric tons and, by 2006, the United States already had accumulated more than 65,000 metric tons of commercial reactor waste and at least 7,000 metric tons of weapons waste.[5] Had Yucca Mountain opened, it immediately would have been filled to capacity. To continue operating current reactors (let alone trying to build new reactors), the United States would have had to locate a new site and start building another Yucca Mountain–size repository.

Without long-term storage facilities, spent fuel rods must be stored in pools or in dry-cask facilities that are secure only for less than a century. The Vermont Yankee reactor, for instance, has accumulated 1,911 radioactive waste bundles in its spent fuel pool and has another 340 bundles stashed on-site in a dry-cask storage facility.

Spent Fuel Pools

The nuclear industry has removed around 270,000 tons of spent fuel rods from nuclear reactors worldwide, and these highly radioactive encasements continue to pile up at the rate of 12,000 tons per year.[6] Fuel rods containing enriched uranium need to be replaced every 5.5 years, if not sooner.[7] Like the euphemism "depleted uranium," "spent fuel" is an intentionally deceptive term. The radioactivity levels in used rods leaving a reactor can be nearly a million times higher than when they were first installed. Spent fuel rods are stored in 45-foot-deep, 100,000-gallon concrete cooling ponds lined with lead and/or steel. Diesel-powered pumps must continuously circulate the water in these ponds. As the Fukushima disaster showed all too clearly, if the oil-fueled pumps fail and backup systems are dysfunctional, the water can boil away, leaving the fuel assemblies to overheat, melt, and ignite.

The NRC considered the implications of a loss-of-coolant incident for spent fuel pools in 2000 and concluded that it would take 140 hours for the water to completely boil off. But the fuel pool at Fukushima's Unit 4 reactor ignited after just 100 hours. (This should be a matter of concern for the residents living near the Pilgrim nuclear plant in Massachusetts. The facility's spent fuel pool contains more radioactive cesium than was released by Fukushima, Chernobyl, and all nuclear bomb testing combined.)[8]

A 2006 study by the National Research Council of the National Academies pointed out what the NRC had overlooked: a dangerous situation occurs as soon as the water drops enough to expose the tops of the fuel assemblies. Quickly rising temperatures accelerate the oxidation of the exposed zirconium alloy cladding that encases the uranium fuel pellets. Exposure to air and steam further increases temperatures while generating explosive hydrogen gas, and this can lead to a self-sustaining "runaway" zirconium-cladding fire with a moving "burn front" that crackles like a fireworks sparkler.

As temperatures increase further, the captive fuel begins to expand until it forces the cladding to bulge and burst, releasing superheated isotopes into the air. At 3,300°F (1,816°C), the contents and the containment merge to form a growing molten mass of zirconium-uranium-oxide. At this point, the exposed fuel rods continue to reignite like "trick birthday candles," posing the risk of a chain-reaction fire jumping from one assembly to the next.[9] Nuclear engineer Arnie Gundersen calls the prospect of a runaway combustion of fuel assemblies "Chernobyl on steroids."[10]

Fearing a spent fuel fire following the multiple meltdowns, the Japanese government began secret preparations to order the evacuation of Tokyo's 13 million residents. The plans were not revealed in order "to avoid panic." As one senior government official told the *Japan Times*, "The content [of the emergency response plan] was so shocking that we decided to treat it as if it didn't exist."[11]

No Plans for Permanent Storage

After 53 years of commercial nuclear power (and a half century of government bailouts), the United States still has no sound plan for long-term storage of high-level radioactive wastes—a fact that leads to another costly, but little known, nuclear subsidy. Between 1983 and 1987, the Department of Energy (DOE) signed contracts with US power companies promising to start accepting the stored wastes from more than 100 commercial reactors beginning January 31, 1998. When the deadline passed and the DOE still had no functioning storage facility, the power companies were able to sue for "breach of contract." Reactor operators filed 71 lawsuits demanding compensation for the continued costs of storing their wastes on-site. By July 2009, DOE had paid out $565 million in damages to five nuclear operators. (Instead of the payments coming from the $23 billion that power companies were required to contribute to the Nuclear Waste Fund, the costs are being borne directly by US taxpayers.) By 2020, taxpayer liability for the DOE's failure to solve the industry's waste problems could top $50 billion.[12]

Despite this boondoggle, in its last three months, the Bush administration quietly signed deals with more than a dozen electric utilities that obliged the government to assume responsibility for long-term storage for nuclear wastes from 21 proposed *new* reactors. When the agreement was signed, it was already clear that the Yucca Mountain facility—even if it were to open—would not have the capacity to store the additional 21,000 metric tons of toxic wastes anticipated from the 21 new reactors. The Bush White House once again gave industry the right to sue if the government was unable to deliver on its promise to provide long-term storage. The beneficiaries were Duke Energy, Southern Nuclear, UniStar Nuclear, Florida Power & Light, Pennsylvania Power & Light, South Carolina Electric & Gas, Progress Energy, Ameren, Luminant, and the South Texas Project.

With the future of a storage facility in Germany in doubt, only two storage sites remain in the works—the Fosmark facility planned for Sweden and the Onkalo site in Finland. Nuclear wastes must be secured for 40-60 years before they are cool enough to be placed in underground storage. Finland's timing could not be better since Onkalo is set to start accepting nuclear waste from the country's six existing reactors by 2020. Three Finnish power companies have announced plans to build three new reactors but this may not be possible since Finland's Nuclear Energy Act revision of 1994 requires all new plant operators to guarantee long-term waste storage and the Onkalo site can only handle the wastes from seven reactors. Meanwhile, there is a daunting long-term problem facing Onkalo: since the repository is located on Olkiluoto island, rising seas may swamp the site as the planet warms.

Lacking a credible disposal plan, the global nuclear industry has tried to promote "reprocessing" as a solution to the growing piles of used fuel rods.[13] But because only 1 percent of the "recycled" material (the plutonium) is actually usable, this creates large volumes of even more toxic leftovers.[14] In the event of an accident, reactors powered by reprocessed plutonium are harder to control and could prove more dangerous than uranium-fueled reactors.[15] (It is the weapons-grade plutonium component that makes the fuel inside Fukushima's Unit 3 reactor the focus of so much concern.)

Instead of solving the waste problem, reprocessing actually increases the volume of radioactive garbage that needs to be stored—safely and forever. Supposedly "safely stored" radioactive wastes have repeatedly contaminated the land and sea around Britain's accident-prone Sellafield reprocessing plant and France's Le Havre reprocessing facilities.[16]

Since waste disposal poses a crippling problem for any nuclear "renaissance," various alternatives to the abandoned Yucca Mountain repository are being considered. One involves new drilling techniques that could bore hundreds of barrel-size holes five kilometers deep into the stable bedrock that exists in many parts of the United States. In theory, this would avoid the leakage issues posed by Yucca Mountain's geology and circumvent the need to provide leak-proof canisters. Whether such schemes could be carried out successfully and economically remains speculative.

Is Nuclear Entombment Possible?

Time is the critical issue in addressing the waste-storage challenge. Approximately 5,000 natural and artificial radionuclides have been identified, each with a different half-life (that is, the time required for the initial amount of radioactivity to decrease by one-half). The half-life of a radionuclide can range from 1.07 seconds (for sodium-26) to 4.47 billion years (for uranium-238). The isotopes left behind in a reactor's nuclear waste include cesium-137 (with a half-live of 30 years), plutonium-239 (with a half-life of 22,000 years), and iodine-129 (with a half-life of 15.7 million years).[17]

The Yucca Mountain nuclear waste disposal site was to have secured 77,000 tons of hazardous radioactive materials from 110 US commercial power plants, but the facility was only designed to operate until 2133—at a cost topping $99 billion (in 2011 dollars). Looking to the long term, the EPA's "radiation protection standard" for people in the vicinity of Yucca Mountain was set at 15 millirem per year for the first 10,000 years (and 100 millirem per year thereafter). The EPA estimated that the 100-millirem exposure (equal to about 1,700 chest X-rays) would produce an extra cancer in roughly every hundredth person exposed—a cancer risk 100 to 10,000 times greater than EPA currently permits.[18] This is an unconscionable legacy to inflict upon future generations.

For nuclear entombment to be even moderately successful, a storage facility must remain intact for at least 100,000 years (the time it would take for most of the radioactivity to degrade to safe levels). No man-made structure has ever survived for 10,000 years (Egypt's pyramids were built around 4,500 years ago). With Yucca Mountain's fate uncertain, the world's only permanent nuclear waste repository still under construction is located 186 miles northwest of Helsinki.

Work on Onkalo (its name is Finnish for "hidden") began in 1972 and will not be completed until sometime in the twenty-second century. When (and if) it is completed a century from now, Onkalo would only have room for Finland's nuclear wastes—about 1 percent of the world's growing stockpile of radioactive garbage, now estimated at between 250,000 and 300,000 tons. The waste would be stored in two-inch-thick copper containers interred 1,600 feet deep in bedrock that has remained stable for 1.8 billion years. The facility was engineered to withstand the weight of a two-mile-thick layer of ice that is expected to accumulate during the next Ice Age.

Onkalo is intended to remain secure for 100,000 years. That's about as long as humans have walked the earth—3,000 generations. This radioactive graveyard may become the single lasting legacy of our times. In 10,000 years (let alone 100,000), Beethoven's sonatas, Picasso's paintings, Rumi's poems, and most modern languages may likely be long forgotten. The only artifact of our civilization that is guaranteed to outlast the centuries is a graveyard of toxic trash buried in Finland's crystalline gneiss bedrock—and around 300,000 tons of radioactive wastes left untended in crumbling surface storage sites around the world.

THE BLUE RIBBON COMMISSION ON AMERICA'S NUCLEAR FUTURE

Nuclear plants were designed to produce electricity for 40 years. They were also designed to produce radioactive wastes that last for thousands of years. More than half a century after President Dwight Eisenhower opened Pennsylvania's Shippingport PWR reactor, the country's first wholly commercial civilian nuclear power plant, the United States finds itself confronted with tens of thousands of tons of radioactive spent fuel rods stacked in "temporary" storage sites at plants across the country.

On January 29, 2010, the White House and the DOE created the 15-member Blue Ribbon Commission on America's Nuclear Future (BRC) to assess "reactor and fuel cycle technologies, storage and transportation, and disposal." On January 26, 2012, the BRC released its findings.[19] Unfortunately, while the panel's criticisms were scathing, its final recommendations were merely tepid.

"This nation's failure to come to grips with the nuclear waste issue has already proved damaging and costly," the BRC's report begins.[20] "The approach laid out under the 1987 Amendments to the 1982 Nuclear Waste Policy Act (NWPA)—which tied the entire US high-level waste management program to the fact of the Yucca Mountain site—has not worked to produce a timely solution." The BRC calls the decision to halt work on the Yucca Mountain repository "the latest indicator of a policy that has been troubled for decades and has now all but completely broken down."

But, after the tough talk came the caveats: "There are several questions the Commission was not chartered to address. We have not . . . proposed any specific site (or sites) for any component of the waste management system [or] offered a judgment about the appropriate role of nuclear power in the nation's (or the world's) future energy supply mix."

The BRC takes careful note of the failures of federally funded waste management. The NWPA proposed a "polluter pays" system that was supposed to hold utilities—not the taxpayers—accountable for the costs of disposing of nuclear waste. (When it comes to the Pentagon's weapons waste, however, taxpayers pay the costs in full.) In exchange for Washington's pledge to take ultimate responsibility for waste disposal, nuclear operators were allowed to bill customers a surcharge on every kilowatt-hour of electricity produced by a reactor. The money went to the federal Nuclear Waste Fund.

On April 5, 2010, following the government's decision to suspend work on the Yucca Mountain storage site, the Nuclear Energy Institute and 16 utilities filed suit against the DOE, claiming that the government no longer had a right to collect "waste storage fees." (Of course, the utilities had never paid these fees; they had simply passed the responsibility on to their customers in the form of a surcharge of one-tenth of a cent per kilowatt-hour. At the time of the lawsuit, this "tiny" surcharge had generated $24 billion.)

The BRC notes that the Nuclear Waste Fund "does not work as intended." Thanks to a number of White House and congressional initiatives, neither the $27 billion that has accumulated in the Nuclear Waste Fund nor the $750 million in annual fee revenues collected by utilities is currently available for the waste-handling program. "Instead," the BRC complains, "the waste program must compete for federal funding each year and is therefore subject to exactly the budget constraints and uncertainties that the Fund was created to avoid."

On one hand, the BRC report acknowledges, "many people fear the transportation of nuclear materials," but it then concedes that addressing the current toxic inventory will call for "greater transport demands." In a sign that the BRC still sees Native American lands as prime dumping sites, the report recommends that "tribal and local officials" should be given the funds and training "necessary to discharge their roles and obligations in this arena."

The tone of the BRC's concluding section is so at odds with the caveats and criticisms of the balance of the report that it sounds as if it was written by a nuclear lobby public-relations team. "Globally, some 60 new reactors are under construction and more than 60 countries that do not have nuclear power plants have expressed interest in acquiring them," the report states. Although none of these nuclear wannabees has yet come up with a plan "for safe storage and disposition of spent nuclear fuel," the BRC recommends streamlining the regulatory process to "guide the design of new systems and lower barriers to commercial investment by increasing confidence that new systems can be successfully licensed." Another recommendation: expand federal spending on "nuclear workforce development" to assure "a viable domestic nuclear industry."

Looking ahead to an imagined global renaissance, the BRC report concludes with a call for "multi-national fuel-cycle facilities . . . to give more countries reliable access to the benefits of nuclear power." US sponsorship of a "global nuclear fuel bank" under the auspices of the International Atomic Energy Agency is applauded as "a driver for some states to engage in uranium enrichment." The BRC proposes that US taxpayers shoulder the burden of creating "one or more multi-national spent fuel storage or disposal facilities."

Although it still hasn't mastered the trick of long-term storage of its own wastes, the United States currently offers to store highly enriched uranium fuel from foreign research reactors. The BRC suggests that the United States should embark on a "spent-fuel take-away" program to help smaller, poorer nations "avoid the costly and politically difficult step of providing for spent fuel disposal on their soil." This would make the nuclear renaissance more marketable in the world's less-industrialized nations—at considerable cost and risk to US taxpayers.

11

The Decommissioning Dilemma

These reactors produce 50 years of electricity and half a million years of waste. It's not a particularly good deal.

—DANIEL HIRSCH, COMMITTEE TO BRIDGE THE GAP

AS OF APRIL 2011, the NRC listed 23 US reactors as permanently closed with "work underway at 13 of these plants that are currently in some phase of the decommissioning process." In fact, only five of the 23 reactors have been fully decommissioned, the most recent being Illinois's Zion Unit 1, which entered DECON status on February 21, 1997. Of the nearly two dozen closed reactors, the NRC lists more than half (14) as still hosting spent fuel on-site (and this includes four of the five sites classified as DECON).[1]

DECON is the NRC's code for a reactor that has been "immediately dismantled." The more common status is SAFSTOR (a.k.a. "delayed DECON"), meaning a closed reactor has been placed in "safe storage" and is being monitored to allow its radioactivity to abate before any dismantling work commences. The most extreme state is ENTOMB, in which a reactor and its wastes are permanently isolated beneath a thick shield of cement (à la Chernobyl).

One-third of America's shuttered reactors currently survive as independent spent fuel storage installations (ISFSI). The ISFSI status was created after Washington failed to come up with a safe means of storing radioactive spent fuel. Instead, the NRC came up with the ISFSI plan, which allows utility owners to "sell off part of their land . . . while maintaining a small parcel under license for storing the spent fuel." The NRC apparently sees no fundamental need to push the process of decommissioning reactors, since it gives plants 60 years from the time they cease operations to complete decommissioning.[2] (Of course, many businesses routinely go out of business or are consumed by mergers in much shorter spans of time.)

The NRC requires all potential plant operators to "establish or obtain a financial mechanism . . . to ensure that there will be sufficient money to pay for the ultimate decommissioning of the facility," but under federal law, it is up to the private corporations running the reactors to "estimate the minimum amount needed for decommissioning." Not the maximum, mind you, but the *minimum*. Making it even easier, the NRC has ruled that 70 percent of the "traditional, rate-regulated electric utilities" currently operating nuclear reactors "are not required today to have all of the funds needed for decommissioning."[3]

Nuclear plant operators are allowed two years from the time of filing a notice to shut down a reactor to come up with a plan to decommission the radioactive remains. The operators then inform the NRC how long they believe the work will take and how much they believe it will cost. As far as "legacy concerns" (an elegant verbal evasion that refers to the challenge of safely containing radioactive poisons for up to 250,000 years), the NRC requires only that the operators "discuss the reasons for concluding that environmental impacts . . . have already been addressed."

According to NRC guidelines, plant owners can spend up to 3 percent of the decommissioning funds on initial "planning." Another 20 percent can be spent on planning within 90 days of filing a "post-shutdown decommissioning activities report." The remaining 77 percent of the trust fund is left to cover the cost of completing the demolition and decontamination of the site.

And what is the NRC's long-term view for the disposition of these reclaimed reactor sites? "Most plans envision releasing the site to the public for *unrestricted use*, meaning any residual radiation would be below NRC's limits of 25 millirem annual exposure and there would be no further regulatory controls by the NRC" (emphasis in the original text).[4]

The story of the Rocky Flats nuclear weapons faciity offers a cautionary example of the practical limits of the decommissioning process. From 1952 to 1992, Dow Chemical (and later Rockwell International) worked under DOE supervision to produce plutonium "triggers" for the Pentagon's Cold War nuclear arsenal. In June 1989, Rocky Flats, a 175-acre site located 15 miles northwest of Denver, Colorado, was the target of an FBI raid that exposed criminal negligence on the part of the DOE and Rockwell International, the site's manager. Subsequent investigations revealed a long history of cover-ups. Large plutonium fires in the 1950s and 1960s had sent radioactive plumes downwind to Denver while poor storage of radioactive wastes had caused widespread contamination throughout the site, including patches of land where spilled plutonium was found to be generating 120 times more radiation than expected. When Jon Lipsky, one of the FBI agents involved in the investigation, was ordered to keep quiet about the dangers, he quit the agency in disgust. Calling the DOE cleanup "woefully inadequate," Lipsky said that the Justice Department had "covered up the truth [and] . . . some dangerous decisions are now being made based on that government cover-up." The case was settled with a plea bargain and a grand jury's extensive evidence of the radioactive danger was put under seal by the Department of Justice.[5] (The disturbing details of the government's complicity in denying and downplaying the danger is recounted in Kristen Iverson's first-person account, *Full Body Burden: Growing Up in the Nuclear Shadow of Rocky Flats* [Crown; June 5, 2012].)

Weapons production at the site ended in 1992 and the DOE began a long process of decontaminating the buildings and surrounding land. In 2001, Congress voted to declare the decontaminated land the Rocky Flats National Wildlife Refuge. In 2007, the land was transfered to the US Fish and Wildlife Service, which set to work promoting the 6,420-acre site as a tourist destination—a vast, natural playground where schoolchildren could enjoy field trips on 16 miles of hiking paths.

While the cleanup of the "refuge" had been declared complete in 2005, a 2011 survey concluded Rocky Flats remains heavily contaminated with plutonium and chemical wastes—at levels unchanged from 40 years ago. "The material is still there; it's still on the surface," according to Marco Kaltofen, president of Boston Chemical Data Corp., the company contracted to run soil tests for the Rocky Mountain Peace and Justice Center.[6] In some areas, levels of plutonium in a single gram of soil still were radiating 1.579 picocuries—nearly 159 times higher than normal.

The Pace of Decommissioning to Date

So far, the track record for decommissioning could best be described as plodding. An April 2011 NRC report[7] provides the following updates:

Dresden Unit 1, Morris, Ill.: Shut down October 1978. "No significant dismantlement activities are underway." Licensee plans to focus on dismantling the reactor in 2020, when the site's other two reactors are due to be retired.

Humboldt Bay Unit 3, Eureka, Calif.: Shut down July 1976. Decommissioning plan approved July 1988. Spent fuel transferred to ISFSI storage in December 2008. License termination expected in 2015.

Indian Point Unit 1, Buchanan, N.Y.: Shut down October 1974. "There is no significant dismantlement underway."

Peach Bottom Unit 1, York County, Pa.: Shut down October 1974. Currently in SAFSTOR status "with no significant dismantlement underway." Decommissioning not expected until 2034, when two remaining reactors are due to close.

San Onofre Unit 1, San Clemente, Calif.: Shut down November 1992. Fuel removed and placed in ISFSI. All structures removed "down to the minus-8-foot building level."

Three Mile Island Unit 2, Middletown, Pa.: Shut down by accident March 1979. De-fueling completed April 1990. "There is no significant dismantlement underway." The DOE came to the rescue of the plant owners by taking "title and possession of the spent fuel" and trucking it across the country for storage at the Idaho National Laboratory.

Zion Units 1 and 2, Zion, Ill.: Shut down February 1998. Fuel transferred to spent fuel pool. Placed in SAFSTOR status until "about 2013 when the decommissioning trust fund will be sufficient to conduct DECON activities." On September 1, 2010, Exelon sold the two reactors to ZionSolutions, which plans to employ a "rip and ship" approach to "reduce the labor-intensive separation of contaminated materials." The radioactive debris will be shipped to the EnergySolutions disposal site in Utah. License termination expected in 2020.

Insufficient Funding

According to an October 2011 report from the NRC's Office of Nuclear Reactor Regulation (ONRR), 99 of the nation's 104 reactors had successfully put aside funds for decommissioning, though 27 showed "shortfalls." For the remaining 5 reactors, the licensees "were not providing the full amount of decommissioning funding assurance." In the case of Prairie Island Unit 1, one of the reactors that had lacked decommissioning funding assurance, the plant owner regained compliance with the funding mandate by winning a request to relicense the aging plant for an extended period of use. According to the ONRR report, "The shortfall for Prairie 1 was resolved in June 2011 when it received approval of its application for license renewal, the effect of which was to extend the period of time for the fund to grow."[8]

The NRC estimates that it can cost from $500 million to more than $1 billion and take a decade (or longer) to properly decommission a nuclear reactor. Given these numbers, it's understandable why plant operators might want to see their aging relicts relicensed for an additional 20 years. It's a convenient way of putting off the inevitable and costly burden of decommissioning—a way of kicking the nuclear canister down the road.

And that road looks to be a dead end. There is no existing government fund to handle the costs of decommissioning. Instead, each reactor operator is required to assemble sufficient funds to dismantle and decontaminate their nuclear sites. According to the ONRR October 2011 report, a total of $40.4 billion had accumulated in these various decommissioning funds as of December 31, 2010. But $40.4 billion would provide only enough funding to dismantle 80 of the country's 104 reactors—and that's assuming each reactor would cost only $500 million to dismantle. Using the NRC's higher estimates, the existing fund would cover the costs of dismantling and decontaminating fewer than 40 percent of the country's aging plants.

But what if the $1 billion decommissioning estimate turns out to be too low? As many US plants approach their inevitable end, the costs of decommissioning are starting to soar beyond previous projections. In August 2011, for example, Florida Power & Light and Florida's Public Service Commission estimated that dismantling four Florida reactors (two at Turkey Point and two at St. Lucie) would cost $7.8 billion. That's almost $2 billion each.

How Do You "Decommission" Radioactive Waste?

On average, a nuclear reactor burns through 20–30 metric tons of fuel each year. Over an extended 60-year operating life, that adds up to 1,200 metric tons per plant. According to the Nuclear Energy Institute, US reactors had generated around 65,000 metric tons of spent fuel by 2011, and 75 percent of that has wound up simmering away in vast hot tubs (that is, cooling pools) installed alongside the reactors.

When a reactor finally keeps its date with the wrecking ball, the demolition process is far from elegant. Deconstructing a nuclear plant generates mountains of rubble, tons of toxic and radioactive wastes, and clouds of global-warming gases puffed into the sky by fleets of cranes, bulldozers, freight haulers, and dump trucks. And here's another fuel-burning, global-warming problem: all of the proposed "permanent storage" sites for nuclear waste have been located in the far west—Nevada, Utah, and Washington State—while 77 of the country's 104 reactors are situated east of the Mississippi. Moving nuclear waste to these sites would entail spending vast sums of money (and generating vast plumes of diesel smoke). It would also increase the miles driven and the opportunities for road accidents that could result in long-lasting, no-go zones of regional contamination.

The pending takedown of Exelon Corporation's Zion nuclear plant near Chicago will be the largest decommissioning project yet undertaken in the United States, but the project appears somewhat lacking in sophistication. In August 2011, the *Baltimore Sun* praised Exelon for "setting several remarkable precedents." As the *Sun* explained, "Instead of separating the radioactive debris from the nonradioactive, the usual method, workers will ship most of the rubble to Utah and dump it in the desert. (Spent fuel will be encased in concrete and stay in Illinois.)"[9]

The Spent Fuel Challenge

In March 1992, Northeast Utility (NU), owner of the Millstone nuclear power plant in Waterford, Connecticut, had to shut down the reactor for refueling. NRC regulations require that the fuel rods from the reactor core be allowed to cool for 65 days before being transferred to a cooling pool, so that they don't overheat the pool and cause its cooling equipment to fail. Millstone's spent fuel rods were so radioactive that, if left exposed, they could deliver a lethal dose of radiation within seconds, and they were so hot that, in a previous incident, they had melted the boots off a Millstone worker. But NU executives, loath to purchase replacement power at $500,000 a day for two months, decided to take a shortcut. They ordered the thousands of spent fuel rods dumped into the plant's already full spent fuel pool.

As former DOE official Robert Alvarez has explained, NU had taken this shortcut before "while the NRC deliberately looked the other way. . . . The corporations that owned the nation's nuclear reactors were stuffing about four times more spent fuel into storage pools than the pools were designed to accommodate, with the NRC's blessing."[10]

The NRC finally was forced to crack down on NU, and the Millstone Unit 1 reactor was permanently shut in 1998. But that wasn't the end of the story. The Republican-controlled Congress took offense at the NRC's regulatory intrusion into the free market and responded by cutting 700 jobs at the NRC. The NRC's "impact on the price of nuclear energy" was unconscionable, New Mexico senator Pete Domenici explained, saying, "This 'tough love' approach was necessary." The NRC got the message. As Alvarez recounts, "By 2000, the NRC sharply curtailed its oversight activities and became more of an enabler of nuclear power than a regulator."

The NRC's regulatory focus may have changed, but the problem of spent fuel storage remains. In the United States, as much as 40 percent of the material in spent nuclear fuel is radioactive cesium-137. The spent fuel pools at US reactors contain as much as 30 times the amount of cesium-137 that was released by the Chernobyl explosion and fire. And in the United States, as in Japan, spent fuel pools were not designed with multiple emergency backup systems to keep the fuel rods cool in the event of a loss of electricity. In the United States, as in Japan, this lack of backup emergency cooling puts tightly packed stored fuel rods at risk of overheating, ignition, and catastrophic explosions.

Spent fuel pools in the United States now hold as much as ten times more radioactive material than is contained the reactors' cores—amounting to tens of thousands of tons of stored spent fuel nationwide. And, unlike the active fuel rods in the reactor's core, the used fuel in the storage pools is not completely encased by a protective shell. While a 45-foot-deep spent fuel storage pool is typically surrounded by five-foot-thick walls of steel-reinforced concrete, the overall structure is often covered by little more than the kind of metal siding found in a Sears backyard storage shed. In order to accommodate the nuclear industry, the NRC has rewritten its rules to allow operators to fill spent fuel pools with five times the number of fuel assemblies the units were designed to contain. The Union of Concerned Scientists has warned that "release of ten percent of the cesium in a typical US spent fuel pool could result in far more extensive land contamination than what has occurred at Fukushima."[11]

A 2003 report by Robert Alvarez emphasized the risks. Alvarez concluded that a spent fuel fire at the Indian Point reactors upwind of New York City could cause up to 5,600 cancer deaths and $461 billion in damages.[12]

High and Dry: A Spent Fuel Solution

Fortunately, there is a partial remedy to the spent fuel dilemma. After an initial five-year cooling-off period, fuel stored in pools can be moved into secure dry-cask storage. This is a proven technology. Germany has been using dry-cask storage for more than a quarter century. Robert Alvarez has proposed tapping the $24 billion-plus trust fund established by the 1982 Nuclear Waste Policy Act (NWPA) to finance the transfer.

But there's a hitch. According to the NWPA, this money can only be spent on a "permanent solution" to the waste problem. Unfortunately, a "permanent" solution has not been found, though a "temporary" solution clearly exists. But even the dry-storage option has its downsides. Many of the dry-cask containers in the United States are stacked like bowling pins under open skies, clearly visible from nearby roads, rivers, and the air—a potentially tempting target for would-be terrorists.

Many US power plant owners have rejected the dry-cask solution because moving the fuel rods to new, safer storage would cost at least $3.5 billion. So, instead of backing Nevada senator Harry Reid's bid to release NWPA money to fund a dry-cask solution, reactor owners have chosen to sue Washington for its failure to provide a permanent resting place for the industry's radioactive excrement. By 2011, reactor owners had filed dozens of suits asking for $6.4 billion in damages and had won judgments totaling $956 million. Just defending itself in court has cost the government $170 million.[13]

12

Proliferation Dangers

"There is no technical demarcation between the military and civilian reactor and there never was one. What has persisted over the decades is just the misconception that such a linkage does not exist."

–"SOME POLITICAL ISSUES RELATED TO FUTURE SPECIAL NUCLEAR FUELS PRODUCTION," LOS ALAMOS NATIONAL LABORATORY, AUGUST, 1981

DESPITE GENERATIONS of glowing publicity about "safe, clean electricity" and the "peaceful atom," the nuclear industry remains fatally intertwined with nuclear weapons proliferation. Since its earliest days, the "civilian" nuclear energy program helped *provide* plutonium for America's nuclear arsenal. In 2000 alone, civilian reactors produced enough plutonium to make more than 34,000 nuclear bombs.

Back in the 1950s, the Pentagon called for the creation of a "civilian" nuclear energy program to fill a plutonium shortage that was slowing the production of nuclear weapons.[1] Dwight Eisenhower's Atoms for Peace program directly promoted nuclear weapons programs in India, Iran, Israel, and Pakistan.[2] Iran signed on to Ike's Atoms for Peace plan on March 5, 1957, and Washington gave Tehran a 5 MW US-built "research reactor" in 1967. With US encouragement, the despotic shah of Iran envisioned building as many as 23 nuclear reactors by the 1990s.[3] During Gerald Ford's presidency Dick Cheney and Donald Rumsfeld went even further, pushing for "a multi-billion-dollar deal that would have given Tehran control of large quantities of plutonium and enriched uranium—the two pathways to a nuclear bomb."[4]

As nuclear critic Dr. Helen Caldicott has observed, "Nuclear power has always been the nefarious Trojan horse for the weapons industry, and effective publicity campaigns are a hallmark of both industries." It was Stefan Possony, a consultant with the Pentagon's Psychological Strategy Board, who helped come up with the Atoms for Peace campaign. As Possony explained: "The atomic bomb will be accepted far more readily if, at the same time, atomic energy is being used for constructive ends."[5] This good cop–bad cop approach was given a boost when Walt Disney and his team were recruited to produce *Our Friend the Atom*, an animated children's film and accompanying book, featuring a benign and beaming genie who emerges from a magic lamp ready to bring light and warmth to the world.

The linked nature of nuclear energy and nuclear weapons was underscored with the revelation that Abdul Qadeer Khan, the "father of Pakistan's atomic bomb," obtained his blueprints for a uranium enrichment centrifuge from URENCO, a Dutch nuclear power company. (Khan later sold his bomb designs to other countries, including Libya.) The enduring unity of the military-nuclear-industrial complex was further underscored when AREVA, France's cash-strapped nuclear firm, announced plans to build reactor equipment in Virginia with a new partner—US arms manufacturer Northrop Grumman.[6]

Every country with a nuclear power reactor has access to the materials needed to produce nuclear weapons. Every country now possessing nuclear weapons (the United States, Russia, the United Kingdom, France, China, India, Pakistan, Israel, and North Korea) got its start operating "peaceful" nuclear reactors.

Of the 42 countries now possessing fissionable material, 22 have "civilian" facilities that can separate highly enriched uranium or produce plutonium. International treaties safeguard only about 1 percent of the world's highly enriched uranium and only 35 percent of the world's plutonium. While Russia and the United States account for 95 percent of the 30,000 nuclear warheads in existence, enough refined and unrefined fissionable material exists to build another 100,000 such weapons.[7]

An average nuclear reactor produces 20 to 30 tons of highly radioactive spent fuel each year. Each ton of spent reactor fuel typically contains around 10 kilograms of plutonium, enough to build a primitive nuclear bomb. Any country with minimal industrial skills can build a small "quick and dirty" reprocessing plant capable of extracting a bomb's-worth of plutonium a day.[8] Spent nuclear fuel can also be turned into "dirty bombs," in which conventional explosives are used to blast radioactive material over a wide area.

How the United States Brought Reactors to Post-War Japan

The post–World War II geopolitics that gave rise to Eisenhower's Atoms for Peace initiatives required convincing Japan to accept nuclear reactors built by US corporations. To help overcome the fears of Japan's atom-bomb-scarred residents, Washington and the CIA enlisted the help of Matsutaro Shoriki, a fascist and militarist who, after the war, had been jailed for three years as a war criminal but was then released without a trial. He had been in charge of government propaganda during the war and was the owner of the right-wing

Yomiuri Shimbun newspaper. Shoriki (whose CIA code name was "Podam") received tens of millions of US dollars to expand his newspaper and TV empire in order to broadcast the message of "a new Industrial Revolution by means of nuclear power."[9]

Shoriki went on to found Japan's Atomic Energy Commission (and became its first chairman). He had dreams of becoming the leader of a new, militarized Japan, and part of his plan involved "research into weapons technology, including nuclear weapons."[10]

Shoriki wasn't the only militarist with nuclear dreams. In 1940, long before the end of World War II, the Japanese army decided to pursue the creation of an atomic bomb and initiated work at the Institute of Physical and Chemical Research under the supervision of Yoshio Nishina. And in 1942, Japan's navy initiated its own program to create a "superbomb" as part of its F-Go Project ("F" stood for "fission").[11]

In the closing months of World War II, Tokyo's leaders—like their rivals in Washington and Berlin—were desperately racing to build the first A-bomb. Japan's secret weapons team was led by Dr. Nishina, a pioneering physicist who built Japan's first cyclotrons in the 1930s. In 1948, a US journalist who served as an investigator with the 24th Criminal Investigation Detachment in Korea claimed that Japan had succeeded in detonating a small nuclear device on August 12, 1945 in Hungnam, North Korea—three days before Japan's surrender.[12] (The claim was repeated in Robert K. Wilcox's 1985 book, *Japan's Secret War: Japan's Race Against Time to Build Its Own Atomic Bomb,* but critics have dismissed the story as lacking sufficient substantiation.)

After the war—and after US soldiers had demolished Japan's atomic cyclotrons—US corporations brought the Atoms for Peace program to Asia, selling Westinghouse and General Electric reactors to Japan. But every civilian reactor program carries an inherent weapons potential—an axiom that was acknowledged in a 1969 policy paper coauthored by Yasuhiro Nakasone, the director of Japan's Defense Agency: "For the time being, Japan's policy will be not to possess nuclear weapons. But it will always maintain the economic and technical potential to manufacture nuclear weapons and will see to it that Japan won't accept outside interference on the matter."[13] With US technical support (and US supplies of enriched uranium), Japan began to accumulate a growing stockpile of plutonium for use in its breeder reactor program. (Breeder reactors "breed" plutonium, the stuff of nuclear weapons.)

Japan's "Latent" Nuclear Weapons Capacity

Although Japan is a signatory to the Nuclear Nonproliferation Treaty, it has quietly amassed all the materials and technical know-how needed to produce nuclear weapons. As an investigative report by the National Security News Service (NSNS) concluded, "[Japan] has used its electrical utility companies as a cover to allow the country to amass enough nuclear weapons materials to build a nuclear arsenal larger than [that of] China, India, and Pakistan combined." The NSNS investigation revealed that Japan's stealth nuclear weapons potential was quietly abetted by "tens of billions of dollars worth of American tax-paid research that has allowed Japan to amass 70 tons of weapons grade plutonium since the 1980s."[14]

Three days before the Fukushima disaster, Tokyo governor Shintaro Ishihara openly called for Japan to begin turning its plutonium into nuclear weapons. "The fact is that diplomatic bargaining power means nuclear weapons," Shintaro told a reporter from Britain's *Independent* newspaper. "All the [permanent] members of the [United Nations] Security Council have them. . . . We should develop sophisticated weapons and sell them abroad," the governor insisted. "Japan made the best [jet aircraft] fighters in the world before America crushed the industry. We could get that back."[15]

On October 29, 2011, Japan's *Mainichi Daily News* electrified the nation with a report that confirmed Japan was, in fact, "a 'latent' nuclear weapons state." Since 1988, Japan has been permitted to operate a nuclear fuel reprocessing plant at Rokkasho. "Using its own enrichment technology," *Mainichi* reported, Japan has the ability, "in theory, to produce the raw materials necessary to build nuclear bombs." The newspaper noted that, for all the post-Fukushima talk about closing Japan's civilian power plants, "our leaders have failed to make any mention of the country's latent nuclear weapons capacity. The true elimination of our dependence on nuclear power," *Mainichi* concluded, "must include our abandonment of nuclear weapons possession." (The inconsistency between Washington's response to the theoretical risks posed by Iran's civilian nuclear energy program and those of Japan's well-established fuel-enrichment operations cannot pass without mention.)[16]

The IAEA: Promotion vs. Proliferation

The International Atomic Energy Agency (IAEA) was created to promote nuclear energy and control the proliferation of nuclear weapons. In practice, these proved to be contradictory goals since the spread of nuclear power inevitably opens the door to weapons programs.[17] One of the reasons there has been so little progress toward the disarmament goals articulated by the United Nations' Nuclear Nonproliferation Treaty is that the five permanent members of the UN Security Council—the United States, France, Great Britain, Russia, and China—happen to be the world's first five declared atomic powers.

The IAEA's shortcomings are evidenced by the fact that India, Israel, North Korea, and Pakistan have all found the means to join the club of nuclear-armed nations—without agreeing to the restrictions of the Nuclear Non-Proliferation Treaty. Every country with a nuclear energy program—31 operate more than 430 commercial power reactors and 56 operate around 240 research reactors[18]—has the potential to build nuclear weapons. Countries with present or past nuclear power ambitions include Algeria, the Arabic Gulf states, Australia, Azerbaijan, Bangladesh, Belarus, Brazil, Chile, Egypt, Estonia, Georgia, Ghana, Indonesia, Iran, Iraq, Ireland, Italy, Jordan, Kazakhstan, Latvia, Libya, Malaysia, Morocco, Namibia, New Zealand, Nigeria, Norway, the Philippines, Portugal, Syria, Thailand, Tunisia, Turkey, Venezuela, Vietnam, and Yemen.

More than 40 bilateral treaties allow the United States to sell nuclear fuel around the world—so long as all the nuclear waste is returned to the United States (ostensibly a weapons-proliferation-prevention strategy). The 2008 US–India civilian nuclear agreement—a long-term agreement of cooperation between India, the United States, and other global nuclear technology providers—envisions constructing dozens of atom-powered plants in India, a country that has refused to sign the Nuclear Nonproliferation Treaty (NPT) and, in 1998, exploded five "nuclear devices" at its Pokhran test site. While this scheme will generate a lot of global cash flow for the nuclear marketers and their government boosters, it could deal a deathblow to nonproliferation hopes by allowing India to become the first country to buy nuclear materials without being a party to the NPT.[19]

The Nuclear Powers vs. Disarmament

The five original nuclear powers—the United States, France, Great Britain, Russia, and China—encouraged the spread of nuclear-fueled power plants. However, to protect the nuclear status quo, Article 4 of the 1967 Nuclear Nonproliferation Treaty (NPT) declared that, while all signatories to the treaty would be allowed to pursue the peaceful use of nuclear energy, only the Big Five could possess nuclear bombs.

Under Article 6, the NPT's signatories pledged to work toward complete nuclear disarmament. Forty-five years later, the nuclear powers still have not disarmed (and the United States has actually embarked on new nuclear weapons programs). The only countries ever to have voluntarily abandoned their nuclear weapons ambitions are Argentina, Brazil, Libya, South Africa, and Taiwan.[20]

The United States could do more to promote the nonproliferation agenda but Washington's economic ties to the corporate weapons industry have undermined the country's NPT pledge to to "cease the nuclear arms race." At the same time, Washington's foreign policy also sends mixed messages that further undercut the NPT.

In 2005, Washington ended a 30-year moratorium on nuclear sales to India (imposed because the New Delhi government refused to become a signatory to the NPT). In June 2012, US-based Westinghouse secured a deal to build a 1,000-MW nuclear reactor in Gujarat. While Washington has expressed little public concern over Saudi Arabia's plan to construct 16 nuclear power plants,[21] the United States has threatened Iran for pursuing a civilian energy program. Both Saudi Arabia and Iran are signatories to the NPT and thus equally entitled to build and operate nuclear power plants. Meanwhile, Israel—which has refused to sign the NPT and refuses to permit international inspections of its nuclear site at Dimona—is known to possess the only nuclear arsenal in the region. Washington's inconsistency on this issue sends a destabilzing message to the world community.

Enrichment and reprocessing make nuclear weapons possible by generating plutonium. To control weapons proliferation, Presidents Ford and Carter suspended support for reprocessing plants. Ronald Reagan reversed the ban on commercial reprocessing, but Bill Clinton then reinstated the ban when he took office. And on February 6, 2006, George W. Bush announced the creation of the Global Nuclear Energy Partnership (GNEP) to address the problem of nuclear waste by reviving the US reprocessing program (while preventing other countries from building their own enrichment and reprocessing plants).[22]

Far from being proliferation-resistant, the GNEP actually threatened to boost stockpiles of enriched uranium and plutonium because it allowed some of the 25 GNEP members (including the United States and France) to retain the "right" to reprocess uranium to supply other GNEP countries. The problem is that reprocessed plutonium is easily weaponized. The 250 metric tons of plutonium already produced worldwide are sufficient to build 40,000 atomic bombs. Under GNEP, the amount of additional plutonium from reprocessing US spent fuel would top more than 500 metric tons.[23]

The Bush administration announced plans to build a new $80 billion reprocessing plant to handle at least 2,000 tons of spent fuel a year. However, the United States has tried, and failed, to build reprocessing plants on three previous occasions. The first facility in West Valley, New York, was shut after six years (taxpayers have so far spent $4.5 billion to clean up its contamination), and the other two plants were declared inoperable.[24]

In June 2009, the DOE cancelled critical funding for GNEP, explaining that the Obama White House was "no longer pursuing domestic commercial reprocessing, which was the primary focus of the prior Administration's domestic GNEP program."[25] In the meantime, however, in April 2010, Washington signed off on a deal that permits India to reprocess its own nuclear fuel. The arrangement has raised fears in neighboring Pakistan, which is now expected to embark on a "significant nuclear military buildup."

Ending the threat of nuclear proliferation requires the closure of all enrichment and reprocessing facilities. And nuclear disarmament would be a long-overdue step toward a more peaceful and secure future. It would also liberate a wealth of national treasure and scientific expertise that could be redirected toward the promotion of smart and sustainable carbon-free energy technologies. However, as long as the United States continues to plan for "winning" a nuclear war, maintains plans for pre-emptive nuclear strikes, and works to expand its nuclear arsenal, other countries will have an incentive to pursue their own nuclear options.[26]

13
Reactors, War, and Terrorism

Nuclear terrorism is still often treated as science fiction—I wish it were.

–FORMER UN GENERAL SECRETARY KOFI ANNAN

FOR NEARLY HALF A CENTURY, civilian nuclear reactors have posed a clear and present danger to national security. The government has been largely silent on the danger and the public has been unaware of the peril. But on June 7, 1981, an Israeli jet raid on the French-built Osirak nuclear reactor situated on the outskirts of Baghdad blew the lid off Pandora's box. The secret was out: "peaceful, civilian" nuclear reactors are strategic targets.

This last dirty secret of the Atomic Age, once it is understood, places the promoters of nuclear power in an ironic no-win position. Their dilemma is simply this: *in an age of nuclear technology, nuclear weapons, and nuclear reactors can no longer coexist.*

The danger was spelled out in a *Current Affairs Bulletin* article in 1982. In the event of a major war, author Brian Martin wrote: "It is possible that nuclear power reactors would be nuclear targets, because of their high economic value, because of their capability of producing plutonium for making nuclear weapons, or because of the devastating radioactivity that would be spread about. . . . The main concentrations of large nuclear reactors are found in the United States, Europe, the Soviet Union, and Japan, that is, those areas most likely to be involved in nuclear war."[1]

"Reactors are seductive targets," said Bennett Ramberg, then a research associate at the former Center for International and Strategic Affairs at the University of California at Los Angeles, shortly after the Israeli attack. "They represent one of the greatest concentrations of capital investment a nation is likely to possess."[2] The danger is not merely theoretical. In addition to the 1981 attack, Israel destroyed a suspected nuclear site in Syria in 2008 and has threatened to bomb nuclear research facilities in Iran. The United States has also implicitly threatened to bomb Iranian nuclear sites.

In the United States, 108 million people—nearly a third of the population—now live within 50 miles of a nuclear plant.[3] More than four million now reside within 10 miles (a 16.9 percent increase over the previous decade).[4] In the 1960s and early 1970s, reactors were not thought to be attractive targets, since they were small (1,000 MW and less) and missiles were only accurate to within a half mile of a distant target. But as Oak Ridge National Laboratory researchers Conrad and Rowena Chester correctly predicted in 1976, "From the 1980s through the end of the century, the situation will be quite different."[5]

A Nuclear Blind Spot

A disturbing conspiracy of silence seems to surround any recognition of this pervasive danger. The Atomic Energy Commission's 1962 study *The Effects of Nuclear Weapons* contains the first reference to the effect of atomic war on electric power generation. While admitting that Hiroshima's electrical system had "suffered severely" from the atomic blast that marked the end of World War II, the study concluded with the reassurance that "although the pole lines would have required some rebuilding, the general damage was such that it could have been repaired within a day or so with materials normally carried in stock by utility companies."[6]

In 1962, of course, nuclear power stations had not yet become a part of the American landscape. Yet when the Department of Energy (DOE) issued an updated version of *The Effects of Nuclear Weapons* in 1977, the chapter on nuclear threats to the American power system was reprinted without elaboration.

In 1975, then secretary of defense James Schlesinger unveiled a new nuclear strategy of "flexible response" (a more limited alternative to all-out "mass assured destruction") and the National Academy of Sciences issued a report that, for the first time, acknowledged that civilian reactors "may pose special problems" in a nuclear war. But the report failed to address these problems.

In 1979, the Congressional Office of Technology Assessment (OTA) published a major study on the impacts of nuclear conflict, which warned that, in the event of an enemy missile attack, "the most vulnerable element of the US economy was judged to be the energy supply system." Remarkably, America's nuclear power facilities were not mentioned. The OTA report equated the "energy supply system" with the nation's 300 oil refineries.[7] The OTA's hypothetical attack scenario for Philadelphia targeted four conventional power plants—Delaware, Schuylkill, Southwark, and Richmond—but said nothing of the state's nuclear plants at Shippingport, Three Mile Island, and Peach Bottom—the latter located about 70 miles west of Philadelphia.

The consistent omission of references to nuclear reactors in these assessments of the vulnerability of the United States to attack is exceedingly odd, given the strategic importance of the electric power grid. But the silence could have been intended to avoid drawing undue attention to the reason civilian reactors have a special role in the thinking of nuclear war-gaming strategists.[8]

The conspiracy of silence is even more remarkable give that Sir Brian Flowers, a prominent British nuclear physicist, described the problem back in 1976.[9] Flowers pointed out that if nuclear power plants had been built and deployed in Europe before World War II, large parts of Europe would be uninhabitable today because of conventional warfare and conventional sabotage directed against those nuclear plants.

In 2011, the international organization, Mayors for Peace, drew new attention to one of the chilling conclusions from Flowers's 35-year-old report: "If your enemy has nuclear power plants then you don't need nuclear weapons—you can achieve much the same effect by tripping up the power plants and making them melt down."[10] In most cases, this can be done just by disabling both on-site and off-site power, and then compromising the containment system for good measure. The spent fuel bay is an even easier target, and as a bonus, it contains even more radioactivity than is present in the reactor core.

What Happens If . . .

While billions of dollars have been spent to "harden" military bases against nuclear attack, large nuclear power complexes remain strategic sitting ducks. If a single one-megaton nuclear warhead can unleash the blast damage that decapitated Mount St. Helens, what could happen if a nuclear missile were to crash near the containment vessel of a nuclear reactor?

In the early 1980s, two Massachusetts Institute of Technology scientists, Kosta Tsipis and Steve Fetter, set out to investigate just such an eventuality, exploring what would happen if a single one-megaton weapon were exploded at a 1,000-MW reactor.[11] (To put this in perspective, the payload of a nuclear-armed Minuteman missile is approximately one megaton, equal to about 50 times the explosive power of the Hiroshima bomb).

"The lethal zone for the detonation of the weapons on the reactor would be more than 500 square miles, an area a third larger than the lethal zone created by the detonation of the weapons alone," Tsipis and Fetter estimated. A fireball rising 12 miles into the sky would carve a crater 400 feet deep and spread a lethal blanket of molten rock and white ash over 4,500 square miles. In addition to the millions who would die immediately, the reactor-bomb detonation would result in a 30 percent increase in delayed casualties, over what might be expected from detonation of the weapon alone. Anyone who could not relocate swiftly could be dead within two months.

While the area contaminated by the reactor-bomb combination would not be "significantly" larger, Tsipis and Fetter state, "the length of time a given area is made uninhabitable . . . [is] ten times larger in the case of the reactor-bomb combination." This is because a bomb yields generally short-lived fallout, while blasting open a reactor core would scatter long-lived isotopes over the land. The total loss would be "of the order of 4,000 square-mile-years; consequently it would result in vast capital losses that would dwarf losses from any other single natural disaster in the history of the modern world," the MIT study concluded. Hundreds of square miles would remain "inaccessible for over a hundred years, creating a permanent monument to the event. It is doubtful that decontamination procedures could mitigate these losses, even if decontamination proved cost-effective." All this, remember, from *one* bomb hitting only *one* reactor.

Tsipis and Fetter cautioned that these estimates rested on a "best case" assumption that "there will be no neutron activation or fission of reactor materials by the fast neutrons of the weapon." If that were to happen, there might be a *second* explosion *within* the first explosion.

Tsipis and Fetter's conclusion is one that has certainly not been lost on the Pentagon war gamers. Striking a reactor with a missile renders a "dirty explosion" even dirtier and, in so doing, it becomes an effective way to devastate large portions of an entire nation with minimal effort. If World War III were to break out, a missile attack on reactors in the Northern Hemisphere would envelop the entire United States in fallout and eradicate western Europe. Only Russia—with fewer reactors and a greater landmass—would escape total obliteration.

Fortified Fallout

Tsipis and Fetter also looked at the probable results of a single nuclear weapon exploding over a single civilian nuclear waste storage site at West Valley, New York. At the time of their study, in the early 1980s, this site (located near Buffalo) included a single steel tank containing 600,000 gallons of high-level radioactive waste that had been generated by the defunct Nuclear Fuel Services reprocessing plant. (Compare that to the 37 million gallons of radioactive wastes that are now stored in 49 leaking tanks at the DOE's Savannah River Site, 60 miles upwind of Charleston, South Carolina.)[12]

"If a one-megaton weapon were exploded directly above these steel tanks," Tsipis and Fetter predicted, "their radioactive contents would rise with the fireball and join the weapons debris" to form the deadliest fallout imaginable. It would extinguish all life in an area the size of Utah (85,000 square miles). One year later, this no-man's-land would still be the size of the state of Washington. After 10 years, the dead land would equal the acreage of West Virginia. After 100 years, a tract of land the size of Rhode Island would remain permanently poisoned with radioactivity. All the legacy of a single one-megaton bomb.

Tsipis and Fetter concluded that it would be "unrealistic to expect that, in a nuclear exchange, only one or just a few ground bursts of nuclear weapons will occur." More likely, the US continent would reel and buckle under "the near simultaneous detonation of hundreds and even thousands of megaton-sized nuclear warheads."

Underground Zero?

In 1980, when vice president George H. W. Bush spoke of a "winnable" nuclear war with 20 million "acceptable" dead, his glib optimism was based on the narrow focus of war gamers who are not overly concerned about the loss of lives. As the Reagan-era wargamers understood it, the "game" would be won if they could assure the "survivability" of their nuclear response. But what about the "survivability" of nuclear reactors and the civilian populations? As Thomas K. Jones, President Reagan's Deputy Under Secretary of Defense for Strategic and Theater Nuclear Forces blithely observed: "Everybody's going to make it if there are enough shovels to go around. Dig a hole, cover it with a couple of doors and then throw three feet of dirt on top."[13]

In the dawn of the Atomic Age (at least as early as 1958), there were plans to place reactors underground. As physicist Edward Teller (the "father of the atomic bomb") was prone to remark, nuclear reactors are not foolproof. Given this reality, it was argued that "undergrounding" would provide the best protection against a catastrophic reactor mishap. Since undergrounding would also enhance the survival of nuclear complexes in the event of military or terrorist attack, why was it that all of America's plants wound up being built on the surface?

An answer can be found in an obscure report in a 1974 issue of *Nuclear Safety* magazine.[14] "Siting of nuclear plants underground has not proliferated for various reasons," the author explained, including the lack of suitable sites, licensing uncertainties, and "apparent economic penalty." The report projected that the cost of excavating 770,000 cubic yards of earth to safely bury the Indian Point 3 reactor would have cost the private utility company an additional $21 million (nearly $97 million in 2012 dollars), adding 10 percent to the cost and 2.5 years of additional construction to the completion of the facility.[15] The reason why reactors were placed in plain sight was simply because plant safety and national security were not cost-effective for utility investors.

What Are the Options?

MIT researchers Tsipis and Fetter called for a formal scientific investigation of the dangers disclosed in their report. They also challenged government leaders to begin an international public education campaign on the danger of nuclear weapons. While the dangers of a nuclear accident are apparent and appalling, they noted (and were borne out by the events at Chernobyl and Fukushima), an explosion or meltdown at a nuclear reactor is nothing compared to the detonation of the smallest tactical nuclear weapon. Further, they charged, "the nuclear industry, the scientific authorities, and the government in this and other countries must be faulted for maintaining this state of public ignorance."

Meanwhile, the potential for nuclear terrorism continues to grow. From a terrorist perspective, a nuclear power plant is a ready-made weapon, already positioned to do the most damage. Some 53,000 metric tons of highly radioactive spent fuel sit in storage pools upwind of major US cities, and 90 percent have no protection against a terrorist attack. Blowing up a spent fuel storage site could unleash a devastating plume of fallout over civilian populations. Benjamin Sovacool, a public policy professor at the National University of Singapore, estimates that such an attack could kill 43,700 people instantly, cause 518,000 subsequent cancer deaths, and require around $2 trillion in decontamination and restoration costs.[16] These facts have not gone unnoticed; the mastermind of the 9/11 attacks reportedly claimed that Al Qaeda had planned to strike US reactor sites.

Although it has been little publicized, there have been a number of terrorist attacks directed at nuclear installations. In 2001, the energy information organization WISE-Paris, published a report documenting 13 such attacks between 1977 and 1999. One of the more remarkable assualts occurred in France on January 18, 1982, when a nighttime raid sent five RPG-7 rocket-propelled grenades streaking over the Rhone in the direction of the unfinished Superphénix breeder reactor. Two of the rockets struck the facility, causing minor damage. (The attack was initially blamed on the international terrorist known as "Carlos the Jackal" but, in a bizarre twist, Chaim Nissim, an elected member of the Swiss Green Party, ultimately claimed responsibility in 2003.)[17]

Given this past history of attacks targeting nuclear facilities, America's—and indeed the world's—close links to the nuclear power industry need to be radically rethought. Each of the civilian reactors operating in the United States today constitutes a potential homegrown doomsday machine. In an age of nuclear-armed ballistic weapons, each new reactor placed in operation becomes a new target—with a 40-to-1 payoff in death for any incoming missile that strikes it. (And it wouldn't take a missile strike. A land-based terrorist attack could potentially disable a nuclear plant's power systems, leading to a meltdown and/or spent fuel ignition.)

The conclusion is as inescapable as it is ironic: surface-based nuclear reactors are not consistent with national security. Not ours, not anyone's. Nuclear reactors *endanger* national security.

It follows that any sincere proponent of nuclear preparedness would have to press for the elimination of nuclear reactors as inexcusably tempting targets of opportunity. On the other hand, advocates of nuclear-generated electric power will be compelled to oppose the continued existence of nuclear weapons—one of the few forces capable of breaching the protective shield of a power reactor. It appears the nuclear fraternity may have come to loggerheads. Atomic power? Atomic bombs? We can't have it both ways.

14

"New, Improved" Reactors Won't Save Us

The tools of the academic designer are a piece of paper and a pencil with an eraser. If a mistake is made, it can always be erased and changed. If the practical-reactor designer errs, he wears the mistake around his neck; it cannot be erased.

–ADMIRAL HYMAN RICKOVER, THE "FATHER" OF THE US NUCLEAR NAVY

NUCLEAR PROPONENTS frequently talk about the promise of a new generation of reactors that will overcome the cost, safety, waste, and other problems that have plagued the industry so far. Thirteen countries (led by France, Japan, and the United States) have joined the Generation IV International Forum (GIF), which is committed to invest an estimated $5 billion over 15 years to design the next generation of nuclear reactor. GIF has identified six new designs planned for deployment between 2020 and 2030. While today's light-water reactors reach operating temperatures of 330 degrees Celsius, the new reactors are designed to operate at temperatures ranging from 510 to more than 1,000 degrees Celsius (1,832 Fahrenheit). The most important thing to grasp about these so-called fourth-generation reactors is that these new, untried designs are neither risk-free nor waste-free.[1]

We have already discussed the inherent design flaws of the "third-plus-generation" AP1000 reactors. (See "'Renaissance' Reactors: Plagued with Problems," page 3.) As for the fourth-generation designs, it would be decades (if ever) before any working prototypes could be built. While it is unwise to write off new technologies entirely, the possibility that fourth-generation reactors could offer a timely solution to the world's climate or energy problems remains extremely remote. A prototype of a very-high-temperature reactor might be completed by 2021, but (according to estimates by the French government, which has the greatest experience with state-supported nuclear power) none of the other designs would be available even to begin commercial construction until 2045.[2] Here, then, is a rundown of some of the leading alternatives.

The very-high-temperature reactor (VHTR) is a helium-cooled reactor that has received major support from the US Department of Energy (DOE). Built to run at fuel temperatures of 2,300°F (1,260°C), the VHTR is designed to produce industrial process heat, commercial electricity, and hydrogen. Because this design is best suited for generating heat and hydrogen, the VHTR would not be expected to play a leading role in filling the country's electric power needs—even if the nation shifted to a "hydrogen economy."

Pebble bed modular reactors (PBMRs) were marketed to the government of South Africa as "safe, clean, cost-competitive, versatile, and adaptable." Westinghouse officials boldly predicted that South Africa's PBMR technology would become "the world's first successful commercial generation IV reactor." Work on this helium-cooled, high-temperature reactor (HTR) began in 1999, but after more than a decade, the government cancelled work in February 2010, explaining that it "could no longer justify putting more money into the project."[3]

The gas-cooled fast reactor is helium-cooled and uses novel core configurations aimed at increasing efficiency through very high operating temperatures. It presents control problems, durability problems, and heavy waste impacts after decommissioning.

The lead-cooled fast reactor could be built in various sizes, with modular components having a long refueling interval. It would supposedly be cooled by natural convection. It has component compatibility problems, extreme pressure and temperature characteristics, uncertain maintenance, and the possibility of environmental contamination with lead.

The molten salt reactor uses nuclear fuel dissolved in molten fluoride salt flowing into a graphite core. It requires super-durable structural materials and poses proliferation risks.

The sodium-cooled fast reactor would breed plutonium and retain it within the reactor. Overheating would cause the chain reaction to slow down. Sodium is highly corrosive and reacts explosively in contact with water or air, making this design very demanding. Sodium coolant problems, combined with uncontrolled costs and recurrent shutdowns, led to the abandonment of France's Superphénix liquid-sodium-cooled fast-breeder reactor.

The super-critical water-cooled reactor. Operating at much higher temperatures and pressures than current reactors, this design would theoretically offer modestly better thermal efficiency (albeit with greater accident risks) and lower electricity costs—although not to the point of being competitive with renewable sources, much less gas or coal.

Traveling wave reactors (TWRs) are being promoted as "a financially and socially attractive emission-free energy that is safe and sustainable."[4] Unlike traditional reactors that require enriched uranium to operate, a TWR needs only a small initial amount of enriched uranium as a trigger, after which a chain reaction begins to slowly consume a core of depleted uranium. TWRs are "breeder" reactors capable of making and consuming their own fuel. Since the major feedstock would be the "nuclear garbage" created by traditional reactors, TWRs are presented as offering a solution to the problem of stored nuclear waste.

The technology received a major boost in 2010 when Microsoft chairman Bill Gates publicly praised "traveling wave" breeder reactors and became a major investor in TerraPower, a TWR pioneer. TerraPower's backers claim that computer simulations demonstrate that "a wave of fission moving slowly through a fuel core could generate electricity continuously for well over 50 to 100 years without enrichment or reprocessing."[5] The proposed benefits of TWRs include the reduction of stockpiles of stored nuclear wastes; the ability to tap a freely available fuel that could supply energy "for thousands of years"; the elimination of enrichment and reprocessing; a reduction of the proliferation risks associated with uranium enrichment; and extended reactor service life.

But TWRs still carry many of the basic flaws of traditional and new, fourth-generation (or Gen IV) reactors: they produce nuclear wastes that need to be removed at the end of the plants' working life; the designs have not been built or tested; it will take a decade before the first TWR could be built; and a single reactor could cost $4 billion. Given these impediments, it's not surprising that billionaire Gates and his other TerraPower partners are lobbying the DOE for billions of dollars in R&D subsidies.[6]

Mini-Reactors to the Rescue?

With hopes for a nuclear renaissance waning, the American Nuclear Society (ANS) has proposed a mini-renaissance of small modular reactors (SMRs). SMRs (small enough to fit in a garage) could power homes, factories, and military bases. SMRs would be cheaper to build and easier to site than large reactors. Some would have no "footprint" because they could be installed underground. After a working life of 7 to 10 years, they could be dug up and relocated for reuse elsewhere (after a suitable cooling-off period). The retired units would still remain a disposal risk, though.

In February 2010, the NRC invited companies to bid for $39 million in federal funds set aside for mass-producing "mini-nukes." Energy secretary Steven Chu followed up with an op-ed in the *Wall Street Journal* praising SMRs.[7] The DOE's Savannah River National Laboratory (SRNL) in South Carolina is now working to see that the state becomes a pioneer in SMR technology and has announced plans to collaborate with New Mexico's Gen4 Energy, Inc. (previously known as Hyperion Power Generation) to build 10 mini-reactors (700 MW each) at a plant employing as many as 500 workers. SRNL has announced that it is taking a leadership role in this effort by establishing a demonstration project for a small reactor to be built by GE-Hitachi.

An international SMR conference held in Columbia, South Carolina, in April 2011, attracted scores of companies and agencies ranging from Westinghouse, AREVA, and GE to the International Atomic Energy Agency, the China National Nuclear Corp., and the Iraq Energy Institute. Also attending were senator Lindsey Graham, three members of Congress, and representatives from the DOE, NRC, US Army, and many US utilities. Conference organizers hoped the event would "cement the Midlands and South Carolina as a center for the development and production of the mini-reactors."[8]

Mini-reactor proponents envision new markets in Third World countries that cannot afford traditional nuclear plants. But mini-reactor detractors note that SMRs would still depend on a costly, inefficient, and hazardous fuel cycle that generates intense heat loads, employs dangerous materials (like helium and highly reactive sodium), and produces nuclear waste. Building mini-nukes would decentralize and scatter all the operational risks of supplying, maintaining, safeguarding, and dismantling nuclear reactors. A mini-nuke would still require its own control room operators and security personnel, and because of economies of scale, smaller reactors could wind up costing more than large plants. Tom Clements, a nuclear power critic with the environmental advocacy group Friends of the Earth, charged that the proposal to develop SMRs at the DOE's Savannah River site was simply a ploy by the nuclear industry to avoid licensing oversight from the NRC. (As the Savannah River Site Citizen's Advisory Board pointed out back in 2000, by tradition, "DOE self-regulates its nuclear facilities."[9]) The DOE denies the charge.

Clements's charges are given credence, however, by the remarks of John "Grizz" Deal, the co-founder and CEO of the company now known as Gen4 Energy. While attending a gathering of engineers and defense contractors in Britain, Deal volunteered the following comments to a reporter for *Inc.* magazine: "We have great connections with the US military, and they're not bound by regulatory agencies," Deal began. "We have political connections who see us as having a chance to be the ones who lead a US nuclear renaissance. We're overcommitted on orders. We keep hearing, 'When will you be able to ship us one?' And we hear it in a lot of different accents. And we're not worried about the approval process. The reception from the agencies has been great, and we expect the process to take less than two and a half years. This thing was designed to be approved. We've got $200 million of US National Lab research behind us."[10]

PART TWO

Cover-ups and Consequences

15

Japan: Living with the Consequences

This time no one dropped a bomb on us. . . . We set the stage, we committed the crime with our own hands, we are destroying our own lands, and we are destroying our own lives.

—**HARUKI MURAKAMI, JAPANESE NOVELIST**

IN JAPAN, as in every other nuclear nation, the unresolved problems and unthinkable consequences of nuclear energy were ignored, minimized, or hidden. The former residents of Fukushima now are paying the price for the hubris and denial that fueled the Atomic Age. Tragically, the fallout damage from the Fukushima disaster might have been minimized had Japan's nuclear watchdogs placed a greater priority on public health than private profit. For example, as the *Kyodo News* newspaper revealed, Kenkichi Hirose, head of Japan's Nuclear and Industrial Safety Agency (NISA), successfully asked the country's Nuclear Safety Commission (NSC) to cancel a May 2006 safety study that called for a larger "disaster mitigation zone" around nuclear plants "for fear of spreading concerns about nuclear power."[1]

For the most part, the 100,000 people who fled their homes (and the 1.5 million affected by dangerous fallout levels well beyond the official 12.5-mile evacuation zone) were left to their own devices. Worried parents were forced to buy their own Geiger counters and dosimeters. In the early days of the crisis, government officials failed to alert the populace to the mounting radiation dangers, and when Tokyo *did* respond to the problem, it was simply to announce that the official "acceptable" level of radiation in food was being increased by a factor of five.[2]

When troubling levels of radioactivity showed up in tap water at Tokyo schoolyards, the government responded by increasing the level of "permissible" radiation exposure for children 20-fold. This proved too much for Japan's radiation safety advisor, who resigned, calling the new rules "inexcusable."

By July 2011, it was clear the radiation had entered the human food chain. Radioactive cesium was found in the breast milk of one-third of 27 women tested near Fukushima Prefecture in May and June of 2011. Tea leaves picked by Tokyo schoolchildren were found to contain 6,000 becquerels per kilogram (Bc/kg) of cesium-137. Straw collected 45 miles from Fukushima was contaminated with radioactive cesium at levels equal to those found around Chernobyl. In July, after recalculating its estimates of radioactive releases, TEPCO warned of a dramatic increase in "hot particles"—the potentially deadly particles that cause a "metallic taste" when inhaled or swallowed.

In October 2011, radioactive strontium was detected in Yokohama, 155 miles from Fukushima, and mushrooms in Yokohama (165 mies from the striken reactors) were found to be seriously contaminated with radioactive cesium. The news came too late to protect the 794 people (including 258 children) who picked and ate the mushrooms.

Eight months after the disaster, a joint study by Tokyo University and the University of Tsukuba revealed that water from the Abukumagawa river watershed was still pouring 50 billion becquerels of cesium-134 and cesium-137 into the ocean every day. Cesium-137 was even detected in whales swimming offshore.[3] Rice in five locations 35 miles from Fukushima registered unsafe levels of cesium, forcing more than 150 farms supplying 192 tons of rice to halt operations. Across Japan, beef cows were contaminated with cesium-137. Between August and December 1, cesium-contaminated beef reportedly was served to 433 schools and 26 kindergartens in 18 prefectures and 46 cities. School lunches in Miyagi Prefecture included beef spiced with 1,293 Bq/Kg of radioactive cesium. After revelations that 180,000 students had consumed cesium-contaminated beef, the government finally agreed to purchase radiation monitors for use at schools.[4] A year after the Fukushima meltdowns, downwind residents were still complaining about the government's failure to test their children for radiation exposure or thyroid cancer. In wasn't until February 2012, that the Education Ministry finally began monitoring radiation levels on a real-time basis at schools, parks, and playgrounds.[5]

In March 2012, Tokyo announced a 20-fold reduction in the amount of cesium isotopes allowed in children's food. In April 2012, Tokyo finally bowed to public outrage and lowered the "safe" level of cesium radiation in food from 500 to 100 Bc/kg. Levels of "permissible" cesium in milk were lowered from 200 to 50 Bc/kg, while levels in drinking water were reduced from 50 to 10 Bc/kg. By this point, however, distrust of the government had grown so pervasive that Aeon, the country's largest supermarket chain, had begun its own independent program to monitor food for radiation.

A Legacy of Poisoned Land

The fallout over soil, forests, and waterways has left the government facing an unprecedented (and, quite likely, impossible) cleanup challenge. In Fukushima Prefecture alone, one-third of the land—an area the size of Rhode Island—is contaminated by fallout. The region's $3.2-billion-a-year agricultural sector has been wiped out. While some land clearly will remain uninhabitable for generations, the government expects it may be neccesary to remove 100 million cubic acres of poisoned soil from 2,000 square kilometers (772 square miles). With much of the fallout absorbed by trees, vast tracts of forests face clear-cutting and the arrival of spring now brings a new risk—radioactive pollen blown through streets from city trees.

Japan's Environment Ministry believes radiation levels in the evacuation zone will fall 40 percent within two years but, even with a cleanup, the health risks of long-term exposure to remaining low-levels of radiation are unclear. The International Commission on Radiological Protection insists the general public should not be exposed to more than 1 to 20 mSv a year. (Tests conducted in December 2011, showed Fukushima residents were being exposed to annual doses exceeding 9 mSv.) Even with a massive decontamination effort, cesium-137 that has migrated into the soil will continue to resurface in plants and crops. The deeper you dig, the more contamination you remove—but this comes at a cost. As Atomic Energy engineer Shinichi Nakayama observes: "You take away the deeper layers and [radiation levels] fall more. But you take it all away and the ecosystem is destroyed.[6]

Tokyo spent $3 billion on decontamination work in 2011 and expected to spend twice that amount in 2012. The cost for decontaminating the area around the damaged reactors is projected to exceed $13 billion and take 40 years. Professor Tatsuhiko Kodama of the University of Tokyo's Radioisotope Center projects that a responsible cleanup of all the land poisoned by TEPCO's fallout could cost nearly $10 trillion.

Although the government utterly failed to accurately inform the public about the radioactivity contained in food and certain consumer goods, Tokyo's leaders now insist that it is safe for some refugees to leave their relocation squats and begin returning home. To justify this, Tokyo has determined that radiation levels 10 times greater than pre-accident background radiation can now be considered "safe." Fukushima governor Yuhei Sato urged displaced residents to return to their homes and even offered returning evacuees the promise of job opportunities. These include "decontamination jobs."

Fallout in the Waves

While the ocean's impact on Fukushima was sudden and specific, Fukushima's environmental impacts on the ocean will be widespread and long lasting. In the first, desperate days of the reactor calamity, the failure of emergency cooling systems on the General Electric Mark I reactors forced TEPCO to cool the seething reactor cores with seawater. With no place to store the irradiated coolant, TEPCO dumped a million gallons (11,500 tons) of seawater back into the Pacific Ocean—with radiation levels 7.5 million times the legal limit. TEPCO told the public that no more than 15,000 terabecquerels had been released into the ocean, but the French Institute for Radiological Protection and Nuclear Safety (IRSN) subsequently placed the estimate closer to 27,000 terabecquerels.

China's State Oceanic Administration reported finding nearly 100,000 square miles of the Pacific tainted with radioactive iodine, strontium, and cesium at levels 300 times above normal. (Cesium-137 is absorbed by phytoplankton, zooplankton, and kelp that are ingested by fish, marine mammals, and humans.) On December 1, 2011, the IRSN reported that the Fukushima disaster had caused the worst ocean contamination in world history—100 times greater that Chernobyl's pollution of the Black Sea.

In the first days following the accident, nearly 13,500 terabecquerels of seaborne cesium-137 were expected to pass the Philippines before turning north and heading east along the Kuroshio current.[7] Computer models indicated that the huge swirl of radioactive water was heading for Hawaii and could reach the West Coast of North America by early 2013. But it took only a month for *airborne* radioactive iodine-131 to show up in kelp beds off the West Coast; California State University scientists found levels 250 times normal in kelp sampled in the waters off the southern California.[8]

Fukushima: One Year Later

A year after the meltdowns, high levels of radioactivity continued to be found throughout Japan. In March 2012, dangerous levels of alpha radiation were reported in Iwaki City, 31 miles from Fukushima. Concentrations of cesium isotopes were rising in Tokyo Bay, and one-third of the fish caught in Onuma Lake contained dangerous levels of radioactive cesium. Meanwhile, the simmering remains of Fukushima Daiichi continued to pose a threat of airborne iodine-131 and strontium-90 exposure in the United States.

Despite Tokyo's December 2011 announcement of a "cold shutdown" of the three damaged reactors at Fukushima, University of California researchers testing milk in the San Francisco Bay Area on January 16, 2012, continued to record rising levels of Fukushima fallout—including the highest cesium levels in six months. At the same time, the Australian Radiation Protection and Nuclear Safety Agency was reporting a "radiation cloud" over the country's east coast with radiation levels "eight times normal."[9] University of California tests released on February 7, 2012, showed even higher fallout levels, with radioactive cesium in Bay Area milk registering 150 percent of the EPA's safety limit, and the April 2012 readings were nearly double the EPA limit. Researchers explained that the increase was due to the long half-lives of cesium isotopes and the fact that the milk was obtained "from a Bay Area organic dairy where the farmers are encouraged to feed their cows local grass."[10]

On January 24, 2012, TEPCO confirmed that radioactive cesium was still leaking from Units 1 and 3 at the rate of 70 million becquerels per hour, a 20 percent increase over December levels. Possible explanations included "agitation" stemming from the January 1 earthquake and—a more troubling prospect—leaks from broken containment vessels.

The Human Impact

A year after being driven from their homes, farms, and businesses, *Japan Today* reported, "tens of thousand of evacuees are still in limbo, unable to return." More than 150,000 evacuees were forced to survive crammed inside government-run relocation camps. "Some say we can go home after 20 or 40 years," one displaced resident complained, "but what are we going to live on until then?"[11]

With nearly two million people demanding compensation, TEPCO offered an initial payment of 120,000 yen ($1,519) per month for "mental suffering" but required evacuees to reapply for the payment every three months. TEPCO discharged some of its "obligations" with a single onetime payment of $13,045. To apply for additional reparations, survivors needed to wait six months and then fill out a 60-page TEPCO application that came with 150 pages of instructions. TEPCO offered the 1.5 million affected people living outside the exclusion zone a lump-sum payment of 80,000 yen ($1,012) (400,000 yen, or $5,062, for children and pregnant women), but these payments were only in effect until December 31, 2011. TEPCO insisted that anyone accepting payments had to agree not to seek additional compensation in the future—that is, when the inevitable onslaught of cancers would begin to emerge.

Meanwhile, a growing number of Japanese citizens are starting to show early symptoms of radiation poisoning, including lethargy, nausea, hair loss, skin rashes, and loss of teeth.[12] Five months after the meltdowns, physicians at the Funabashi Futawa Hospital (located in Chiba Prefecture, 124 miles from Fukushima) were reporting "increased nosebleeds, stubborn cases of diarrhea, and flu-like symptoms in children."[13] On February 28, 2012, doctors reported that one-third of Fukushima's children were found to have developed "lumps" in their thyroids.

Looking to the Future

While Japan's new prime minister Yoshihiko Noda has pledged to pursue alternative energy, he also reserved the option to resurrect a "safe" nuclear power program. In September 2011, only six months after the Fukushima meltdowns, the *Yomiuri Shimbun* reported that Noda was "insisting that a stable electric power supply utilizing nuclear power plants is essential for economic growth."[14] On March 11, 2012, the first anniversary of the earthquake and tsunami, Noda delivered a national address that made no reference to the ongoing nuclear crisis. It was another story when 78-year-old Emperor Akihito (who also is a world-renowned biologist) delivered a national address on the same day and proclaimed:

> As this earthquake and tsunami caused the nuclear power plant accident, those living in areas designated as the danger zone lost their homes and livelihoods. . . . In order for them to live there again safely, we have to overcome the problems of radioactive contamination, which is a formidable task.[15]

What happened next unleashed a tsunami of public anger. Japan's corporate and state-run TV stations censored the emperor's speech, deleting his every mention of the "nuclear power accident" and the "formidable task" of dealing with "radioactive contamination." On March 20, demonstrators descended on the Tokyo offices of NHK (Japan Broadcasting Corporation), accusing Japan's state-sponsored TV of silencing the emperor in order to protect the interests of the powerful nuclear industry. The censorship controversy served to further deepen the average citizen's distrust of government, industry, and the mainstream media.

Undeterred by growing public opposition, on June 16, 2012, Noda ordered the restart of two reactors at the Ohi nuclear facility in western Japan. The fate of Japan's other 48 idled reactors will be determined once Japan's discredited nuclear regulatory agency is replaced by what the public hopes with be a "more independent" successor.[16]

There are encouraging signs that the nuclear-industrial clampdown on the media is finally starting to bend under the pressure of mounting public anger. In the past, the power of the nuclear establishment has been so entrenched that, prior to Fukushima, the Japanese government legally required school textbooks to extol the virtues of "safe, clean" nuclear power. Now, as Greenpeace International blogger Christine McCann has noted, "numerous publishers have submitted requests for updates to high school textbooks . . . in order to add information about last year's Fukushima nuclear disaster and the 'myth of safety' pertaining to nuclear power."[17]

16

Why You Can't Believe the Official Story

Run! Run as fast as you can. Don't believe the government. The government will lie to you.

—**NATALIA MIRONOVA, RUSSIAN NUCLEAR ENGINEER AND CHERNOBYL "LIQUIDATOR"**

ONE CONSISTENT LESSON from Three Mile Island, Chernobyl, and Fukushima is clear: when the public is at risk, plant operators and government officials inevitably cover up and lie. "They lied to us," physicist Michio Kaku memorably told CNN on June 22, 2011. "TEPCO isn't in cold shutdown and won't be for another year. And, if there's another quake, it could start all over again."[1] Behind the official reassurances from Tokyo, full meltdowns were under way at three of the six reactors, 600,000 spent fuel rods were at risk of burning off into the atmosphere, and the fallout burdens turned out to be 10 times greater than officially reported. Plutonium rained down 28 miles from the plant and strontium-90 turned up 155 miles away—well outside the official 12.5-mile "evacuation zone."[2]

Immediately after the Fukushima accident, the Health Physics Society (HPS) joined the American Nuclear Society (ANS) and the Nuclear Energy Institute (NEI) in downplaying the danger. The HPS stated that "loss of life and future cancer risk are small, particularly in contrast with those resulting from the Chernobyl accident,"[3] the ANS claimed that "no public ill health effects are expected,"[4] and the NEI echoed the view that "no health effects are expected among the Japanese people."[5]

London's *Guardian* newspaper subsequently revealed that just two days after the Fukushima quake (about the time the first wave of fallout hit the US West Coast), British officials "approached nuclear companies" to fashion a PR strategy "to play down" the accident lest it undermine "public support for nuclear power." The government-nuclear complex worked closely "with the multinational companies EDF Energy, AREVA, and Westinghouse."[6]

Infrastructures of Deceit

For more than 50 years, the creation, activation, and expansion of nuclear power has been accelerated, promoted, and protected by a vast infrastructure of deceit. From the earliest days of atomic power, the military-industrial-nuclear complex engaged in untruths and propaganda, from the false promises of electricity "too cheap to meter" to the depiction of atomic energy as "clean" when its proponents knew full well there was no effective long-term plan for safely dealing with tons of toxic nuclear wastes. The deceit continued with the creation of supposedly safe "threshold" exposures that ignored medical science and perpetuated the false notion that nuclear risks are manageable. The deception continued in 2011 as plumes of fallout blew across the Pacific, dusted Hawaii, and headed for the West Coast.

On March 15, 2011, President Barack Obama assured viewers of Pittsburgh's KDKA-TV that "the nuclear release from Japan will dissipate by the time it gets to Hawaii, much less the US mainland."[7] By reassuring his audience (many of whom were still haunted by memories of Three Mile Island), the president was engaging in the kind of public relations subterfuge that has long been a staple of the nuclear industry and its government enablers.

In fact, the swirling cloud of radioactive gases that crossed the Pacific Ocean *did* reach the US mainland. Independent scientists monitoring the plume detected xenon-133 and "high concentrations" of cesium-137 in the United States and Canada. Nuclear expert Arnie Gundersen estimates that, within four days of the Fukushima explosions, Seattle was exposed to xenon and krypton levels 40,000 times above normal, followed by slower-moving clouds of iodine, cesium, and strontium isotopes.[8] From March 15 to 19, radiation swept over Hawaii and the West Coast, covering all of western North America. While the Rocky Mountains blocked much of the fallout, rainwater was contaminated across New England. Two of three East Coast monitoring stations (including one in Boston) detected the presence of dangerous "hot particles."[9] Air sampling traced the fallout all the way to Stockholm.[10]

Although the Norwegian Institute for Air Research was reporting that fallout had blanketed most of the United States and Canada, little official alarm was demonstrated in either country. Three weeks after Fukushima's reactors began to overheat and explode, the US government still refused to publish any official data on radiation levels in the United States.

Nuclear expert Gundersen issued a call to form a citizen's brigade of radiation trackers. The procedure involved swiping a one-square-meter surface with a cloth following a rainstorm and placing the cloth next to a Geiger counter. Hundreds responded with photos and videos of Geiger counters clicking away from California to Connecticut. One volunteer in St. Louis monitored "rainout" levels 178 higher than normal following a storm.

The importance of monitoring fallout patterns would be dramatically underscored by a December 2011 report linking the Fukushima fallout to 14,000 "excess deaths" in the United States in the 14 weeks following the meltdowns.[11] The authors, Joseph Mangano and Janette D. Sherman, reviewed weekly mortality records for 122 cities collected by the Centers for Disease and Control and reported that the greatest mortalities in excess of what would be expected were seen among children under the age of one. The findings echoed the 16,500 "excess deaths" recorded in the 17 weeks following the Chernobyl explosion. On February 23, 2012, Mangano updated the report, raising the excess death count to nearly 22,000.[12] Industry pundits and pro-nuclear bloggers immediately dismissed the findings as "bogus."

The Government's Response

"It's not that government agencies aren't tracking the spread of radioactive material," the *Seattle Times* reported on April 5, 2011; the problem was that "they have so far released very little actual data on isotopes of concern to human health, including iodine-131 and cesium-137."[13] The Environmental Protection Agency (EPA) maintains a coast-to-coast, border-to-border network of stations, called RADNET, that monitor radiation in the nation's air, rain, and milk, but the agency failed to offer a full accounting of its findings. Similarly, the Department of Energy (DOE) Pacific Northwest National Laboratory declined to make its findings public. The EPA didn't come forth with news that it had found radionuclides in US drinking water until six days after a team at the University of California at Berkeley alerted the nation to the situation.

Robert Alvarez, a former DOE deputy assistant secretary for national security, was aghast. "If a university professor and his students can collect samples and turn them around in a reasonable amount of time and report it, you would think government officials could do the same," he declared.[14]

At UC Berkeley, where nuclear engineering professor Kai Vetter and his students had been busy monitoring Fukushima fallout, there was also criticism of the government's failure. Referring to an array of monitors set up on the roof of a campus building, Vetter expressed amazement: "This is the only source of hard data out there—which is a surprise to me." Researchers at the university's nuclear energy department complained that the EPA had "rigged" its RadNet monitors to show lower radiation readings and called the system "severely flawed."[15] In any event, RadNet tests only for iodine-131 and ignores radioactive isotopes from cesium, uranium, and plutonium.

On May 3, 2011, the EPA announced it was suspending its weekly radiation monitoring and would, henceforth, test milk and drinking water for radiation only once every three months. On February 7, 2012, the *Washington Post* confirmed that the NRC knew there was a good chance radioactive iodine would reach Alaska but chose to tell the public there was no health risk.[16] The EPA refused to test the Gulf of Alaska for fallout.

Vetter's rainwater monitors began to detect the first traces of fallout on March 17, a week after the quake and tsunami. Levels of iodine-131 peaked on March 20 at 4.5 millibecquerels. This spike appeared to be linked to the hydrogen explosions that destroyed the reactor containment buildings. The monitors also detected radioactive isotopes of cesium and tellurium. On March 23, Vetter's monitors recorded iodine-131 levels at 181 times the EPA's maximum contaminant level (MCL) for drinking water. UC San Diego researchers reported similar readings as Fukushima's radioactive cloud swept over southern California. Meanwhile, milk and vegetables on sale in Bay Area markets during this period were found to be mildly radioactive.

Although some of the rainwater Vetter and his students harvested on the Berkeley campus contained iodine-131 concentrations 181 times greater than the EPA's MCL of three picocuries per liter, university scientists agreed with the government that short exposure to 543 picocuries per liter of iodine-131 would too low to pose any serious health threat to the public. (The MCL standard refers to a constant level of exposure for over a year.) Data subsequently released by EPA's RADNET laboratory analysis indicated that rainwater falling on Richmond, California, on March 22, 2011, contained 138 picocuries of iodine-131 and 5.96 picocuries of tellurium-132. On April 12, the same monitors recorded a deluge of stormwater laced with iodine-131, cesium-134, and cesium-137. The high readings lasted until April 28, with iodine-131 levels peaking at 8.9 picocuries per liter—nearly triple the MCL's "safe" limit.

Hot Particles over Seattle

On October 31, 2011, Marco Kaltofen, a fallout specialist with the Worcester Polytechnic Institute in Massachusetts, informed the annual meeting of the American Public Health Association that his research had found "isolated US soil samples [containing] up to 8 nanocuries per kg of radiocesium." Kaltofen reported detecting cesium fallout in the soil of five California cities—Alameda, Oakland, Sacramento, San Diego, and Sonoma.

After Fukushima's radiation levels peaked on March 20, garage operators in Seattle were warned that automobile air filters now qualified as "radioactive waste" that required special handling for disposal. But it wasn't just air filters that posed a risk. As nuclear energy expert Arnie Gundersen explained during a videotaped interview with Dr. Helen Caldicott on June 17, 2011, the air over Seattle during April and May was "loaded with hot particles. . . . We're seeing plutonium and americium . . . strontium and cesium." Gundersen extrapolated that, if the average resident of Tokyo was inhaling "ten hot particles a day . . . the average person in Seattle breathed in six."[17]

Responding to evidence of widespread contamination, the EPA announced on April 3, 2011, "We did not expect to see radiation at harmful levels reaching the US from damaged Japanese nuclear power plants."[18] While acknowledging that the danger was small, Robert Alvarez took issue with this argument. The nuclear industry "likens [radiation] to everyday life and it is not like everyday life," Alvarez stressed. "You shouldn't have radioactive iodine, even in tiny quantities, finding its way into your milk supplies."[19] The EPA promised to move quickly to release its tests for radioactivity in rain and snow but failed to do so. Within a week of the EPA's reassuring advisory, independently tested milk samples in Phoenix and Los Angeles were registering iodine-131 at levels roughly equal to the EPA's MCL. (Note: While the EPA's MCL allows for one death per million Americans, the FDA's more lenient "safe exposure" level permits 2,000 deaths per million.)[20]

As Fukushima's hot breath blew across North America—contaminating strawberries in California[21] and milk in Vermont[22]—word began to circulate that Washington was preparing to follow Tokyo's example by simply increasing "permissible" exposures. As it turned out, the nuclear-industry-backed plan to raise the "permissible" levels of radiation exposure started back in 2009. In the closing days of the Bush Administration, the EPA agreed to update the EPA's 1992 Protective Action Guides (PGA) to subject the public to vastly increased radiation exposures. The Obama Administration has not closed the door on raising exposure guidelines. On May 29, 2012, the EPA's Radiation Protection website reported that revisions to the PAG Manual "are under review."

According to investigators at Public Employees for Environmental Responsibility (PEER), the new guidelines would "significantly increase allowable public exposure to radioactivity." The new guidelines would include a nearly 1,000-fold increase for exposure to strontium-90 and a potential 100,000-fold jump in exposure to iodine-131. "With the Japanese nuclear situation still out of control and expected to continue that way for months," PEER executive director Jeff Ruch observed, "this is the worst possible time for the EPA to roll back radiological protections for Americans."[23] (At least nobody in Washington went as far as one cavalier Japanese politician who advised, "Smile and the radiation won't harm you.")

The industry benefits from radiation's long lag time. While radiation-linked leukemia can manifest in 5 to 10 years, solid cancers do not appear until 15 to 60 years after exposure. And since radiation-induced genetic mutations tend to be recessive, many generations may pass before the damage from Chernobyl and Fukushima eventually resurfaces in the misshapen form of stillborn fetuses or deformed and disease-ridden children. As Dr. Helen Caldicott emphasizes, there is no such thing as a "safe dose" of radiation, and all internal bodily exposures are cumulative.[24]

In March 2011, Friends of the Earth, Physicians for Social Responsibility, and the Nuclear Information and Resource Service filed a Freedom of Information Act request in an attempt to discover the basis for the NRC chair Gregory Jaczko's remarkable recommendation that US citizens in Japan evacuate from locations within 50 miles of the Fukushima reactors. When the documents were finally released, nearly a year later, they further confirmed the government's practice of "safeguarding the truth" rather than safeguarding the public. Among the documents was a 507-page transcript of NRC phone conversations revealing a multiparty discussion on March 17, 2011, about "the [radiation] doses they saw all the way out in California." One speaker states, "They were calculating doses, particularly for children—thyroid doses [that] . . . are showing millirem range doses, like one to 10 millirem." Later in the conversation, a speaker identified only as "Mr. Lewis" mentions a "dose estimate that was done for California . . . estimating what we believe to be very high doses to children." Referring to the accident in Japan, another NRC staffer notes, "The public doesn't know what percentage of core damage [inaudible]. We did not on purpose put that in the press release, because it's a little alarming."[25]

Institutional Failures

On the first anniversary of the Japanese triple meltdown, Greenpeace International released a major report that identified the most critical lessons from Fukushima. Greenpeace concluded that the leading causes of the growing calamity were "institutional failures of political influence and industry-led regulation." In short, it was "a failure of human institutions to acknowledge real reactor risks, a failure to establish and enforce appropriate nuclear safety standards, and a failure to ultimately protect the public and the environment."[26]

In page after page of analysis, the Greenpeace postmortem describes a pandemic of symptoms in Japan's nuclear cabal—symptoms that eerily echo patterns in the US government-nuclear complex. "The [Japanese] nuclear industry kept saying that the probability of a major accident . . . was very low.

With more than 400 reactors operating worldwide, the probability of a reactor core meltdown would be in the order of one in 250 years." Instead, over the past 30 years, "a significant nuclear accident has occurred approximately once ever decade."

In Japan, emergency planning failed. It was impossible to coordinate the evacuation of so many people. People were told to "shelter in place," but their food and water ran out before the cesium-137 stopped falling from the skies. The 20-kilometer mandatory evacuation zone proved to be too small. A 50-kilometer zone would have been more advisable from a health standpoint (except for the fact that a larger evacuation would have been even more problematic). Evacuation plans in the United States are no better. The challenges range from the impossibility of evacuating Manhattan to the plight of rural Americans living downwind from US reactors who would be expected to flee down unpaved dirt roads.

Greenpeace estimates that the total costs to compensate Japan's citizens for their losses, pay for the decontamination of the land, and cover the expense of decommissioning all six reactors at the Fukushima Daiichi site could soar to $650 billion. The Greenpeace report states, "It is staggering to witness how the [Japanese] nuclear industry managed to build up a system whereby polluters harvest large profits, while the moment things go wrong, they throw the responsibility to deal with losses and damages to the impacted citizen." Sound familiar?

Greenpeace found that TEPCO and government agencies shared an "attitude of allowed deception" that was reinforced by "undue political influence" on the regulatory process. This "self-regulatory environment" was a key factor in Fukushima's failure.

Even on those occasions when regulators finally demanded modifications in the interests of public safety, TEPCO was permitted to go years without actually implementing the changes. (This echoes the situation in the United States, where the NRC spent a year devising a list of twelve post-Fukushima safety improvements but chose to demand only three and then gave the industry up to five years to comply.)

Ending on a positive note, the Greenpeace assessment celebrates the fact that the new renewable power plants that came online worldwide in 2011 were now churning out as much electric power as 16 large nuclear power stations. If that's not a renaissance, what is?

17

The Regulatory-Industrial Complex

[The NRC is] a moribund agency. It's become captive of the industries that it regulates, and I think that's a problem.

—PRESIDENTIAL CANDIDATE BARACK OBAMA, 2008

UNFORTUNATELY, it is not uncommon to find government regulators working on behalf of industry rather than in the interests of the public. A corporation that is beholden to maximizing profits for its shareholders naturally views any regulatory intrusion as a potential obstacle to profit sharing. Regulatory bodies are just another form of "Big Government"—another obstacle to maximizing profit that must be subverted or overcome.

Consequently, corporations have learned to deal with regulatory agencies in the same way they deal with government agencies—through lobbying, influence peddling, revolving-door job offers, camaraderie, junkets, and gifts. Regulatory agencies, whether at the state or federal level, have always been fair game for industry flacks. But while the public knows how corporate lobbyists work to game the system by cozying up to politicians on the Hill or inside state capitols, less attention is paid to how the same techniques are used to influence the people who head the agencies entrusted with regulating the behavior of Big Coal, Big Oil, and Big Nuclear.

Back in 1974, Congress became so alarmed by the Atomic Energy Commission's atomic boosterism that it replaced the AEC with the NRC. Today, the new agency's clear failure to fulfill its role as an independent watchdog has prompted critics to call for reform or replacement of the NRC.

The NRC had become such a handmaiden to the industry that it has routinely handed out operating permits before plants were even built, let alone inspected. This practice of "preconstruction licensing" was first proposed in 1989 after the Shoreham, Seabrook, and Zimmer reactors were built, only to be refused operating permits when critics turned out at public hearings and demonstrated that the plants were unsafe to operate. Previously, operators needed to negotiate a three-stage process of (1) applying for a construction permit, (2) building the nuclear plant, and (3) requesting permission to operate the reactor following an inspection. "Preconstruction licensing" became part of the NRC's consolidated "one-stop" licensing policy. The NRC argued this would speed the introduction of a new generation of reactors, but a US appeals court ruled in 1990 that such a shortcut procedure appeared to violate the safeguards of the Atomic Energy Act.[1] Meanwhile, the NRC has continued to hand out 20-year operating extensions to every nuclear operator that has filed a request.[2]

In July 2011, in response to the Fukushima nuclear disaster, the NRC released a 96-page guide advising nuclear plant operators on how to improve reactor safety. In preparing the report, the NRC met with members of the nuclear industry but not with members of the public. The Union of Concerned Scientists (UCS) faulted the report for recommending safety reevaluations every 10 years rather than demanding immediate and sweeping overhauls. The UCS also noted that the report failed to fully address hazards resulting from "unlikely" extreme events (like the quake and flood that hit Fukushima). Instead, the NRC suggested that, in an emergency, operators should simply be prepared to run their emergency cooling systems "for at least eight hours and should have procedures to keep the reactor and spent-fuel pool cool for 72 hours."[3]

Inside the NRC: Regulator or Enabler?

In 2011, the Associated Press published an extraordinary series of reports revealing the extent to which the NRC had become a tool of the very industry it was supposed to be regulating. After spending a year poring over more than 11,000 NRC documents obtained through a Freedom of Information Act request, the AP came to a chilling conclusion: "Federal regulators have been working closely with the nuclear power industry to keep the nation's aging reactors operating within safety standards by repeatedly weakening those standards, or simply failing to enforce them."[4] (Enforcement is not as rigorous as most near-nuclear neighbors might wish. Because of current budget limitations, there are only enough on-site NRC inspectors to check 5 to 10 percent of each plant on an annual basis.)[5]

The AP uncovered extensive evidence of NRC collusion with industry. The record showed that the NRC had misled the public, abetted industry cover-ups, and repeatedly responded to evidence of operational risks by loosening regulatory restraints. If reactors were running too hot or leaking too much coolant, the AP found, the NRC would simply rewrite regulations to keep aging power plants in compliance:

> When valves leaked, more leakage was allowed—up to 20 times the original limit. When rampant cracking caused radioactive leaks from steam generator tubing, an easier test of the tubes was devised, so plants could meet standards. Failed cables. Busted seals. Broken nozzles, clogged screens, cracked concrete, dented containers, corroded metals and rusty underground pipes—all of these and thousands of other problems linked to aging were uncovered in the AP's yearlong investigation.[6]

To the AP's amazement, "not a single official body in government or industry has studied the overall frequency and potential impact on safety" posed by the country's increasingly decrepit reactors. Instead of cracking down on reactors that have been cracking down, "the NRC has extended the licenses of dozens of reactors."

The reason is clear. "With billions of dollars and 19 percent of America's electricity supply at stake, a cozy relationship prevails between the industry and its regulator, the NRC. Records show a recurring pattern: Reactor parts or systems fall out of compliance with the rules. Studies are conducted by the industry and government, and all agree that existing standards are 'unnecessarily conservative.' Regulations are loosened, and the reactors are back in compliance."

The Fine Art of "Pencil Engineering"

In the course of their investigation, AP reporters encountered a number of engineers and "former regulators" who used the very same phrase to identify this process: "pencil engineering," or "the fudging of calculations and assumptions to yield answers that enable plants with deteriorating conditions to remain in compliance."[7]

When the first US reactors went into operation in the 1960s, it was assumed that these early models would be replaced by more improved designs at the end of their 40-year lives. But with the collapse of the nuclear bubble, there has been no second generation of "beta reactors." The industry's only recourse has been to beg the NRC to allow existing reactors to operate far beyond their planned retirement. The NRC helpfully responded with the "Master Curve" policy, which was designed to redefine troubled reactors as fit and able.

The Master Curve permitted more lenient "embrittlement calculations" for analyzing the toughness and durability of metals, such as those used in reactor containment vessels. As the AP investigation put it, the goal was to "put questionable reactor vessels back into the safe zone." The AP went on to explain:

> A 1999 NRC review of the Master Curve, used to analyze metal toughness, noted that energy deregulation had put financial pressure on nuclear plants. It went on: "So utility executives are considering new operational scenarios, some of which were unheard of as little as five years ago: extending the licensed life of the plant beyond 40 years." As a result, it said, the industry and the NRC were considering "refinements" of embrittlement calculations "with an eye to reducing known *over-conservatisms*" (emphasis added).[8]

As the AP discovered, another tool the NRC has employed to make it easier for industry to avoid more rigorous safety practices is "risk-informed analysis" (RIA). Broadly applied since the 1990s, RIA has allowed regulators to ignore a more comprehensive set of safety concerns and concentrate only on areas of highest risk. Union of Concerned Scientists physicist Ed Lyman argues that RIA generally works "to weaken regulations, rather than strengthen them." This approach also assumes all reactors to be equal and ignores the problems associated with older plants.[9]

In 2010, the AP report noted, "the NRC weakened the safety margin for acceptable radiation damage to reactor vessels—for the second time. . . . The minimum standard was relaxed first by raising the reference temperature 50 percent, and then 78 percent above the original—even though a broken vessel could spill its radioactive contents into the environment."

In the 1990s, the NRC was faced with reports of cracking in the steam generator coolant tubing that keep reactors from overheating. The nuclear industry complained about the government's safety standard, which said that no cracks could extend through more than 40 percent of the tube wall. In response, the NRC came up with an alternative standard: a positive outcome on remote "eddy-current tests" of the tubing. This allowed reactors to continue running despite evidence of cracked tubing.

In a 2001 report, NRC staff confessed "there would be some possibility that cracks of objectionable depth might be overlooked and left in the steam generator for an additional operating cycle" (that is, 18 to 24 months). "They did not want to say [cracked tubing] was a problem," former NRC engineer Joe Hopenfeld told the AP, because "if they really said it was a problem, they would have to shut down a lot of reactors."[10]

A prime example of the NRC's lax oversight was provided by the Byron plant in Illinois, where a critical pipe in the 26-year-old facility was allowed to corrode to the point that it ruptured, spraying a large jet of water and forcing a plant shutdown. It turned out that plant operators had been aware of the corrosion of the originally 9.52-millimeter-thick pipe but responded to the corrosion problem by repeatedly lowering the definition of what thickness was "safe." In June 2007, the plant operators redefined the "safe" thickness, lowering it to 3.82 mm. By October 2007, the pipe's thickness had been reduced to 1.52 mm, but instead of replacing the pipe, Byron officials *again* revised the guidelines to define this new thickness as "safe." When the pipe eventually burst, its thickness measured only 3/100ths of an inch (0.0762 mm).

As former nuclear regulator and whistleblower George Mulley told a film crew from the Center for Investigative Reporting, "That's like 7 or 8 sheets of paper! You're talking paper-thin and this is carrying water under pressure used to cool a power plant!" Mulley recalled that he had filed an NRC review criticizing the lax oversight at the Byron plant, but his report was rewritten by his NRC supervisors to remove blame from the licensee and was never released to the public.[11]

Why didn't the NRC spot the problem? One reason is that the two-person inspection teams assigned to each facility can only examine around 10 percent of plant systems each year. But there's another reason, Mulley explained: a "trust the licensee" attitude. An inspector once told Mulley that he "wouldn't check to see if a licensee lied." And why was that? "Because it's against the law for a licensee to lie to me," the inspector replied.

This regulatory policy of "don't ask, don't tell" extends to the relicensing process. As watchdog group Beyond Nuclear has noted, the NRC has "rubberstamped 59 of 59 license 'renewals' sought by industry, including at the oldest operating reactors in the US, despite . . . very serious, documented safety risks due to age-related degradation."[12] Even the NRC's inspector general reported, in a September 6, 2007 audit, that relicensing often involves little more than a "cut and paste" job, where plant operators' reports are adopted verbatim and misrepresented as "independent safety analysis."[13] The inspector general's report went on to charge that once a final license renewal was approved, the incriminating drafts were destroyed.[14]

Decades of Deceit

In its role as a booster of nuclear power, the NRC has spent decades assuring the media that America's aging reactors pose no danger to the public. However, the AP investigation revealed that, within the confines of its own chambers, the NRC has become increasingly alarmed by a steady decline of reactor safety.

One internal report from 1984 estimated that corrosion, wear, and fatigue were responsible for more than a third of 3,098 parts or systems failures that occurred over just the first dozen years of reactor operations. In 1994, the NRC admitted to Congress that critical "shrouds" used to line reactor cores were found to have cracked at 11 units, with five suffering extensive damage. According to the AP, a 2010 NRC report warned that "cracking of internal core components—spurred by radiation—remains 'a major concern' in boiling water reactors."[15]

The AP exposé series also revealed that "radioactive tritium has leaked from three-quarters of US commercial nuclear power sites, often into groundwater from corroded, buried piping." Leaks were confirmed at 48 of 65 sites, including Braidwood, Browns Ferry, Byron, Dresden, Fort Calhoun, Indian Point, LaSalle, Oyster Creek, Palo Verde, Peach Bottom, Prairie Island, Quad Cities, Salem, Vermont Yankee, and Watts Bar. At 37 plants, tritium concentrations in leaked discharges exceeded those deemed safe by federal drinking water standards—in some cases by as much as 450 percent.[16]

Tritium leaks can be a sign of worse problems. Migrations of tritium, cesium-137, and strontium-90 can be an indication that a reactor's mile-long networks of buried pipes and electrical cables are beginning to fail. Tritium leaking from pipes built to carry cooling water should be a matter of the greatest concern to the NRC, yet despite clear evidence that the problem is rapidly increasing as reactors age, the commission has been accused of looking the other way. Exelon, the country's largest nuclear operator, insists that "100 percent verification of piping integrity is not practical." And the company gets what it wants; most buried pipes and cables at nuclear reactors have not been inspected for 30 to 40 years, and even when a tritium leak signals a problem, the NRC allows operators to postpone repairs for up to two years.[17]

The media may have been deceived and the public may have been endangered, but the nuclear industry could not be more pleased, as can be seen from a 2008 position paper prepared by the pro-nuclear Electric Power Research Institute (EPRI). In its report, EPRI praised the NRC's regulators for creating "a more tractable enforcement process and a significant reduction in the number of cited violations."[18]

The Practice of "Tombstone Regulation"

"More tractable enforcement" and fewer citations: what more could any industry ask of its federal regulators? For the rest of us, it starts to look more like what some industry insiders have called "tombstone regulation." As former nuclear engineer and whistleblower Paul Blanch told the AP investigators, "Until there are tombstones, they don't regulate."[19]

But the tombstones on the horizon have become more numerous, and the latest stand along the eastern shore of Japan. Mindful of the date, two days before the anniversary of the Fukushima disaster, the NRC announced that it would order significant safety changes in the hope of heading off a local disaster at any of the 23 Fukushima-style reactors operating in the United States. The changes included upgrades to equipment and training to handle simultaneous threats at multiple reactors. (Fukushima demonstrated how a failure at one reactor could trigger failure in a second, third, and fourth unit. In the United States, 36 nuclear power sites have at least two reactors sitting side by side. Three sites have three reactors, and the Vogtle plant in Georgia is set to expand to four reactors.)

NRC's Post-Fukushima Response

On February 22, 2012, NRC staff released three safety orders intended to defend against a "beyond-design-basis external event" like what happened at Fukushima: (1) request all US reactor operators to "develop plans to deal with extreme situations"; (2) propose that all plants improve spent fuel pool instrumentation to monitor water levels and temperatures; and (3) address the problem of vents on the roofs of containment structures. (Twenty years after the Mark 1 reactors went into service, vents had to be added to the flawed design to avoid containment structure damage from the buildup of internal pressures. To prevent an exposion, radioactive gases are vented into the air. The NRC has been under pressure from environmental critics and from members of Congress to require that filters be added to the vents to prevent massive releases of radiation into the atmosphere.)

Although these minimal actions could cost millions, Exelon, Entergy, Southern Company, and other energy corporations had no immediate cause for concern since the operators would not need to submit compliance plans until February 28, 2013, and need not make any actual changes until December 31, 2016. The NRC task force also offered other post-Fukushima safety suggestions but the press was given to understand that such recommendations would first require "more study" and "industry review." [20]

In March 2012, the Union of Concerned Scientists (UCS) published a detailed critique of the NRC's post-Fukushima response. In a 26-page report, UCS staff scientists David Lochbaum and Edwin Lyman concluded, "A major flaw in the NRC's approach is that it has relegated the task force's first and primary recommendation to last in line. . . . At present, it is only required that reactors be designed to handle some types of accidents—so-called 'design-basis' accidents—but not most 'beyond-design-basis' accidents such as the one at Fukushima."[21]

The UCS had earlier faulted the NRC for lax enforcement, noting that four of the five commissioners had refused to enforce a deadline for fire protection safeguards that had been ignored since 1980. Three dozen US reactors remain out of compliance with first safety regulations, still—with the consent of the NRC majority—the operators won't have to act until at least 2016. The UCS report noted that the NRC (acting on a policy dating from 1982) does not even "require new reactors to be safer than existing ones." In other words, any "improved" fourth-generation reactors need only be as safe as those built "at least 30 years ago."[22]

Business before Safety

"When the NRC revises regulations or adopts new ones," the UCS report explains, "it sometimes 'grandfathers' (or exempts) existing reactors from these regulations." These safety waivers "continue to apply even when a reactor receives a 20-year license extension."[23]

The UCS has repeatedly faulted the NRC for placing "business ahead of safety," with the result that numerous "generic" safety issues affecting nearly 20 reactors have remained unresolved since 1996. A case in point: emergency pumps at all 60 PWR units in the United States remain susceptible to pipe breakages and steam blasts that could potentially damage the reactor core. The problem was uncovered in 1979, but the NRC did not require operators to address the danger until 2007, and even now, the UCS complains, "some 20 reactors are still not in compliance." The group accuses the NRC of "responding slowly due to industry resistance."[24]

The UCS report also notes a lack of attention to US spent fuel pools, which are more densely packed than those at the Fukushima facility. As of December 31, 2010, more than 49 metric tons of spent fuel remained stored in pools alongside (or, worse, on top of) US reactor buildings.[25] Despite the risk inherent to such an arrangement, the NRC has failed to require operators to move spent fuel assemblies to dry-cask storage units that would provide safe and affordable storage for at least 50 years.

The NRC has also allowed operators to lower operating costs by switching to "high-burn-up fuels" that permit reactors to generate power for twice as long between refueling. The UCS report warns that these fuels "are more vulnerable to damage" during routine accidents.

The NRC also pampers the industry's bottom line by not requiring operators to include the dangers of spent fuel accidents into cost-benefit calculations. Nor does the NRC require operators to prepare realistic risk assessments for seismic threats. And, best of all, the NRC (unlike every other branch of government) has its own definition for the worth of a human life. While other agencies place the value of a human life at between $5 million and $9 million, the NRC's price tag is a mere $3 million. (That figure, adopted in 1995, has not been adjusted for inflation.) In 2011, the Office of Management and Budget warned the NRC that it was hard to justify this figure. In the meantime, the nuclear lobby continues to reap immense cost-benefit savings on its "safety" expenditures.

On April 6, 2012, the UCS sent an urgent request asking the NRC to (1) develop emergency planning rules to cover people outside the immediate 10-mile zone around nuclear reactors; and (2) in the absence of any long-term storage solution, begin construction of interim storage facilities for high-level radioactive waste currently packed into spent fuel pools. These two critical steps are part of the UCS's 23-part plan to improve reactor safety in the United States.[26] Intense public pressure will be needed before the NRC acts on these key requests.

Meanwhile, hundreds of aging reactors continue to operate in the United States and around the world with inherent problems that remain unaddressed and unresolved. (In the next chapter, we profile a "rogue's gallery" of some of the nation's worst nuclear plants.)

The Political Assassination of Gregory Jaczko

On May 21, 2012, Gregory Jaczko abruptly announced his resignation after five years as the head of the NRC. Over the preceding months, Jaczko's fellow pro-nuclear commissioners had relentlessly accused him of "bullying," Republicans had berated him for "inappropriate behavior," and the nuclear industry had castigated him for imposing a "chilled working environment." The attacks continued despite a June 2011 NRC inspector general's report that dismissed charges leveled at Jaczko. During a contentious December 2011 hearing, senator Barbara Boxer charged Jaczko's accusers of resorting to "McCarthyism" to fuel a political "witch hunt."

But Jaczko had his defenders. Congressman Edward Markey (D-Mass.) praised the NRC chief for leading "a Sisyphean fight against some of the nuclear industry's most entrenched opponents of strong, lasting safety regulations, often serving as the lone vote in support of much-needed safety upgrades recommended by the Commission's safety staff."[27] Senator Bernie Sanders (I-Vt.) hailed Jaczko for his "consistent voting record supporting the swift implementation of strong safety reforms." It was because of his efforts to "hold the nuclear industry accountable," Sanders maintained, that "Chairman Jaczko was subjected to repeated personal attacks made by some of his colleagues and pro-industry advocates in Congress. I am extremely disappointed he is leaving the Commission."[28] The lamentations were not limited to the denizens of Washington. On the western edge of the country, Friends of the Earth activist Damon Moglen stood in the shadow of the shuttered San Onofre nuclear plant and lamented Jaczko's departure. "He's the only NRC commissioner to come out to visit the plant, to meet with citizens," Moglen told a reporter from KPBS-TV, on the day of Jaczko's announcement. "He was making a real commitment and I think people should be very concerned and disturbed that this real leader in the NRC has now been forced to announce his resignation."[29]

In a Huffington Post commentary published on the day of Jaczko's announcement, Ryan Grim nailed the political dynamic that led to Jaczko's departure. "The battle against Jaczko reveals the flip-side of corporate influence on politics," Grim wrote. "While the traditional understanding of money in politics has to do with favors and rewards, it works in reverse, too: Act against corporate interests, and become a target for personal destruction." According to Grim, the anti-Jaczko "coup" was orchestrated by fellow NRC commissioner Bill Magwood, a pro-industry nuclear advocate who had once done consulting work for the Japanese firm that owned the Fukushima power plants.[30]

Jaczko was vilified for being too critical, for being too quick to demand fundamental reforms, for insisting on more rigorous (and costly) safety protocols. It was Jaczko who pulled the plug on the Yucca Mountain waste storage boondoggle. It was Jaczko who warned overseas Americans to put 50 miles between their bodies and the fuming remains of Fukushima. It was Jaczko who refused to cast a vote to approve the construction of questionable new reactor designs.

In the wake of Jaczko's resignation, the odds that the NRC will change direction and start making public safety a priority are not promising. Following the Fukushima disaster, Jaczko put the NRC to work revising its emergency planning protocols (for the first time since the Three Mile Island meltdown 30 years earlier). Finally, on May 17, 2012 (four days before Jaczko announced his resignation), the Associated Press published a report on the long-awaited recommendations of the NRC's "quietly overhauled" report. Community watchdogs were astonished to read that the NRC majority had actually voted to *reduce* protections in the event of a radiation accident.

The AP revealed that the new regulations would shrink nuclear evacuation zones from a 10-mile radius to a mere 2-mile radius, "even in a severe accident." The revised rules also called for a reduction in emergency training exercises. While some nuclear critics branded the new guidelines as "insane," the AP reported that the nuclear power industry "praised the changes," citing the NRC's argument that "the revised standards introduce more variability into planning exercises."[31]

18

Five of the Worst US Reactors

> *Hopefully, nations will refuse to accept a situation in which nuclear accidents actually do occur, and, if at all possible, they will do something to correct a system which makes them likely.*
>
> —HERMAN KAHN, US NUCLEAR STRATEGIST

THE CONSEQUENCES of poor regulatory oversight can be seen in the operating histories of the country's nuclear reactors. The following five facilities are representative. Many other nuclear power sites around the country have equally disturbing records of poor performance, emergency shutdowns, and close calls.

Davis-Besse: Beset by Holes, Cracks, Close Calls

First licensed to operate in 1977, Ohio's Davis-Besse nuclear power station was supposed to be decommissioned when its 40-year license expired in 2017. But operator FirstEnergy (like every other owner of one of the rickety reactors in America's creaky fleet) applied to the NRC for a 20-year life extension. The potential extension would be a concern even if the aging reactor (located 20 miles east of Toledo) had racked up a perfect history of safe operation. Davis-Besse has not. This plant is not just "an accident waiting to happen," it is a place where extremely serious accidents *have* happened . . . repeatedly.

The NRC's own records recognize Davis-Besse as one of the most dangerous reactors in the United States. Between 1969 and 2005, this single plant experienced six out of the 34 reported "significant accident sequence precursors"—triple the rate reported at any other US nuclear plant. The problems began on September 24, 1977, after the plant had been operating for only six months. A relief valve became stuck and failed to close, leading to a "7 percent core damage probability" that ranked as the fourth most serious US accident then on record. (A nearly identical problem would occur two years later, resulting in a 50 percent core meltdown at Pennsylvania's Three Mile Island reactor.) It was a good thing for the people of Toledo that the new reactor was running at only 9 percent power and producing less heat and pressure when the accident occurred, but they would not have been comforted by the scene inside the reactor complex, were the control room operators were thrown into a state of panic as they scrambled frantically for 20 minutes in a chaotic attempt to regain control of the runaway reactor.

A Beyond Nuclear report reconstructed the scene inside Davis-Besse that day:

> Over 300 bells and flashing lights were simultaneously signaling alarm as a water column displaced the steam bubble "shock absorber" and filled the pressurizer on the very top of the reactor, risking any sudden jolt fracturing safety-significant pipes. . . . [T]he Number 2 Steam Generator risked boiling dry, which could cause dangerous overheating and even a "loss-of-coolant-accident" in the hellishly hot reactor core. Operators "grasped at straws," rashly deciding to chuck emergency manual procedures that only seemed to be making matters worse. Luckily for the unsuspecting cities just to the east and west, an operator spotted a gauge reading that resolved the perplexing puzzle, and corrective action was taken at the 26th minute of the crisis.[1]

Neither the NRC nor the Babcock and Wilcox Co., the reactor's designer, seemed inclined to look too deeply into the incident. It all would have remained a closely held secret but for James Creswell, a principled regional NRC inspector who broke ranks and took his concerns directly to two top NRC officials. Unfortunately, Creswell's warnings were not heeded until March 22, 1979, six days before an identical series of events led to the destruction of the reactor at Three Mile Island.

Davis-Besse's first year of operation was marked by a second serious accident when both emergency feedwater pumps failed, posing the risk of damage to the reactor core. Electrical problems led to further damage on April 19, 1980, and again on June 24, 1981. During the attempt to recover from the latter incident, a feedwater pump refused to start due to a "maladjusted" clutch and a bent "speed stop pin." Adding to the difficulties, a safety valve failed to "reseat."

A "Scram" and a Near Meltdown

The plant's next near miss occurred on June 9, 1985, following another loss of feedwater coolant. Even though the reactor was successfully "scrammed" (that is, placed in a quick emergency shutdown), the reactor came close to initiating a core meltdown.

Beyond Nuclear, drawing on NRC documents, offered this reconstruction of what transpired inside the plant:

> Personnel had to sprint through the darkened corridors with bolt cutters, not knowing if they had the proper keys or access cards to open locked security doors, in order to cut through chains securing valves, so they could manually open them to restore water flow to steam generators in order to cool the reactor core, with each passing minute increasing the risk of a loss-of-coolant-accident, nuclear fuel damage, and even a meltdown.[2]

Nuclear expert Dave Lochbaum, of the Union of Concerned Scientists (UCS), estimates that this loss of cooling water put Davis-Besse within 41 minutes of having the reactor core completely uncovered. (At Three Mile Island, the core was never completely uncovered, yet half the fuel rods went into a meltdown.) The 12-minute lapse in the flow of water to the steam generators caused enough damage to shut Davis-Besse for a year.

This incident was dubbed the worst accident since Three Mile Island, and the US House Subcommittee on Energy Conservation and Power followed up with a report revealing that Davis-Besse had recorded 48 feedwater problems since July 1979. The reactor also had accidentally shut down 40 times between 1980 and 1985 as a result of equipment failures and human error.[3]

In what proved to be a misguided attempt to improve discipline at the plant, a former nuclear navy commander was hired as plant manager from the mid-to-late 1980s. This plan ran aground when the new manager subsequently showed up drunk during a Christmas holiday, began cursing his employees, and started throwing punches before plant security dragged him from the building. But that wasn't the last of Davis-Besse's problems.

Nature threw the next blow on June 24, 1998. A tornado with wind speeds topping 156 miles per hour hit the plant dead center, crossing between the containment building and cooling tower and damaging the plant's electric transmission lines. The twister hit without warning while the reactor was running at 99 percent power. The operators were able to plunge the reactor into emergency shutdown, but it was still necessary to cool the dangerously hot core. Unfortunately, the tornado had cut access to all off-site electric power (the blackout would last for 27 hours). When the operators tried to start one of the plant's two emergency diesel generators, it failed to start. (It was eventually declared "inoperable" owing to the extreme heat inside the building where it was housed.) The remaining generator also failed due to "an apparent problem with the governor control." With outside power lines down, plant operators were unable to contact state or local authorities, and the emergency system—designed to sound an alarm to warn Ottawa County residents of a potential meltdown, explosion, or fire at the plant—was useless. Without power for cooling, temperatures in the spent fuel pools began to rise. Fortunately power was restored and operators regained control of the reactor before there was any release of radiation.

But the plant's most notorious brush with disaster still lay ahead.

Ohio's "Hole-in-the-Head" Reactor

In 2002, Davis-Besse became the subject of media ridicule for its "hole-in-the-head" reactor. Over the years, corrosive acids had been allowed to spill over the lid of the reactor's pressure vessel, causing cracking in the mechanisms used to drive the control rods that manage temperatures in the core. The massive metal cap had become so degraded that it was close to bursting. The only thing preventing a massive radioactive steam explosion was a slim layer of stainless steel, which itself was beginning to crack and swell.

Repairs undertaken to fix the "hole in the head" forced the reactor to be closed for two years. Two plant engineers were convicted of covering up the situation, which the Government Accountability Office called "the most serious safety issue . . . since Three Mile Island,"[4] and the DOE fined FirstEnergy (the plant's new owner) a record $33.5 million.

Then it was the NRC's turn to take some heat: The NRC's own inspector general accused the commissioners of promoting corporate profit over public safety. Ohio congressman Dennis J. Kucinich declared that the NRC's response to David-Besse was "inadequate, irresponsible, and left the public at grave risk."[5]

FirstEnergy replaced the "hold-in-the-head" lid with a 25-year-old lid. The reactor was put back online in 2004 but shut down again on March 12, 2010, after the metal nozzles atop the "new" reactor lid also showed signs of cracking. An inspection found that 24 of the 69 nozzles were damaged, and two had "through-wall cracks" that were leaking water onto the reactor vessel's carbon-steel lid. The replacement reactor head, which was supposed to be good for 15 years, somehow failed after only six years.

An investigation showed that the FirstEnergy had purchased the lid from the owners of an uncompleted Michigan plant. It turned out the mothballed lid was made from a substandard alloy that had been banned for use years earlier. Although the use of a critical substandard part marked a serious violation of the NRC's quality assurance criteria, the NRC permitted Davis-Besse to resume operations.

In July 2006, FirstEnergy (a.k.a. FirstEnergy Nuclear Operating Company or FENOC) confessed to four "inadvertent releases of radioactive liquids that had the potential to reach groundwater," including a spill of tritium that was double the EPA's "permissible" level.[6]

In June 2009, an explosion rocked the plant's electrical switchyard. A year later, the NRC was still uncertain of the cause and FENOC was unable to explain how it would prevent a future explosion.

By 2010, Davis-Besse had excreted nearly 557 tons of hot and toxic fuel-rod wastes. If the plant is allowed to operate for the full term of a 20-year operating-life extension, that amount would more than double. And ever since the plant's indoor storage pool filled up in the 1990s, the reactor's used fuel rods have been parked aboveground in unfortified concrete-and-steel bunkers. In 1994, the Toledo Coalition for Safe Energy warned the NRC that the bunkers failed to meet technical specifications for safety, but the NRC dismissed their findings. If a power outage, earthquake, extreme weather event, or terrorist attack were to uncover the stored fuel, a resulting explosion and fire could send deadly fallout 500 miles downwind. According to the NRC's own estimates from 1982 (and the population around the plant has expanded significantly over the past 30 years), a radioactive accident at Davis-Besse could cause 1,400 immediate fatalities, 25,000 cancer deaths, and $185 billion (in 2010 dollars) in damages.[7]

A Cracked Containment Building

The latest development in the ongoing Davis-Besse saga erupted on February 8, 2012, when Congressman Kucinich produced documents showing that the upper 20 feet of the reactor's steel rebar reinforcement had been rendered "functionally ineffective" by advanced and extensive cracking. An NRC inspector's report revealed that FENOC had used brittle, corroded rebar to patch a hole in the containment building. If the NRC inspectors had not been alert, the faulty repair would have been cemented into the containment wall and no one would have been the wiser. FENOC actively resisted the inspection, but the NRC insisted on taking a closer look—perhaps recalling that FENOC was suspected of using defective parts to repair its Beaver Valley reactor in Pennsylvania. (The Beaver Valley plant also suffers from containment corrosion, a potentially defective replacement vessel head, and a reactor pressure vessel that has become so embrittled by radiation bombardment that it could shatter "like a hot glass under cold water" in the event of an emergency core cooling.)

Kucinich's broadside came at a bad time for FENOC, which (despite having one of the worst safety and performance records of any US nuclear utility) was expecting NRC approval to extend its operating license for the Davis-Besse plant. Faced with evidence of extensive cracking of the concrete containment building, FENOC argued that the problem was the result not of "long-term exposure" to moisture (as the NRC alleged) but of a particularly powerful blizzard back in 1978.

"If the Blizzard of 1978 is responsible," Kucinich shot back, "the NRC should shut down all reactors in the Midwest and demand inspections." Kucinich also drew attention to FENOC's initial claim that cracking occurred only in the building's "decorative" and "architectural elements." Kucinich observed that FENOC "made those statements even though they knew the cracking was located at the main outer reinforcing steel in the wall, which is clearly 'structural.'"[8]

FENOC also argued that the moisture seepage occurred only because its employees had neglected to treat the walls with sealant—for 42 years. Kucinich was incredulous. "How can we expect FirstEnergy to operate a nuclear reactor without dangerous consequences when they can't even think far enough ahead to paint the wall of their containment structure?" he declared.

While FENOC was awaiting the NRC's decision on Davis-Besse's future, it announced plans to disassemble the plant's containment structure to replace the facility's aging steam generators. (This would mark the fourth time the plant's containment has been breached, a record for any US reactor.) This radical repair is expected to further exacerbate the existing degradation of the containment building.

Meanwhile, millions of Great Lakes residents were quietly hoping the NRC would listen to the pleas of Toledo resident and legal counsel Terry Lodge, who had this message for the commissioners: "We contend that FENOC's current lack of quality assurance and control, its historic and notorious lack of safety culture, as well as its severely degraded containment structure, call into question Davis-Besse's operational safety during the proposed 20-year license extension."[9]

Diablo Canyon: Balanced on a Fault Line

Diablo Canyon's 27-year-old twin reactors overlook the Pacific Ocean from Point Buchon, a coastal bluff 12 miles southwest of San Luis Obispo. This is the plant that state senator Sam Blakeslee (whose district includes the reactor station) grimly predicts could become "our Fukushima." The plant's demise probably would not come from a tsunami (the plant is perched atop an 85-foot-tall cliff). A catastrophic earthquake is the greater threat.

In the event of a loss of outside electric power (as happened at Fukushima), Diablo Canyon's emergency generators are supposed to kick in within 10 seconds, and (assuming the 50,000-gallon underground fuel tanks survived the quake) there would be enough diesel fuel on hand to cool the two reactor cores for seven days. However, if the emergency generators fail to start (or if the emergency persists for more than a week), the only remaining backup is a set of 125-volt batteries. According to a San Luis Obispo Country press release, the batteries would provide "enough power to shut the reactors and provide emergency core cooling and other necessary safety measures for two hours." Plant engineer Rudy Ortega explained what that would mean in practical terms: "We would have two hours to get one of the six diesel generators started."[10]

In 2011, the Union of Concerned Scientists bestowed a "Near-Miss" award on Diablo Canyon after NRC inspectors reported that plant engineers had unwittingly disabled critical valves controlling the emergency cooling system. The problem, which could have lead to a partial meltdown, had gone undiscovered for 18 months.[11]

Following the Fukushima disaster, an NRC inspection at Diablo Canyon revealed some stunning safety lapses. Doors designed to self-latch in the event of flooding proved unworkable. Four of the 30-foot cables used to power fans needed to cool the plant's six 18-cylinder diesel generators were not installed and could not be found. All six emergency generators shared the same central location, leaving them open to a shared, "common mode" failure. And, worst of all, the NRC appeared surprised by the discovery that the two reactors had only *one* emergency cooling pump between them.[12] (Design flaws are not new to Diablo Canyon. In 1981, in one of the industry's most embarrassing engineering flubs, plant operator Pacific Gas & Electric spent four years constructing a reactor dome before a young engineer double-checked the blueprints and pointed out that critical parts had been installed upside down.)

Calls to Close Diablo Canyon

Following the Fukushima meltdowns, both of California's Democratic senators, Dianne Feinstein and Barbara Boxer, contacted the NRC to express their concern for public safety, given that "roughly 424,000 live within 50 miles of the Diablo Canyon and 7.4 million live within 50 miles of the San Onofre Nuclear Generating Station."[13] The group San Luis Obispo Mothers for Peace joined 25 national anti-nuclear organizations to petition the NRC to suspend all relicensing of reactors until there had been a thorough investigation of all safety issues raised by Japan's misfortune. The NRC rejected the petition in September 2011.

The NRC continues to insist the plant is earthquake-safe. "The seismology around Diablo Canyon has been thoroughly studied," NRC administrator Elmo Collins assured the people of San Luis Obispo. But Jeanne Hardebeck, a US Geological Survery (USGS) seismologist, wasn't so certain.[14]

In 2008, Hardebeck discovered a previously unknown earthquake fault, the Shoreline Fault, located offshore less than 2,000 feet from the plant. Pacific Gas & Electric (PG&E) dismissed the new evidence as flawed and said it would be perfectly safe to continue operating the reactors. In a 500-page report, PG&E informed the NRC that it viewed the new fault as inconsequential. Since it was only 15 miles long, PG&E reasoned, the Shoreline Fault could produce nothing stronger than a 6.5-magnitude quake.[15]

However, if the Shoreline were to connect with the longer, nearby Hosgri Fault, it would more than double the assumed length of the Hosgri, extending the zone of its potential impact over 250 miles from Point Conception (about 120 miles northwest of Los Angeles) to the coastal town of Bolinas, north of San Francisco.

Hardebeck questioned PG&E's conclusions: "An interpretation that says the two faults don't connect doesn't seem to fit with the observations that we have." More often than not, Hardebeck explained, earthquakes that began on one fault "have actually jumped to another fault," over distances of up to three miles. If the two faults were to move as one, Hardebeck reasoned, an offshore slippage could produce the equivalent of a 7.7-magnitude earthquake striking directly below the Diablo Canyon site. While reluctant to predict the faults might be connected, USGS scientist Sam Johnson did confide to colleagues at a Spring 2011 USGS meeting in Menlo Park, California, that, having looked at the evidence, it was fair to conclude that the potential force of this compound fault "would be close to an 8.0. That would be a big concern."[16]

New Fault Raises Megaquake Concerns

There is a lot of talk at the NRC about responding to "the lessons of Fukushima," but so far the NRC continues to ignore the fact that the unexpectedly violent quake that triggered the Japanese tsunami occurred when several faults assumed to be "unconnected" suddenly surged at the same time. When the Fukushima quake hit, Hardebeck emphasized, "it ruptured through all of those fault segments."[17]

State senator Sam Blakeslee has a PhD in earthquake studies, so his words carry extra weight when he faults the NRC for not taking the risk seriously enough. Blakeslee was astonished that PG&E sped up its bid to win its relicensing approval *before* the new Shoreline fault earthquake data could be properly assessed. "I could not understand the utility racing to relicense before the seismic information came forward," Blakeslee told the Center for Investigative Reporting (CIR). "It was almost as if they were afraid of what they would find."[18]

With Fukushima in the rearview mirror, Blakeslee called a hearing and grilled NRC officials. "There is a new fault, in my district, next to my constituents, and you're telling me you're just going to continue business as usual and not delay to get the information before you do your safety review?" Blakeslee fumed. "That's unacceptable!"

Asked to justify its decision to relicense, an NRC official told Blakeslee that the commission had relied on safety evaluations submitted by PG&E. "We expect licensees to do those studies," the NRC official testified.

Documents obtained by the CIR revealed that PG&E's scientists had, in fact, looked into the probability of a 7.2 quake occurring along the combined faults and even produced a graph showing that the potential shaking could exceed the stresses the plant was built to withstand. But in its public presentations, PG&E provided Blakeslee and the NRC with a different graph, one showing that a serious quake along the Shoreline Fault was impossible. Hardebeck was not convinced. She insisted her geological mapping evidence showed "earthquakes along the Shoreline Fault very clearly go all the way to the Hosgri Fault."

Activist and author Norman Solomon (co-author with Harvey Wasserman of the 1982 anti-nuclear classic, *Killing Our Own: The Disaster of America's Experience with Radiation*) was so concerned about the danger of California's two coast-sited nukes that he decided to run for a US congressional seat on an anti-nuclear platform. Although he served as an "Obama delegate" to the 2008 Democratic National Convention, Solomon now calls Obama's nuclear policy "fundamentally mistaken." Solomon was particularly critical of the plan to triple US loan guarantee handouts for nuclear plant operators from $18 billion to $54 billion. "The NRC is a nuclear-friendly fox guarding the radioactive chicken house," Solomon declared on the campaign trail. "The federal government has no business promoting this dangerous industry while safe and sustainable energy resources are readily available."[19]

Solomon has called for an immediate shutdown of both Diablo Canyon and San Onofre. As to the NRC's call for more studies, Solomon responded, "I reject the notion that we should wait for such nuclear-enthralled agencies to tell us whether nuclear power is an acceptable risk for Californians." Solomon praised Germany's bold decision to abandon nuclear power—which means replacing 23 percent of the country's power needs with new renewable energy. California, by comparison, produces only 15 percent of its electricity by frying atoms. "Effective conservation options are readily available, and widespread use of renewables like solar is in reach," Soloman wisely concluded.

The Sunshine State has the right political climate to go nuclear-free. In 1976, a citizens' group succeeded in placing an initiative on the June ballot.

Proposition 15 called for a ban on new reactors in the state. After the nuclear industry spent millions of dollars to defeat the proposal, the state legislature took a stand against a nuclear renaissance by passing a law banning further construction of nuclear power plants until the NRC could provide a proven means of safely disposing of nuclear waste. (As of December 2010, 13 states had either banned or placed restrictions on the construction of new reactors.)[20] More recently, anti-nuclear campaigners circulating a petition for a statewide initiative calling for the closure of California's existing plants got an unexpected boost from Mother Nature. On April 26, 2012, Diablo Canyon was forced to shut its Unit 2 reactor when seawater intake pipes became clogged by a swarm of salps—jellyfish-like sea creatures. With Diablo's other reactor down for maintenance and San Onofre's reactors ordered shut for safety reasons, California had become a de facto nuclear-free zone.

Indian Point: The Country's "Most Dangerous" Nuclear Plant

The 2,062 MW Indian Point twin-reactor complex on the Hudson River in New York has a long history of operational problems. Since 2007, Indian Point has experienced nine unplanned shutdowns due to a wide range of problems, including a steam boiler rupture, a transformer explosion, a loss of generator power, a failed generator relay, the failure of a main feedwater pump, and blocked cooling-system intake valves. (The rupture of a steam generator tube in 2000 was ranked, at the time, as the worst such accident in reactor history.)

Indian Point sits 24 miles from Manhattan, athwart two intersecting earthquake faults capable of producing a 7.0-magnitude jolt—10 times stronger than what the plant was built to handle. The NRC rates Indian Point as the US reactor most likely to melt down in the aftermath of a quake. And because it contains the radioactive equivalent of 1,000 Hiroshima bombs, the US Geological Survey ranks Indian Point as the most dangerous power plant in the country.

Before the reactors began operating on September 16, 1962, local cancer rates were 11 percent below the national average. By 1997, cancer rates in the four counties adjacent to the plant were 1.1 percent above the national average. Using New York State Cancer Registry data, the Radiation and Public Health Project (RPHP) has revealed that cancer rates in the Indian Point region have continued to climb with each passing year. By 2007, the cancer rate was nearly 7 percent above the national average. RPHP also reported "unexpected rises" in incidences of 19 of 20 major cancers. "The greatest increase was found in the local rate of thyroid cancer," which jumped from 13 percent below the national average to 51 percent above. "There are no known causes of thyroid cancer other than exposure to radioactive iodine," RPHP director Joseph Mangano noted. Mangano added that "rising cancer rates in areas near Indian Point, whose reactors are aging and corroding over time, raises concerns."[21]

Indian Point is operated by Entergy, a "premier provider of nuclear life-cycle services,"[22] whose laissez-faire attitude toward safety issues was on full display in January 2012, when the aging reactor was shut down after a pipe that circulates 90,000 gallons per minute of 540°F (282°C) radioactive water sprang a leak. When a local TV reporter asked about the "leak," Entergy spokesperson Jerry Napp corrected him. "The seal did have an increased flow-rate of water," Napp explained. "Some might refer to it as a 'leak' but it is really just a water-flow through the seal. . . . As designed, actually."[23]

Despite the plant's history as the most dangerously operated plant in the United States and the fact that New York State Attorney General Eric T. Schneiderman targeted Entergy in 2011 for its continued failure to comply with federal fire safety regulations,[24] Entergy has asked the NRC to extend the licenses of the plant's Units 2 and 3 for another 20 years (Unit 1 was shut down in 1974; the licenses for Unit 2 and Unit 3 expire in 2013 and 2015, respectively). Schneiderman scored a major victory on February 1, 2012, when the NRC sided with New York State and rejected Entergy's request that Indian Point be granted exemptions from more than 100 critical fire safety requirements.[25] The NRC promised to "fast-track" new post-Fukushima retrofits to safeguard Indian Point. "These will start soon," NRC spokesperson Neil Sheehan stated, "like what we did with security after 9/11."[26] The NRC commission itself lacked Sheehan's optimism. In an internal comment issued in 2011, NRC chair Jaczko promised only that NRC staff "should strive to complete and implement the lessons learned from the Fukushima accident by 2016."[27]

"No Widespread Health Effects"

Marvin Fertel, president and CEO of the Nuclear Energy Institute, assures Indian Point's critics that there is no cause for alarm. In a February 17, 2012 op-ed in the *New York Post*, Fertel wrote, "Our facilities are tightly regulated by the independent Nuclear Regulatory Commission, which has at least two on-site inspectors at every US nuclear-energy facility every day." Fertel went on to cite the NRC's "multiyear, multimillion-dollar study modeling the effects of potential accidents at US nuclear facilities." The results, Fertel wrote, "affirm that any such event would unfold slowly and cause no widespread health effects; current emergency plans would protect the public."[28] Those plans include a promise to evacuate 20 million people in the event of a nuclear emergency—an expectation the Department of Homeland Security (DHS) has determined to be totally unrealistic. As one jaded New Yorker observed, "People can't get over the *bridge* on a rainy day, let alone if there was a nuclear event."[29]

Meanwhile, neither Fertel, the NRC, nor the DHS seems overly concerned by the 9/11 Commission's discovery that Mohamed Atta, the alleged leader of the attacks on the World Trade Center and the Pentagon, also contemplated flying a hijacked airliner into a nuclear reactor "near New York City." Atta's obvious target? Indian Point.

Concerned that each of Indian Point's reactors takes in about 1.2 billion gallons of river water a day, Riverkeeper lawyer Phillip Musegass attempted to gauge the plant's riverine impacts. However, as he complained to reporters from the Center for Investigative Reporting, under the NRC's bizarre relicensing process he discovered that "we are not allowed to raise concerns about the spent fuel pool, we aren't allowed to raise concerns about the emergency evacuation plan, we're not allowed to raise questions about terrorism or security."[30]

When the NRC issued a draft supplemental environmental statement declaring that Indian Point poses no "significant" public health threat, Indian Point's neighbors, environmental groups (including the Natural Resources Defense Council and Riverkeeper), and New York governor Andrew Cuomo joined forces to challenge this conclusion. Cuomo took the battle to the next stage, suing the NRC for failing to enforce its own safety regulations. "I've had concern about Indian Point for a long time," Governor Cuomo insisted, "As attorney general, I did a lot of work on Indian Point. My position was that it shouldn't be relicensed. My position was that it should be closed." Cuomo's concerns were further raised by the NRC's new seismic studies. "The Indian Point power plant is the most susceptible to earthquake because reactor number three is on a fault," Cuomo noted, adding "this plant—in this proximity to New York City—was never a good risk."[31]

Local residents presented the NRC with New York State Health Department data showing that rates of thyroid cancer in the four counties nearest the reactor site were nearly twice the US average. Over the previous four years, 992 residents had been diagnosed with thyroid cancer. Childhood cancer rates were also above average. Samples of milk from breastfeeding mothers living within 50 miles of the reactor showed significant levels of strontium-90, with levels climbing the closer a resident was to the Indian Point plant. Strontium-90 also had been detected in local fish and crabs.

In the March 28, 2011 edition of *Newsweek*, Helen Caldicott, cofounder of Physicians for Social Responsibility, calculated the consequences of a Fukushima-style incident at Indian Point. At the sound of an on-site alarm, she explained, residents would have about 78 minutes to evacuate the 10-mile zone around the reactor. Caldicott estimated "early fatalities from acute radiation sickness for those within the 10-mile evacuation zone would range from 2,440 to 11,500. Late cancer deaths, which would occur two to 60 years later, could range from 28,100 to a staggering 518,000 people in the 50-mile zone." Meanwhile, New York would be rendered "virtually uninhabitable, with $1 trillion or more in costs from attempts at decontamination, the condemnation of radioactive property, and compensatory payments to people forced to relocate."

"More than Adequate" Power without Indian Point

In January 2012, a week before two major public hearings into the plant's future, one of Indian Point's reactors was shut down by a leak. On January 31, 2012, the New York State Assembly's Committee on Energy and Committee on Corporations, Authorities, and Commissions concluded, "Indian Point can be shut down without unduly burdening New York's ratepayers or the electric system." The New York Independent System Operator (the nonprofit organization that oversees New York City's energy needs) confirmed that the state had "more than adequate" generation capacity due to expected upgrades and the completion of a 250 MW offshore wind power project, which would give southeast New York 2,000 MW of new renewable power by 2015 and 3,000 MW due to transmission improvements by 2016.

Earlier studies by the Natural Resources Defense Council and Riverkeeper had concluded that renewable power from new wind and solar projects, combined with improved transmission, could replace Indian Point's 2,000 MW in 10 years or less while adding only $3 to $5 to the average customer's monthly utility bill.[32]

The state assembly invited Entergy to present records documenting the price and quantity of power produced by its reactors, but the company was a no-show at the hearing. "Entergy failed to provide even the most basic information associated with the plant's operation," assembly member Kevin Cahill fumed. "Entergy's lack of cooperation will require us to revisit the issue in the very near future."[33] Local patience with Entergy and the NRC is wearing thin. In March 2011, Cuomo asked the NRC to close Indian Point because of "structural flaws." When his request was rebuffed, Cuomo responded by signing the Power New York Act, which would replace Indian Point with a bevy of renewable power projects.

The NRC has deemed most of the concerns raised by local residents to be "out of the scope" of the relicensing proceedings. For their part, officials in Westchester County, where Indian Point is located, have announced an ambitious plan to reduce the county's carbon footprint by 20 percent over the next seven years and by 80 percent by midcentury—turning away from the false solution of nuclear power.

San Onofre: Mysterious Leaks Prompt Shutdown by NRC

In 2009, San Onofre's Units 2 and 3 were shut down so that plant operators could replace four steam generators that had been in operation since the early 1980s. (San Onofre's Unit 1 had been shut down in 1992. The 600-ton reactor was to be shipped 15,000 miles around the tip of South America for disposal in South Carolina. When those plans were scuttled, the reactor was entombed on-site, in a massive casket of cement and steel.) After more than three decades, plant officials were concerned that critical tubing in the aging generators might be subject to leaks or clogging—a significant worry given that each generator contained 9,700 tubes. Plant officials promised to replace the old generators with the "safest, most efficient 21st century machinery."[34]

After an investment of $670 million, the two "new, improved" reactors began service in 2011, but on January 31, 2012, plant workers were forced to shut the Unit 3 reactor following a radiation leak into the atmosphere. San Onofre's owner, Southern California Edison (SCE), initially told the public, "There has been no release to the atmosphere." This was followed a day later by the admission that some radioactive tritium "could have" leaked from the damaged plant. An SCE official subsequently conceded, "There might have been an insignificant or extremely small release," but he hastened to add the industry's inevitable assurance that the radiation "would not pose a danger to anyone."[35]

The truth is, no one knows the nature or amount of radiation that was released that day since the plant operators are not equipped to properly monitor off-site emissions. What is known, however, is that radioactive poisons vented into the sky can be carried as far as 15 miles by the winds. So anyone out on a bike ride in the area that day could have unwittingly inhaled a dose of San Onofre's fallout.

The cause of the unexpected leak sent plant operators into a new round of damage control. An investigation traced the leak to one of the nearly 10,000 metal tubes in one of Unit 3's two generators. The 0.75-inch-thick tube was losing around 3.5 gallons of water an hour. These small tubes draw heat away from the zirconium-clad fuel rods in the heart of the nuclear reactor, carrying the superheated, high-pressure, radioactive water from steam generators located inside the reactor's containment dome to a separate building, where the steam spins turbines to produce electricity. By design, San Onofre's adjacent turbine buildings are not sealed, which allowed radiation from the leaked "hot" water to escape into the atmosphere.

Damage on an Unexpected Scale

Further investigation of Units 2 and 3 revealed damage on an unexpected scale—especially given that $680 million had been spent on the new generators and they had only been in operation for 22 months. In Unit 2, investigators discovered more than 871 tubes had lost 10 percent of their thickness, 69 tubes had lost 20 percent of their thickness, and 2 tubes had lost more than one-third of their thickness. This despite the fact that the tubes are made of high-chromium nickel alloy 690, a supposedly corrosion-resistant metal.

Because of the immense size of most reactor architecture, it is surprising to learn that these critical tubes—the vascular system of the nuclear power process—are only the size of the cork in a bottle of zinfandel and only as thick as a plastic credit card. The integrity of these tubes is critical since the potentially explosive high-pressure water flowing in the reactor's metal veins carries a cargo of radioactivity that includes the noble gases (krypton and xenon), tritium (a radioactive form of hydrogen), and smaller amounts of cesium, iodine, plutonium, polonium, and uranium that routinely escape through small fissures in the reactor's fuel rods.

Clearly, the failure of a single tube is cause for concern, but the possible cascading failure of 69 tubes—let alone 871—could easily lead to blocked cooling pipes, burst steam generators, and a catastrophe that could have required the emergency evacuation of densely populated stretches of Southern California.

"I've never heard of anything like that over so short a period of time," Joram Hopenfeld, a retired NRC engineer, told the media. "Usually the concern is in older steam generators, when they have cracks all over the place." NRC spokesperson Victor Dricks called the corrosion spectacle "unusual for a new steam generator" and declared that the reactors would remain shut down until the owners provided "justification for continued operation."[36]

But it turned out that the NRC had been keeping another secret from the American public. On February 2, 2012, SCE spokesperson Gil Alexander offered an unusual defense of the alarming incident, telling the press that "other plants" with new generators had also reported unusual wear.[37] Alexander's statement forced the NRC to confirm that the agency had been aware of "short-period" wear with other new steam generator retrofits, including reports of damage around the critical brackets that support the thousands of metal tubes.

"Our working hypothesis here is that what we're seeing at San Onofre is the same sort of phenomenon that we're investigating at other plants," NRC spokesperson Scott Burnell stated. While Burnell was unable to produce a list of all the US reactors plagued by this unforeseen problem, he did reveal that the NRC was aware of similar failures at the Arkansas Nuclear One reactor,

Florida's St. Lucie nuclear power plant, and the only operating reactor at Three Mile Island in Pennsylvania.[38] (It is both surprising and disturbing that these defects cannot be traced to a single manufacturer. Japan's Mitsubishi Heavy Industries built San Onofre's generators; General Electric provided the generators for the Arkansas plant; the French firm AREVA manufactured the generators used at St. Lucies and Three Mile Island.)

On March 14, three more tubes failed during a pressure test at San Onofre's Unit 3, prompting the NRC to fly a team of inspectors out to the ailing plant. On March 27, the NRC announced that neither reactor would be allowed to restart until they could be proven safe to operate.

Faulty Tubes, Faulty Procedures

In early April, NRC chairman Gregory Jaczko flew in to inspect the plant. By now it had been determined that the tubes had sustained damage from friction caused by excessive vibrations. The cause of the vibrations remained a mystery. Would the NRC finally hold the industry accountable, or would it, once again, opt for accommodation? An April 5 Associated Press report did not raise hopes. It explained the SCE's proposed solution: "321 tubes with excessive wear will be plugged and taken out of service at the two reactors, *well within the margin to allow them to keep operating*" (emphasis added).[39]

A report from Friends of the Earth (FOE) released on May 15, 2012, traced the failure of more than 1,300 tubes in the two generators to another failure: SCE had falsely informed the NRC that its new steam generators were a "like-for-like replacement." Installing "identical" equipment can save time and money since it does not require a rigorous licensing review. SCE failed to inform the NRC that the new generators contained seven significant design changes that should have required public hearings before approval. According to the FOE investigation, SCE failed to inform the NRC that the new generators (1) changed a preexisting design specifically intended to reduce vibration; (2) removed the generators' main "structural stay cylinder"; and (3) crammed an additional 400 tubes into "an already packaged design."

Edison proposed that it be allowed to restart the reactor after plugging the damaged tubes and promised to operate the reactor only between 50 and 80 percent power. But the FOE report noted that "reducing power does not provide a remedy for underlying structural problems" causing the tube-damaging vibrations. FOE noted that power reductions had failed to solve similar problems at other reactors and, worse, could actually exacerbate the vibration problems. If the reactor were allowed to restart and a "steam-line accident" were to occur, the resulting damage "could cause an inordinate amount of radioactivity to be released outside of the containment system, compromising public health and safety in one of the most heavily populated areas in the United States."[40]

What are SCE's options? An attempt to repair the damaged generators could take 18 months and cost $400 million, so the most straightforward solution would be to replace the existing generators—only this time, SCE should be required to proceed through the full NRC license amendment process.

An Unplanned Swim in the Reactor Pool

Another troubling incident had preceded the headline-grabbing news of a radioactive leak. A private contractor had been entrusted with replacing the lid of the reactor vessel. On January 27, an outside worker hired by that contractor dropped a flashlight. When the worker leaned over a railing to retrieve his equipment, he plunged headlong into the 20-foot-deep reactor pool, still aglow with the blue brilliance of the Cerenkov radiation being thrown off by the submerged reactor core.

SCE did not file an NRC report on the incident, claiming that the worker received only 5 rem of radioactive exposure. SCE spokesperson Gil Alexander reassured the public that the unnamed employee "was able to return to work the same day."[41] The CBS radio affiliate in Los Angeles offered a slightly different spin on the story. According to KNX 1070, an NRC official reported the worker actually swallowed "a little bit of water that had some residual contamination in it." Once the worker was examined and declared unharmed by the 5 rem exposure, KNX reported, "he was *ordered* back to work the same day."[42] (Emphasis added.)

A nuclear expert with the Union of Concerned Scientists (UCS) subsequently discovered that statistics published on the NRC's own website showed that radiation levels in Unit 2's cooling system had doubled from January to February 2011 and continued to rise through the end of the year. This suggested to UCS that the reactor had operated for several months with damaged fuel rods "that allowed radioactivity to escape into the water at ever-increasing rates." Such a breach would have allowed radioactive particles to migrate into the water in the pool the San Onofre worker fell into. Additional "fuel fleas" or "hot particles" could have been released into the water during removal and replacement of the reactor core.

The NRC seemed to be content with SCE's assurances that the worker was in perfect health. When Rochelle Becker, executive director of the Alliance for Nuclear Responsibility, contacted the NRC to obtain information about the worker's radiation dose, the NRC simply told her to go ask SCE. "When you're telling the public to go back to the utility," Becker observed, "I think that's an indication that there is too much trust."[43]

Residents to NRC: "Shut Down San Onofre"

With San Onofre shut down, local governments in San Clemente and Laguna Beach have appealed to the NRC to prevent the plant from restarting until the residents are convinced it would be safe to do so. Local opponents point to a November 2008 California Energy Commission report[44] that warned San Onofre was likely to experience "larger and more frequent earthquakes" than it was designed to handle, which has further fueled concerns among local residents and both of California's US Senators.[45]

Since SCE does not share its radiation data with the public, the San Clemente city council voted to install its own radiation monitoring system. With residents demanding epidemiological studies to assess the health of people living near the reactors, an SCE spokesman announced on February 10, 2012, that the company had "not made a decision on whether we'll apply for renewal" to restart the plant. If SCE does decide to close the plant, it will then have to deal with decommissioning and the safe disposal of 4,000 tons of high-level radioactive waste stored at the site.[46]

California is one of 13 states that have either banned or restricted the construction of any nuclear reactors within their borders. In 2012, Californians began circulating petitions for a ballot initiative that would, if supported by voters, close San Onofre and Diablo Canyon forever. California has now joined a rising tide of grassroots campaigns calling for the immediate shutdown of reactors in Florida, New Jersey, New York, Ohio, Texas, and Vermont. California senator Barbara Boxer used this news to deliver a stern message to NRC officials during a hearing before the US Senate Committee on Environment and Public Works, convened on December 15, 2011:

> Let me tell you what happens when people lose confidence in the NRC and the nuclear industry. Right now, there is a petition being circulated for a ballot initiative that would effectively shut down the two nuclear power plants in California. I believe we will see more of that across the country if America doesn't have confidence in the NRC. If the NRC does not do its job, the American people will demand the ultimate protection—the shutdown of old nuclear power plants that have similar characteristics as the Fukushima plant.[47]

Vermont Yankee: The Green Mountain State vs. the NRC

On March 10, 2011, the NRC unanimously approved a 20-year license extension for the troubled Vermont Yankee nuclear power plant. Within hours of the decision, three similar General Electric Mark 1 reactors were knocked off-line by an earthquake in Japan—and all three overheated and exploded. Despite the devastation in Fukushima Prefecture, the NRC stood by its decision to allow the 40-year-old Vermont Yankee plant to continue operating through 2031. Given Vermont Yankee's history of breakdowns and cover-ups—and the fact that a reactor accident here could put more than a million Americans at risk—the watchdog group Beyond Nuclear excoriated the NRC's decision as both "audacious" and "reckless."[48]

Vermonters received another jolt when it was revealed that the NRC had voted to extend Vermont Yankee's license even though its inspectors had discovered that critical electric cables powering the plant's safety systems had been "submerged under water for extended periods of time."[49]

It was not the only maintenance failure of Entergy Corp., which had acquired the plant in 2002. The company has a reputation for "buying reactors cheap and running them into the ground." In 2004, a poorly maintained electrical system set off a large fire in the plant's turbine building that forced an emergency shutdown. In 2007, Vermont Yankee experienced a series of maintenance problems that included the dramatic collapse of a cooling tower. A waterfall of high-pressure water burst from a ruptured cooling pipe and tore a gaping hole in the plant's wall. Entergy was able to hide the damage—but only until a concerned employee leaked a photo of the wreckage to the press. The huge gap in the side of the building was reminiscent of the hole in the side of the Pentagon following the 9/11 attacks.

Tritium + Entergy = Perjury

During state hearings in 2009, Entergy executives were asked if radioactive tritium detected in the soil and groundwater near the reactor could have leaked from the plant. Company officials repeatedly swore under oath that this was impossible since there were no underground pipes at the plant. It was not until January 2010, after a leak of radioactive tritium was traced to a series of subsurface pipes, that Entergy changed its story. While the plant didn't have "underground pipes," Entergy now explained, it did have "buried pipes."[50]

Attorney general William Sorrell began a 17-month investigation during which Entergy's former executive vice president Curtis Hebert admitted that the company's statements about the pipes "could have been more accurate." The state ordered Entergy to remove more than 300,000 gallons of radioactive water fron the soil and ground water at the reactor site, and Vermont governor Peter Shumlin demanded the plant's closure.

There's another waste problem at the plant: a large and potentially lethal stockpile of used fuel rods. While Fukushima's six reactors had between

360 and 500 tons of slowly dying fuel rods on-site, the nuclear graveyard at Vermont Yankee is filled with 690 tons of dangerously radioactive waste. And the storage pools for this spent fuel lack both backup cooling systems and backup generators.

Beyond Nuclear's "Freeze Our Fukushimas" campaign, which aims to close all 23 Mark 1 reactors in the United States, hoped to score its first victory when Vermont Yankee's 40-year operating license expired on March 21, 2012. The odds were improved by the fact that Vermont is the only state that gives lawmakers the authority to veto a nuclear power plant. In February 2010, a month after Entergy's tritium scandal was exposed, the Vermont Senate voted 26–4 against issuing a new "certificate of public good" that would allow Vermont Yankee to continue operating.[51]

Entergy Sues Vermont

In April 2011, Entergy's lawyers responded by suing the governor and the state, claiming, "We have a right to continue operation."[52] On January 19, 2012, federal judge Garvan Murtha ruled that only the NRC could close a nuclear plant, and therefore Entergy was entitled to its new 20-year operating license. Murtha also made it clear that the Green Mountain State was not entitled to raise any questions regarding plant safety or the prices charged for nuclear power—under federal law, only the NRC could raise such matters.

The decision alarmed Beyond Nuclear and other critics who feared the nuclear industry and the federal government were working in concert "to pre-empt a state's right to self determination for an energy future in the public good." Senator Bernie Sanders (I-Vt.) quickly weighed in. "If Vermont wants to move to energy efficiency and sustainable energy," Sanders told the *Burlington Free Press*, "no corporation should have the right to force our state to stay tethered to an aging, problem-ridden nuclear plant."[53]

Eight days after the judge's decision, Entergy refused the state's second request to investigate the source of a tritium leak that had poisoned a drinking well on the plant's property.[54] On July 25, 2011, Entergy further demonstrated its disregard for due process by announcing a $60 million refueling project—an investment that would pay off only if the power plant won its extension.

Entergy's lawyers publicly confirmed their understanding that the company still needs the permission of Vermont's Public Service Board (a quasi-official board that oversees Vermont's utilities) if it is to continue operating its reactors. In a responding press release, however, the state's Department of Public Service (which represents the interests of utility customers in cases brought before the Public Service Board) cautioned, "Past experience shows Entergy cannot be taken at its word."[55]

In a daunting struggle that pits 600,000 Vermonters against the US government, the nuclear industry, and the NRC, the state attorney general vowed to appeal Judge Murtha's decision—all the way to the US Supreme Court, if necessary.

"People don't trust the NRC," Bob Audette, a reporter for the *Brattleboro Reformer*, told a film crew from the Center for Investigative Reporting (CIR). "They think it's the lapdog of the industry. They think it's there basically to affirm everything the industry does. It's too cozy with the industry."

In another interview with the CIR, Anthony Roisman, a legal consultant for New York and Vermont, expressed his concerns with the NRC: "This regulatory agency does not regulate effectively. And until it does, there is no way that the public can have any confidence that plants, whether they are licensed or re-licensed, won't have some catastrophic event. No one will benefit from a post-catastrophic-event hand-wringing that says, 'Oh we should have done this and we'll do better next time.' The consequences are unimaginable."[56]

19

Near Misses and Unbelievable Mishaps

The phrase "It can't happen here" is an invitation to disaster.

–FORMER NRC COMMISSIONER PETER BRADFORD

THERE IS A TENDENCY to view nuclear reactors as shinning engineering marvels technologically on par with the Starship *Enterprise.* But nuclear reactors are relics of a pre-NASA era. (Ground was broken for the first US reactor on September 6, 1954—four years before the creation of NASA, on July 29, 1958.) When the first commercial reactors were being built in the 1950s, the hottest TV fare was *The Frank Sinatra Show*, *Beulah*, and *The George Burns and Gracie Allen Show*—and TV screens were still black-and-white. Nuclear reactors were designed using sharpened pencils and slide rules. Even today, most reactor control rooms are not digital. Instead, many of America's heirloom nuclear plants still rely on arrays of analog dials with jittery needle gauges—hand-me-downs from a pre-Internet era.

Even with dated technology, you would expect (given the extraordinary expense and risks of nuclear technology) that the industry would entrust its equipment only to the best-trained and most experienced employees. A near miss at Progress Energy's Brunswick reactor suggests otherwise.

In November 2011, alarms sounded when steam and radioactive water began bursting from top of the pressure vessel of the Unit 2 reactor. The reactor was promptly shut down and NRC inspectors rushed to investigate this "first of its kind" incident. In January 2012, the NRC released a report attributing the "fluke mishap" to human error. The reactor had recently returned to service after a shutdown for refueling and the work crew responsible for replacing the huge metal lid atop the pressure vessel had made some serious mistakes. The NRC revealed that 9 of the 12 workers assigned to tackle this critical task not only were unqualified to do reactor vessel assembly but had received only "just-in-time" (that is, "last minute") training to do the work. The workers "didn't know how to read the instrumentation and torqued the reactor vessel lid's studs at 1,300 pounds per square inch instead of 13,000 psi. In other words, they were off by one zero and screwed the studs at 1/10th the required pressure." Some of the crucial bolts were left so loose they could still be turned by hand. Fortunately, this "first of its kind" mistake did not lead to a "first of its kind" nuclear catastrophe.[1]

The irregularities that mark the NRC's regulatory history have a troubling parallel in the quotidian world of nuclear power plant management. A review of NRC inspections and incident reports reveals the seamy side of the inner workings of these high-tech atomic furnaces. In addition to policing the familiar panorama of a control room's gleaming desks and glowing screens, the NRC's inspectors have to clamber deep inside the cranky bowels of these aging structures, stepping over tangles of electrical cables, peering into dark, dripping recesses, and poking at metallic scales encrusting rusted pipes and valves.

The combination of failing equipment and worker error can be downright frightening. For example: During a safety test on October 22, 2009, workers at California's Diablo Canyon plant found they couldn't open the valves that released emergency cooling water to prevent a core meltdown. This potentially catastrophic problem had gone undetected for *18 months*.

Sometimes, it is management failure that compounds the problem. On June 8, 2010, an electrical short triggered a shutdown at Virginia's Surry nuclear plant, and 90 minutes later, a fire broke out in the main control room of the Unit 1 reactor. An NRC investigation noted that a similar fire had erupted at the Unit 2 reactor six months earlier. When workers tested the control rooms, they found that the electrical systems were so degraded that "some produced visible sparks during testing." After the first fire, workers had begged technicians to investigate the threat, but according to a report prepared by the Union of Concerned Scientists, "the company closed the report without any investigation or evaluation."[2]

A Chilling Inventory of Reactor Incidents

Acknowledging that no reactor is foolproof, Nils J. Diaz, a former NRC Chairman, spelled out a theory of preventative oversight in a 2004 speech entitled "The Very Best-Laid Plans (the NRC's Defense-in-Depth Philosophy."[3] The gist of the NRC's "defense-in-depth" philosophy is described as follows:

> An approach to designing and operating nuclear facilities that prevents and mitigates accidents that release radiation or hazardous materials. The key is creating multiple independent and redundant layers of defense to compensate for potential human

> and mechanical failures so that no single layer, no matter how robust, is exclusively relied upon. Defense-in-depth includes the use of access controls, physical barriers, redundant and diverse key safety functions, and emergency response measures.[4]

Each year, the NRC's best inspectors probe America's reactors to gauge whether "defense in depth" is a daily practice or a shallow promise. And each year, the Union of Concerned Scientists (UCS) goes over these inspection records and issues a report. The gist of UCS's 2011 survey was suggested in its subtitle: "Living on Borrowed Time."[5] The report identified 15 "near-misses" that occurred over the course of a single year. Here are a few of the most troubling incidents from 2011:

- Workers at Duke Energy's Oconee reactor in South Carolina discovered that an emergency backup cooling system installed in 1983 would have failed to prevent the reactor core from overheating. The problem went unnoticed for more than a quarter century.
- A routine test of the emergency cooling system at the Callaway reactor in Missouri actually wound up damaging the pump.
- Workers at Nebraska's Cooper reactor and Ohio's Pilgrim reactor were exposed to high levels of radiation while attempting to replace detectors to monitor the reactor core.
- A control room test at Connecticut's Millstone 2 reactor triggered "an unexpected and uncontrolled increase in the reactor's power level."
- A test of the fire suppression system at Minnesota's Monticello reactor revealed that the water flow had been blocked "by rust particles inside the system." Nuclear Management Co., the plant's owner, had ignored NRC warnings about the corrosion threat.
- In 2009, an emergency cooling pump failed at Entergy's Palisades plant in Michigan. Workers installed a replacement. In 2011, the replacement failed, disabling the pump.
- Workers at the same Entergy plant, while "troubleshooting faulty indicator lights," accidentally shut off power to half the instruments in the reactor's main control room, causing a loss of control and forcing the reactor to shut down.
- While attempting to restart Entergy's Pilgrim reactor after refueling, plant operators "lost control of the reactor," triggering an automatic shutdown. "Security problems" at the same Massachusetts plant prompted a "special inspection" by the NRC, but

according the UCS, "details of the problems, their causes, and their fixes are not publicly available."

While "Living on Borrowed Time" credited NRC inspectors for spotting a number of problems that could have led to "major accidents," the report identified a number of systemic problems with NRC oversight. Among them:

- The NRC had permitted 47 reactors to continue operating "despite known violations of fire-prevention regulations dating back to 1980."
- The NRC was allowing 27 reactors to remain online "even though their safety systems are not designed to protect them from earthquake-related hazards identified in 1996."
- The NRC had taken no special action to minimize risks at eight reactors that "suffer from both afflictions."
- US reactors "continue to experience problems with safety-related equipment and worker errors that increase the risk of damage to the reactor core."
- When NRC inspectors discover broken parts or improper maintenance, "they all too often focus just on that problem, not its underlying cause."
- The frequency of reports of high-risk problems (more than one a month) was judged to be unusually "high for a mature industry."

The UCS report singled out one owner for particular attention, noting: "Four of the special [NRC] investigations occurred at plants owned by Entergy." UCS suggested a prudent response: "The NRC should formally evaluate whether corporate policies and practices contributed to the shortcomings."

Nebraska's Fort Calhoun nuclear plant offers a prime example of the NRC's failure to exercise due diligence. On February 17, 2010, NRC inspectors toured the plant after a critical turbine-driven auxiliary feedwater pump failed during a monthly test. According to the UCS, the inspectors discovered that the pump had failed "numerous times over many years." Moreover, "the owner had never found the cause of the problem and therefore had never taken steps to prevent it."

Mistakes and Malfeasance: In the NRC's Own Words

In response to the alarm sounded by the meltdowns in Japan, the NRC ordered a thorough inspection of all US reactors. The results were sobering.

NRC inspectors found significant safety problems with reactors at 60 of the country's 65 nuclear power stations. The plants racked up lapses that varied from mundane to monumental. The following notes are drawn from the NRC's inspection records.[6]

- **Braidwood.** "The licensee would have encountered several unplanned challenges in implementing various B.5.b mitigating strategies." ("B.5.b" is the NRC's code for maintaining reactor safety following a large explosion or fire.) Emergency pumps were slow to start, critical emergency flanges proved too heavy to move and install, and "a weld must be removed to open a hatch to gain access to a refueling water storage tank area used to implement B.5.b strategies."
- **Brunswick.** Critical equipment to deal with explosions and fires would not be available "in a severe weather event." In the event of a flood, diesel fire pump batteries "would be underwater." In the event of an earthquake, the building "is not seismically qualified." In either event, the safety equipment "would be adversely affected by a severe natural phenomenon."
- **Comanche Peak.** "Some suction hoses listed in the extreme damage mitigation procedure . . . were not in the designated location or were absent." Critical electrical equipment for dealing with an emergency was also missing. Inspectors reported that "the lube oil storage building is a potential fire/explosion hazard"—and it was positioned near a trailer storing emergency response gear. The alternate service water pumps "were located in a non-seismic building" and the plant operators had "never tested the fire truck pumping from the reservoir."
- **Davis Besse.** The "operational contingency response action plan" calls for "a total of 11 prepared 8-gauge insulated wires" to support emergency power requirements, but "the inspectors only found a total of 10 wires," which would make it impossible to connect the block of emergency batteries to provide power. Because of "a reduction" in its cooling capacity, the diesel generator needed during a station blackout would be rendered "unavailable whenever outside ambient air temperature exceeds 95 degree F."
- **Diablo Canyon.** "Cooling Pump 0-1 would not function when tested." "The licensee was unable to place the necessary temporary hoses from raw water reservoir to the plant due to

obstructions created by recent security modifications." "Licensee did not have the ability to implement" emergency feedwater to cool "both units simultaneously to support mitigation of a postulated earthquake."

- **Hatch.** The operators' procedure to reduce internal pressure through periodic venting was judged "not able to be performed due to the *high temperature conditions* expected to exist within the area." In addition, "some operations on Unit 1 require operators to disconnect and install air lines by climbing and standing on piping directly over a 30-foot drop." And in Unit 2, "temporary electrical connections" were found to be "coated, which may impact the ability to make effective electrical connections."
- **Indian Point 2.** A serious problem cropped up at the Entergy-owned plant in New York when inspectors identified "an unresolved item for Entergy not ensuring the operability of the containment hydrogen recombiners as required." Entergy also failed to make the mandated inspections of the plant's water bays and pumps every five years. Furthermore, the NRC reported, "there was no planned date for this inspection."
- **Indian Point 3.** Entergy admitted its fire prevention plans "would present *potential vulnerabilities*" because critical equipment was stored in underground or in quake-vulnerable cinder-block buildings and could be lost during a severe quake.
- **Limerick.** "In certain SBO [station blackout] scenarios, only two EDGs [emergency diesel generators] from the non-blacked out unit would be available." The current operating license calls for three EDGs to be available. The NRC concluded that this "inability to satisfy the licensing basis . . . requires further review."
- **Millstone.** A "fire main isolation valve would need to be operated to pressurize the fire main to mitigate a fire in Unit 1, but the valve would be under water."
- **Oconee.** "Equipment that would be used to pump water . . . in the event of flooding . . . did not fit any outlet." During tests of sump pumps, "the start capacitor failed on two of four pumps tested." Inspectors found "preventative maintenance activities had been deferred several times and were currently outside of the recommended frequency for performance."
- **Point Beach.** Inspectors noted that the plant had "an insufficient number of power cables," the plant's G-05 gas turbine could fail during an electrical blackout, and there were "no

specific requirements" for training staff to handle blackouts, explosions, or fires.

- **Robinson.** "Up to 8,000 feet of electrical cabling" were required but "could not be located."
- **Sequoyah.** Despite the example of Fukushima, eight maintenance supervisors "had not yet received initial B.5.b training." "Six station drainage valves" essential to respond to flooding "could not be located," manpower and timeline expectations were "unrealistic," and "generator fuel oil replenishment connections were located below the probable maximum flood level."
- **Surry.** "One of the portable generators . . . failed to start when tested." A leak in a PolyPipe tube was spewing 25 gallons per minute. "The licensee repaired the leak and restarted the pump" but the repair "did not completely eliminate the leak."
- **Three Mile Island.** "The assumed operator response time for a circulating water rupture in the turbine building was unrealistic."
- **Vermont Yankee.** The operators "did not include the fire prevention hoses" essential to fighting a fire or explosion into a "preventive maintenance program." Portable nitrogen gas bottles used to power valves following a fire or explosion "were not periodically inspected" to ensure they would work in an emergency. The inspectors noted "a common theme in that the licensee did not maintain equipment in accordance with guidance provided in NEI 06-12." (NEI 06-12 was a regulation prepared in December 2006 by the Nuclear Energy Institute, a lobbying organization that calls itself "the policy arm of US nuclear energy industry." NEI 06-12 laid out "acceptable means for developing and implementing the mitigation strategies . . . intended to maintain or restore core cooling, containment, and SFP [spent fuel pool] cooling capabilities under the circumstances associated with the loss of a large area of the plant due to explosions or fire." The NRC officially accepted the NEI's recommendations in a letter dated December 22, 2006.[7])
- **Vogtle.** The future site of the country's first newly licensed power plants (Vogtle 3 and 4) was found at fault because plans for restoring off-site power "cannot be performed if main control room evacuation is required."
- **Wolf Creek.** "Procedures to refill the refueling water storage tank are not viable" because "the specified connection point is not readily accessible." The operators admitted that the situation

> risked "excessive reactor coolant system depressurization, which could compromise natural circulation core cooling." Put to the test, "station operators failed to promptly locate certain emergency operating components."

Two of the 60 cited plants deserve special mention.

Fitzpatrick

Inspectors reported "an apparent beyond design and licensing basis vulnerability." Specifically, the plant's emergency procedures "do not address hydrogen considerations during primary containment venting." (It was inadequate hydrogen venting that caused the explosions of the reactors at Fukushima.) The NRC reported that the plant's current license did not require a "hardened vent system as part of their Mark I containment improvement program." Instead, Fitzpatrick was relying on "a hard pipe from primary containment." The situation was so worrying that the NRC reported it would order an "agency task force to conduct a near-term evaluation of the need for agency actions." The plant's system for removing "decay heat" was found not to be "workable." The NRC accepted the following fix: the licensee agreed to "fabricate an adaptor to connect a 1.5-inch fire hose to a 1-inch pipe thread fitting in order to fill the decay heat removal system using fire water, if necessary."

Dresden

"The flood strategy is that when water level rises above 517 feet (plant ground level), the doors to the reactor building will be opened and water will be let in." While opening the doors would "flood all safety-related emergency core cooling pumps," this desperate action would be needed because of "the inability of the reactor building walls to support water levels above 517 feet." This would mean "the only equipment capable of removing decay heat expected to remain functional after the PMF [probable maximum flood] are the isolation condensers."

In the event of an expected 517-foot flood, the plant operators were required to surround the diesel-driven isolation condenser pumps with a ring of sandbags 519 feet high. But in order to handle a PMF that could rise to 528 feet, the NRC concluded, "the pump would have to be hoisted into the air using a crane and a chain fall to remain above the flood waters." In other words, "the pump has to be operated and refueled while hanging in the air." "This pump will also provide cooling and makeup water to the spent fuel pools." And where will the pump source its coolant? "The source of water for the pump is the flood water off the reactor building floor."

Dresden's operators told the NRC they expected "as much as 72 hours warning before water level would reach 509 feet," which would be plenty of time, they reasoned, to shut down the reactor, flood the building to the reactor's head flange, and pluck the critical diesel pump out of the path of the onrushing floodwaters. There was only one problem. As the NRC report noted, "The flood pump has never been operated while hoisted into the air."

PART THREE

The Path Forward: Better Options Exist

INTRODUCTION

The "Energy Turn": Germany Leads the Way

We want to end the use of nuclear energy and reach the age of renewable energy as fast as possible.

—GERMAN CHANCELLOR ANGELA MERKEL

ANGELA MERKEL began her career as a quantum chemist studying subatomic reactions, so it was no surprise that, as chancellor, she became an outspoken advocate for a nuclear renaissance. In 2010, Merkel called for extending the operating lives of Germany's aging reactors by an average of 15 years, assuring her critics that nuclear power was both clean and safe.

But like much of the world on March 11, 2011, Merkel's eyes were trained on the video images streaming out of Japan. When a hydrogen explosion tore apart the containment building surrounding the Unit 1 reactor, a shaken Merkel turned to an aide and said, "It's over."

As a scientist, she realized that the nuclear industry's predictions about the probability of a disaster had been fundamentally flawed. Merkel went on to declare that Fukushima had "forever changed the way we define risk in Germany." Merkel's environment minister, Norbert Röttgen, underscored the shift in thinking. Fukushima, he stated, "has swapped a mathematical definition of nuclear energy's residual risk with a terrible real-life experience."[1]

Within weeks of the Fukushima meltdowns, Merkel unveiled a radical plan to speed the decommissioning of all 17 German reactors. Merkel's plan called for reduced use of fossil fuels—coal, petroleum, and natural gas—and the accelerated approval of renewable energy projects. The goal was to make "green power" the major source of energy for all 81 million Germans by 2030. Merkel called the proposal the "energy turn."

With reactors now set to go off-line by 2022 (or sooner), the challenge is to meet the country's baseload demands with energy drawn from the sun, wind, biomass, and geothermal. Merkel's goals are to double renewable power generation to produce 35 percent of Germany's needs by 2020, 50 percent by 2030, 65 percent by 2040, and 80-plus percent by 2050. Merkel's "energy turn" is also committed to reducing CO_2 emissions by phasing out fossil fuels.

At the time Merkel made her stunning announcement, Germany was pulling around half of its 81 gigawatts of peak demand from coal and 23 percent from nuclear. Renewables were in third place, providing 17 percent of Germany's power, ahead of natural gas at 10 percent.

Wasting no time, Merkel showed up on the coast of the Baltic Sea on May 2, 2011, to flip the switch on Baltic 1, the country's inaugural offshore wind farm. As Merkel pressed a button, 21 gargantuan wind turbines stirred to life 16 kilometers out at sea, and enough new electricity to power 50,000 homes began to surge into the German grid. (Notably, Siemens—a German company that is quickly joining the "energy turn" by shifting its manufacturing from nuclear and fossil fuels to renewables—built the turbines.)

Introducing "Energy Democracy"

The Renewable Energy Sources Act (RESA) helped to hasten Merkel's vision of a green Germany. Passed back in the 1990s, RESA requires that power companies and utilities reward anyone who generates and contributes "green electricity" to the grid. Under this system (also known as a "feed-in tariff"), a homeowner with rooftop PV panels or a small wind turbine is paid for any excess power generated—and at rates *higher* than prevailing energy prices. This premium rate for renewable energy (financed by a small surcharge on all energy users) has stimulated impressive growth of green energy design and development. At the same time, annual rebates on the energy from the solar panels atop an average four-person home were generating $220 in "renewable revenue" for German households.

RESA's payments-for-power plan unleashed an unprecedented race for renewables. In just two years, Germany added more than 10,000 MW of distributed photovoltaic power, with households and small businesses accounting for a majority of the new, privately owned power panels. With more than one million solar systems installed, solar PV generated 3 percent of Germany's electric power in 2011—a 60 percent increase over the previous year. Renewables provided nearly 20 percent of Germany's energy needs in 2011. But equally important was the news that Germany's overall electricity consumption for the year was cut by nearly 5 percent.[2]

In 2011, Germany commanded 16 percent of the global green-tech market and had added 300,000 "green collar" jobs over the previous decade. Its application of renewable technology had more than tripled in less than a decade, with wind farms blossoming along the coasts and mountaintops and solar panels proliferating across village rooftops. Rapeseed is now being widely grown as a biofuel stock, and large "biomethane" tanks (filled with fuel brewed from agricultural wastes) are now a common sight at many farms.

The "feed-in" laws (*Stromeinspeisungsgesetz*, in German) set off what can only be called a "feed-in frenzy" that saw homeowners, farmers, and community groups racing to take power generation into their own hands. By 2010, 51 percent of Germany's 50,000 MW of renewable energy was being produced by these *Stormrebellen* ("electricity rebels").[3] Here is a telling comparison: In 2010, the United States installed 890 MW of new solar photovoltaic (PV) panels, raising the national PV energy total to 2,200 MW. Thanks to the *Stromeinspeisungsgesetz* effect, Germany added 7,408 MW of new PV panels in 2010, boosting the national total to 16,516 MW. Individual panel-owners generated 6,648 MW of the nation's PV power (more than 39 percent) while PV-equipped German farms generated another 3,586 MW (more than 21 percent) of the country's PV power.[4] The tiny Bavarian farming village of Wildpoldsried managed to produce 321 times more renewable energy than the village itself needed. In 2011, Wildpoldsried's farmers were able to sell the excess power and reap a financial harvest of $5.7 million.[5]

Across Germany, corporate control of power is being replaced by citizen ownership (*Bürgerbeteiligung*), as "power to the people" swiftly evolves from a wistful cliché to a working reality. Renewable energy advocate Paul Gipe notes "travelers to Germany marvel at how Jeffersonian the Bürgerbeteiligung movement has become in democratizing electricity generation."[6]

Instead of writing billion-euro checks to oil and gas providers in Russia, Germany now spends its euros at home, supporting local green-tech innovators. The green energy choice provides a better economic dynamic since, as environment minister Röttgen observes, "When more people consume oil and coal, the price will go up, but when more people consume renewable energy, the price of it will go down."[7]

This message apparently is turning heads, even in the oil capital of planet Earth. In May 2012, Gipe reported the surprising news that "the conservative, oil-rich Kingdom of Saudi Arabia has proposed one of the most sweeping and massive moves to renewable energy on the planet. . . . Starting from zero, the Kingdom plans to install a staggering 54,000 MW of renewable-generating capacity during the next two decades." By comparison, the United States currently operates around 50,000 MW of solar panels and wind turbines and has a population 10 times the size of Saudi Arabia's. Gipe predicts that the Saudis' decision to "go solar" may lead policymakers in Canada and the United States "to reconsider their recalcitrance toward the renewable revolution sweeping the globe."[8]

COMPARING ENERGY SYSTEMS: BEST TO WORST

How do various energy production options stack up when they are rated not just on the energy and pollution they produce but also on their impact on water supplies, land use, wildlife, resource availability, thermal pollution, chemical pollution, nuclear proliferation, and malnutrition? Stanford researcher Mark Z. Jacobson conducted a detailed "lifetime assessment" survey of current energy choices. The study found that wind, concentrated solar, geothermal, tidal, photovoltaic, wave, and hydro can best power the country's transportation, residential, industrial, and commercial needs while addressing global warming, air pollution, and energy security. Wind is two to six orders of magnitude safer than any other energy source. It causes the least wildlife loss and the lowest human mortality. Replacing oil-burning cars with wind-powered electric vehicles could cut US CO_2 emissions by 32.5 to 32.7 percent and prevent 15,000 annual deaths linked to exhaust pollution.

Life-Cycle Assessments–Best to Worst

(Note: "g CO_2e/kWh-1" refers to grams of carbon dioxide equivalent per kilowatt-hour.)

Wind
Emissions: 2.8–7.4 g CO_2e/kWh-1
Operational lifetime: 30 years
Cents per kWh: 8.55
Time to completion: 2–5 years
Energy payback time: 1.6 months

Concentrated Solar
Emissions: 8.5–11.3 g CO_2e/kWh-1
Operational lifetime: 40 years
Cents per kWh: 12.65
Time to completion: 2–5 years
Energy payback time: 5–6.7 months

Solar Panel
Emissions: 19–59 g CO_2e/kWh-1
Operational lifetime: 30 years
Cents per kWh: 6-14
Time to completion: 2–5 years
Energy payback time: 1–3.5 years

Wave
Emissions: 21.7 g CO_2e/kWh-1
Operational lifetime: 15 years
Cents per kWh: 4.5
Time to completion: 2–5 years
Energy payback time: 1 year

Hydroelectric
Emissions: 17–22 g CO_2e/kWh-1
Operational lifetime: 50–100 years
Cents per kWh: 10.53
Time to completion: 8–16 years
Energy payback time: 5–8 years

Geothermal
Emissions: 15 g CO_2e/kWh-1
Operational lifetime: 35 years
Cents per kWh: 10.18
Time to completion: 2–3 years
Energy payback time: 10-16 years

Nuclear
Emissions: 9–70 g CO_2e/kWh-1
Operational lifetime: 40 years
Cost per kWh: 15.316
Time to completion: 10–19 years
Energy payback time: 15 years

Coal
Emissions: 790–1,020 g CO_2e/kWh-1
Operational lifetime: 5–30 years
Cents per kWh: 10.55
Time to completion: 5–8 years
Energy payback time: 2-4 years

Coal with Carbon Capture and Storage (CCS)
(Note: The process of compressing and transporting carbon dioxide for storage consumes 14–25 percent more energy.)
Emissions: 255–440 g CO_2e/kWh-1
Operational lifetime: 35 years
Time to completion: 6–11 years
Cents per kWh: 17.32
Energy payback time: 20 years

Sources: Mark Z. Jacobson, "Review of Solutions to Global Warming, Air Pollution, and Energy Security," *Energy & Environmental Science* 2 (2009): 148–73. Costs: California Energy Commission, 2008. Solar panel costs: Solar Energy Industries Association, 2012 (http://http://www.seia.org). Wave power costs: "Wave Energy," Ocean Energy Council, 2012 (http://www.oceanenergycouncil.com). Note: costs are approximations and are subject to change depending on such variables as time, technology, resource availability, and economic conditions.

20

The Path Ahead

Humans are inventing a new fire—not dug from below but flowing from above, not scarce but bountiful, not local but everywhere. This new fire is not transient but permanent . . . and grown in ways that sustain and endure. Each of you owns a piece of that $5 trillion pie.

–AMORY LOVINS, *REINVENTING FIRE*

WHERE DO WE GO from here? How do we start the process of change? If nuclear energy is not the answer to the natural limitations of nonrenewable energy and resources, then what is the best course to fashion an economy that provides for the long-term welfare of people and the preservation of nature? Instead of continuing to fantasize about a "nuclear rebirth," it's time to accept the fact that this ailing industry suffers from a long list of incurable maladies. Instead of prolonging the lives of reactors well past their designed age of retirement, we need to pull the plug on these aging artifacts. We need to demand that reactors be closed at the end of their operational lifetime of 35 to 40 years.

Beyond that, we need to accept that we live on a finite planet with limited resources: Earth's oil, water, minerals, soils, and oceans are all showing troubling signs of devastation and depletion. On a finite planet, the pursuit of continuous, ever-expanding economic growth—the primary goal of the prevailing US economic system—is a recipe for disaster. By itself, even the creation of new "clean energy systems" will not enable us to maintain our current unsustainable rates of industrial-economic growth. We need a renewables revolution that supports a steady-state economy driven by a conservation imperative.

The United Nations' $24 million Millennium Ecosystem Assessment (undertaken by the largest body of social and natural scientists ever assembled to evaluate human impacts on the planet's ecosystems) found that 60 percent of Earth's 24 most critical ecosystems have been seriously degraded over the past 50 years, to the point that the planet's ability "to sustain future generations can no longer be taken for granted." The report's 1,360 global experts called for "substantial changes in institutions and governance, economic policies and incentives, social and behavior factors, technology, and knowledge."[1]

As Richard Heinberg eloquently demonstrates in his 2009 report *Searching for a Miracle*,[2] there is no way to sustain an economic system built on a false assumption of endless resources, inexhaustible energy, and infinite expansion. Even if we could manage to switch totally to renewable energy systems—which we certainly advocate—no combination of these systems could sustain an economy built on the mad premise of endless growth—and the projected rise in the human population only compounds the dangers. The good news is that thousands of groups around the world are already far ahead of their governments in planning for a post-oil, post-growth, conservation economy. There is no shortage of good ideas, only a shortage of ability to make them manifest.

Over the years, the Worldwatch Institute has been an astute observer of the planet's energy progress. On the 25th anniversary of the 1986 Chernobyl explosion, Worldwatch published *The World Nuclear Industry Status Report 2010-2011*, a survey of "nuclear power in a post-Fukushima world."[3] Among the report's findings:

- In 2009, nuclear power plants generated 2,558 Terawatt-hours of electricity, about 2 percent less than the previous year—and the fourth year in a row to record a decline.
- In 2010, for the first time, worldwide cumulative installed capacity from wind turbines, biomass, waste-to-energy, and solar power surpassed installed nuclear capacity.
- Annual renewable capacity additions had outpaced nuclear start-ups for the previous 15 years. In the United States, the share of renewables in new capacity additions skyrocketed from 2 percent in 2004 to 55 percent in 2009.

The New Power Paradigm

Even if all of the world's current energy output could be produced by renewables, consuming that much energy would still inflict terrible harm on Earth's damaged ecosystems. In order to survive, we need to relearn how to use less. It is critical that we adopt a conservation imperative.

Faced with the inevitable disappearance of the stockpiles of cheap energy we have used to move and transform matter, we need to identify society's fundamental needs and invest our limited energy resources in those key areas. A post-oil/post-coal/post-nuclear world can no longer sustain the onetime extravagances of luxury goods, designed-to-be-disposable products, and brain-numbing entertainment devices. The long-distance transport of raw materials, food, and manufactured goods will need to decline in favor of local production geared to match local resources and needs. Warfare—the most capital-, resource-, and pollution-intensive human activity—must also be addressed. Neither the costly inventory of nuclear arms nor the Pentagon's imperial network of 700-plus foreign bases is sustainable. There will doubtless still be wars, but in a post-oil world, they will be either waged with solar-powered tanks or fought on horseback.

Modern economies insist on powering ahead like competing steamboats in an upstream race. We have become addicted to overconsumption on a planet that was not designed for limitless exploitation. As the late environmental leader David Brower noted, "In the years since the Industrial Revolution, we humans have been partying pretty hard. We've ransacked most of the Earth for resources. . . . We are living off the natural capital of the planet—the principal, and not the interest. The soil, the seas, the forests, the rivers, and the protective atmospheric cover—all are being depleted. It was a grand binge, but the hang-over is now upon us, and it will soon be throbbing."[4]

On the eve of India's independence, Mahatma Gandhi was asked whether his new nation could expect to attain Britain's level of industrial development. "It took Britain half the resources of this planet to achieve its prosperity," Gandhi noted. Therefore, he famously estimated, raising the rest of the world to British levels of consumption would require "two more planets."

The United Nations Development Program recently reconsidered Gandhi's equation as it applies to "a world edging toward the brink of dangerous climate change." Working from the assumed "sustainable" ceiling of climate-warming gases (14.5 Gt CO_2 per year), UNDP confirmed, "If emissions were frozen at the current level of 29 Gt CO_2, we would need two planets." Unfortunately, UNDP noted, some countries are producing more CO_2 than others. Fifteen percent of the world's richest residents are using 90 percent of the planet's sustainable budget of shared resources. According to UNDP's calculations, if developing countries were to follow the example of Canada or the United States, each would need to tap the resouces of *nine* additional planets.[5]

In this final section, we have bundled some attainable survival strategies into four broad categories: technological efficiencies; alternative-renewable energy systems; public policy options; and, most important, the conservation imperative—powering down. A combination of these efforts, enthusiastically embraced, will have a good chance of saving the world by adopting sustainable lifestyles that eschew the bizarre notion that happiness can only come from never-ending consumption and accumulation. A new level of appreciation for sufficiency and equitability will be the ultimate answer.

21

Efficiency: The "Fifth Fuel"

The goal of life is living in agreement with nature.

–ZENO OF CITIUM (335-264 BC)

WHEN IT COMES to the question of "net energy," the laws of physics are clear and immutable: in order for an organism (or a technology) to succeed, it cannot expend more energy than it consumes. In order to survive, a migrating salmon needs to expend no more than one calorie of energy to acquire three calories of food energy—a 3:1 ratio. Prehistoric hunter-gatherers burned one calorie of energy in the pursuit of every 4–5 calories of berries or deer meat. Preindustrial farmers and the nomads of the Kalahari could subsist successfully with a 10:1 energy return on investment (EROI). However, with the discovery of coal and oil, humans were able—for the first and only time in history—to tap an EROI of 100:1. In order to match the energy potential of a single barrel of oil, a single human worker would have to labor 8.6 years—without pausing to sleep.[1]

This onetime gift of fossil energy (representing 500 million years of captured solar energy) sent human beings into unprecedented overdrive—building, traveling, flying, planting, logging, polluting, and toying with weapons of mass destruction. Human "progress" was fueled by the profligate combustion of fossilized biomass, consumed at a rate of about 5 million year's worth of solidified sunlight every year. But today, as Richard Heinberg has memorably observed, "the party is over." As the most easily accessed reservoirs of cheap oil have been exhausted, the EROI for oil has plummeted toward 10:1, forcing industry to undertake costly, high-risk, deep-ocean drilling and to scrape desperately after tar sands, an unattractive carbon resource with an EROI of only 4:1.

Investing in Efficiency

Energy efficiency has been called "the fifth fuel."[2] According to energy expert Amory Lovins, "Each dollar invested in electric efficiency displaces nearly seven times as much carbon dioxide as a dollar invested in nuclear power, without any nasty side effects."[3] It is clear that a serious international commitment to promote energy efficiencies would yield spectacular paybacks, not only by reducing global warming, energy consumption, and pollution, but also by lowering the costs of economic productivity. But we begin with a daunting task. Consider the current state of affairs:

Each day, on average, human activity pours 16 million tons of CO_2 into the atmosphere. The average citizen of the world contributes about 12 tons of CO_2 a year; the average American contributes 23 tons.[4] The United States burns about 17 million barrels of oil a day, 7 million barrels of which are imported. The United States, with only 5 percent of the world's population, generates 23 percent of all energy-related global carbon emissions. The United States consumes nine times the electricity used by the earth's average resident, while more than 1.6 billion people live without electricity.

Surprisingly, it's not cars but buildings that are the greatest sources of planet-cooking emissions. Homes and offices consume around 50 percent of US energy. Globally, buildings consume 76 percent of the world's electricity.

Since 1980, the size of the average US home has ballooned 45 percent, a trend that must be reversed, since this increase in rooms and windows requires more energy for heating, cooling, lighting, and added electrical appliances.[5] The average US home produces 11 tons of CO_2 annually. An average home uses 26 plug-in appliances, including multiple TVs, TV set-top boxes that consume half as much energy as a refrigerator, and videogame consoles that require twice the power of a fridge.[6]

Since developing countries use only 30 percent of the world's generated energy, it is clear that the ultimate solution to our energy future rests largely in the hands of the industrialized nations. If the United States, for example, were to reduce its industrial energy use by 1 percent, that seemingly small step would save about 55 million barrels of oil (and about $1 billion) a year.

A 2010 wide-ranging assessment by Clean Technica (a world leader in clean technology reporting) concluded US industry—from steel, paper, and cement mills to power plants—wastes nearly 200 gigawatts of potential power per year. The US Energy Information Administration (EIA) estimates coal-fired power plants lose as much as 51 percent of their energy to conversion losses—a scandalous performance characteristic that has remained unchanged since the 1950s. According to Clean Technica, the average US power plant consumes three units of fuel to produce a single unit of electricity. The wasted heat that could be used to produce power is pumped into the sky as superheated steam. Nuclear plants routinely waste 21 percent of their thermal potential.

The Department of Energy has identified 135,000 megawatts (135 gigawatts) of untapped cogeneration potential in the United States while the Lawrence Berkley National Laboratory estimates 64,000 megawatts (64 gigawatts) could be harvested by reclaiming wasted industrial energy.[7]

The energy wasted by America's power plants would be sufficient to power Japan. Our automobiles, industrial motors, and water heaters are less efficient than those in Japan and Europe. Ironically, many of our new "smart" technologies are not energy efficient. Some modern manufacturing systems consume a million times more energy per pound of output than traditional industries. For example, it takes more energy to make a microchip than to forge a manhole cover. Production inefficiencies of current manufacturing processes have become a critical factor in determining the life-cycle energy costs of solar panels.[8] This must change, and an ever-growing number of sustainable business practices are coming online to make this happen.[9]

The "Cheapest, Cleanest, Fastest Energy Source"

On the plus side, since 1973, improved efficiencies have saved more than six times the energy produced by America's nuclear power plants. A global "Green New Deal," like the one proposed by the UN,[10] could create jobs, spur technological development, cut the world's energy demand 20 percent by 2020, and lower US electricity consumption by 75 percent.[11] Strong national efficiency standards could replace the need for 450 power plants by 2020. Calling efficiency "the cheapest, cleanest, fastest energy source," President Obama has vowed to cut the government's energy use by 15 percent.[12] Over the past 30 years, the United States has routinely saved far more power through efficiency improvements than it has generated from nuclear power plants. Between 1997 and 2011, thanks to efficiency gains, the United States consumed 2,300 fewer BTUs per dollar of GDP—a 24 percent decline. The energy saved was approximately equal to 14.5 million barrels of oil per day, easily outpacing the country's increase in oil production. Despite the demonstrated power of energy efficiency, US oil production continues to surge. The first quarter of 2012 witnessed the highest level of US oil production since 1998.[13] And while efficiency investments usually pay for themselves in three years (and thus have been adopted on economic grounds by many US companies), they continue to be dismissed by the corporate interests that stand to profit (in the short term) from the ever-increasing consumption of energy.[14]

The fact is that efficiencies have already had a major impact. In September 2010, a Lawrence Livermore National Laboratory study found that total US energy consumption fell from 99.2 quadrillion BTUs in 2008 to 94.6

quads in 2009 because of changing economic and lifestyle impacts.[15] "The US Energy Information Administration reports that the country's CO_2 emissions dropped nearly 10 percent between 2007 and the end of 2009.[16] While around one-third of the drop was attributed to the recession and lower natural gas prices, most of the reductions came from efficiency gains and new state renewable energy standards boosted by a federal "clean energy" stimulus.[17] (Note: Part of the "dash for gas" that brought down prices involved the exploitation of "unconventional" sources like shale gas. But the hydraulic fracturing needed to extract the gas can pollute groundwater with toxic "fracture fluids" and trigger earthquakes.)

Meanwhile, the US Department of Energy notes that 75 percent of home electricity is wasted, consumed by "standby" appliances that continue to draw power even when "turned off." While modern refrigerators use two-thirds less energy than they did 20 years ago, a proliferation of "vampire appliances" with always-on dials, clocks, and displays are costing US homeowners $1 billion a year in added energy bills.[18]

Redesigning or eliminating the "standby" option can tap significant energy savings. In the meantime, homeowners can drive a stake through these energy vampires by plugging them into surge protectors so they can all be turned off at one time at night or when homeowners leave the house. This would accomplish a significant gain in "negawatts."

Negawatts, not Megawatts

"Negawatts" are watts saved from one application and made available for another. For example, replacing one 75-watt incandescent bulb with one long-lasting, energy-efficient 20-watt compact fluorescent bulb creates 550 kWh of saved energy over the new bulb's lifetime—and keeps 1,300 pounds of CO_2 out of the air.[19] Here's another form of negawatt: when electricity demand rises, large users can be paid for creating "negawatts" by cutting back on consumption.[20] Taking full advantage of off-the-shelf energy-saving devices can cut the costs of heating, cooling, and lighting US homes and offices by as much as 80 percent.[21]

The Zero-CO_2 Economy

Transition to a truly zero-CO_2 industrial economy may be possible, but it won't resemble the kind of extravagant industrial dystopia we now inhabit. This can be achieved with a combined array of old and new technologies, including photovoltaic panels, solar thermal, passive solar, wind turbines, wave power, biomass, geothermal, hydrogen, ultracapacitors, solar light pipes, and other options.[22] Because oil-dependent industrial agriculture generates as much as 30 percent of the world's climate-changing greenhouse gases, a zero-CO_2 economy will also require replacing large-scale, mechanized farming with organic and agroecological practices that actually capture and return excess atmospheric CO_2 to the soil.[23]

A Green Transition by 2050

Greenpeace and the European Renewable Energy Council have published a detailed blueprint for the transition to a sustainable and just energy future by midcentury. Given the recent trend of double-digit growth in renewable energy options, their plan, called *Energy [R]evolution*, would close all nuclear reactors by 2020 while stabilizing both the climate and the economy.[24] As homeowners go off-grid, producing their own rooftop power and backyard food, this transition to "micropower" promises to propel a fundamental efficiency revolution.[25] Taking the lead in North America, the Canadian province of Ontario has passed the Green Energy Act, while the university town of Berkeley, California, has fashioned a climate action plan that the United Nations has praised as "the best in North America."[26] The following sections detail these and other possibilities.

22

The Power of Renewables

Of all the forces of nature, I should think the wind contains the largest amount of motive power. . . . It is applied extensively, and advantageously, to sail-vessels in navigation. Add to this a few windmills and pumps, and you have about all. . . . As yet, the wind is an untamed and unharnessed force; and quite possibly one of the greatest discoveries hereafter to be made, will be the taming and harnessing of it.

—ABRAHAM LINCOLN, 1860

IN 2007, the fast-growing global renewables sector received $71 billion in private capital investment, while the nuclear industry received nothing.[1] At the same time, the world invested $100 billion more in renewables than in fossil fuels, adding 40 billion "green watts" to global output. Nuclear power added zero watts.[2] And the renewable sector continues to surge. In a May 2011 report, the United Nations Intergovernmental Panel on Climate Change concluded that renewable energy could provide 77 percent of global power demands by 2050. In addition, this expansion of clean, renewable energy would avoid the creation of 220 to 560 gigatons of carbon dioxide between 2010 and 2050—equal to about a third of current global CO_2 pollution and significant enough to stabilize atmospheric CO_2 at 450 ppm.[3]

A more conservative International Scientific Congress study predicted that wind and photovoltaic technologies could provide 40 percent of the world's electricity by 2050—if renewable alternatives receive adequate financial and political support. Investing in existing and innovative green-tech energy and efficiencies also would prevent the release of 38 billion tons of greenhouse gases a year.[4] Each month, a single home with a one-kilowatt PV system prevents the release of 300 pounds of CO_2 into the air, keeps 150 pounds of coal from being mined, saves 150 gallons of water, and avoids the release of nitrous oxide and sulfur dioxide into the atmosphere. While solar and wind power provide energy on a variable rather than a constant basis, they can become collectively reliable when integrated with steady-output renewables like geothermal, small hydro, and wave resources.

Biomass, solar, wind, geothermal, and hydropower are all projected to be competitive with "cheap, dirty" coal as early as 2015. There is an added advantage to replacing carbon-combustion energy with renewable power—because wind, water, solar, and geothermal energy (WWSG) is more efficient, meeting the world's projected global demand for 16.9 TW of electricity by 2030 could be accomplished with only 11.5 TW of WWSG.[5] To understand why this is the case, we need only to look at the automobile. Only about one-fifth of the gas burned in a car is actually transformed into motion—most is wasted as heat. By contrast, in an electric-powered vehicle, as much as 86 percent of the energy is converted into motion.

Hopes to replace "dirty coal" (which is responsible for 86 percent of the country's stationary CO_2 emissions) with "clean coal" technology have been dimmed by warnings that this new techno-fix could add as much as 70 percent to the cost of turning coal into electricity.[6] In another sign that "clean coal" is no panacea, major utilities are starting to drop out of the industry's multimillion-dollar lobbying group, the American Coalition for Clean Coal Electricity. The clean-coal cause was further damaged by a January 2010 report condemning the process of using "mountaintop removal" to extract coal. As reported in the journal *Science*, "The preponderance of scientific evidence is that impacts are pervasive and irreversible and that mitigation cannot compensate for losses."[7]

Many renewables are already cheaper than nuclear. US utilities in the Southwest, for example, are signing purchase agreements for wind power at 4.5 to 7.5 cents per kWh—one-third the cost of nuclear power.[8] And according to an 18-page report by John O. Blackburn, PhD, a professor emeritus of economics and former chancellor at Duke University, 2010 marked a "historic crossover" in North Carolina, when electricity from solar PV panels became cheaper than nuclear power.[9] Electricity from the state's nuclear reactors averaged 16 cents per kWh, while electricity from photovoltaic panels was only 14 cents per kWh—and that rate continues to fall. (The data covered only PV power, not concentrated solar systems, which can lower costs even more.) The study factored in the hidden costs of government subsidies for both power sources but concluded that—even if all subsidies were removed—solar power would still be cheaper within a decade.[10]

In 2010, the Energy Information Administration's *Monthly Energy Review* reported, the production of renewable energy in the United States jumped by 5.6 percent, rising to 10.92 percent of total energy production. At the same time, nuclear energy's share of power production fell to 11.26 percent (down from 11.48 percent in 2009).[11]

What would it take to move to a self-sustaining WWSG economy by 2030? According to a blueprint offered by two California researchers, more than 50 percent of the world's power needs could be met by building 3.8 million large wind turbines. (Not an impossible task, given that the world's automakers build more than 50 million cars and light trucks each year.)[12] Another 40 percent could be generated by rooftop solar panels and concentrated solar power stations, with the remainder coming from geothermal energy, wave energy, and 900 hydroelectric stations (70 percent of which already exist).[13]

President Obama's 2009 stimulus package routed at least $70 billion to renewable energy R&D—triple the baseline budget of the US Department of Energy (DOE) and more than the annual budgets of the Labor and Interior Departments combined. The stated goal was to double renewable energy capacity in three years. (Reality check: while this is a lot of money, it's less than the world's three major oil companies spend in a single year. And Obama's stimulus lasted only two years.)

In 2006, for the first time in history, micropower—which embraces small renewables, cogeneration, and other efficiencies—outperformed nuclear reactors, producing one-sixth of all the world's electricity (and one-third of all new electricity).[14] In China, the country with the world's most ambitious nuclear program, electricity from distributed renewables charged ahead of atom-powered electricity by a factor of seven. The amount of new nuclear capacity added between 2000 and 2007 was six times less than the 13,300 MW of wind power added during the same period. In 2011, the world added another 41,000 MW of wind power to the global energy mix. This 21 percent annual increase in wind power brought the total installed global capacity to more than 238,000 MW—equal to the output of 238 average 1,000-MW reactors. By the end of 2011, 22 of the world's 75 wind-powered nations were generating more than 1 GW in clean, renewable wind energy.[15]

Compared to these vibrant new developments, nuclear power is a stodgy and obsolete technology. We need to move past the old "dig-it-up-and-burn-it" energy technologies and focus our attention on fashioning a sustainable green future. Here's a brief rundown of some of the most promising technologies.

Wind Power

Wind power is the world's fastest-growing energy source, and there's good reason for its prodigious growth. The potential for wind power over land exceeds 72 TW—20 times the world's current electric power production. In a major study in 2010, DOE's National Renewable Energy Laboratory

(NREL) estimated the potential for wind power off US coasts could exceed 4,000 GW—four times the entire electricity demand in the United States.[16]

By 2011, the World Wind Energy Association listed 83 different countries that had installed wind power systems capable of generating a combined annual output of 344.8 TWh. The top ten wind-energy nations continue to be (in descending order of installed power): China (63 GW), the United States (47 GW), Germany (29 GW), Spain (22 GW), India (16 GW), France (7 GW), Italy (7 GW), the United Kingdom (6.5 GW), Canada (5 GW), and Portugal (4 GW).[17] Although Denmark is not on the list of top ten generators, it currently stands at the top of the world's wind-powered economies, with about 20 percent of its electricity flowing from the wind.

In 2008, as US coal-powered electricity generation fell by 1.5 percent, wind power generation increased by 60.7 percent.[18] Wind installations worldwide generated a record 27 GW. The United States alone added more than 4 GW—more than 40 percent of all new global electrical capacity—to eclipse Germany as the planet's leader in wind energy. (The United States—with 40.2 GW of turbine capacity—would manage to remain the world's wind-power giant until 2011, when it was passed by China—with 42.3 GW).

In 2009, Europe added 37.5 GW of new wind energy, while electricity production from oil, coal, and nuclear energy fell a collective 32.1 GW.[19] Global revenues for solar, wind, and biofuels reached $139 billion. (And by 2019, revenues from wind power alone are expected to exceed $114 billion.)[20] China doubled its wind power capacity for the fifth year in a row, while the United States increased its wind power capacity 40 percent, bringing it to 35.2 GW.[21]

Surprisingly, 2010 was a bad year for the wind industry. The American Wind Energy Association (AWEA) reported the lowest figures for new capacity, new installations, and new investment in years. Wind power advocates hoped to boost demand through passage of a strong national renewable energy standard and expansion of federal "clean energy" loan guarantees. "We have a historic opportunity to build a major new manufacturing industry," the AWEA cautioned, but without the same federal support enjoyed by the nuclear, oil, and coal industries, the AWEA feared "manufacturing facilities will go idle and lay off workers."[22] (While the attempt to pass a renewable energy standard was defeated, a new attempt at funding a green energy renaissance was mounted on March 1, 2012 with the introduction of the Senate's "Clean Energy Act of 2012.")

Global wind power rebounded in 2011, surging 21 percent with the installation of more than 41 GW of new wind turbines.[23] As European Wind Energy Association CEO Christian Kjaer notes, in 2011, "the world installed

50 percent more new wind-power capacity (41.2 gigawatts) than all new nuclear capacity installed from 2002 to 2011 (27.3 GW). In terms of electricity production, the wind-power industry has installed the equivalent of 1.3 nuclear reactors per month over the past three years." Moreover, Kjaer added, "between 2004 and 2011, more nuclear-power capacity was decommissioned worldwide than was installed."[24] In the first quarter of 2012, the United States added 1,695 MW of new wind energy. Over the previous five years, US wind energy capacity grew 35 percent—more than nuclear and coal combined.[25]

Wind power has the potential to supply a good portion of US electrical needs at an affordable price, since power-friendly wind speeds have been found to exist in 17 percent of US land and coastal regions. The NREL estimates that the wind in North Dakota alone could generate one-third of the country's electricity. Some studies suggest that combining North Dakota's breezes with the wind-energy potential of Kansas and Texas could provide for all our country's needs for clean, affordable electric power.

The NREL projects that offshore wind turbines could supply at least 20 percent of all electricity for US coastal cities, and secretary of the interior Ken Salazar believes that wind turbines off the East Coast could produce 1 million MW—enough power to replace 3,000 coal-fired power plants. New Jersey plans to triple its wind power resources to 3,000 MW by 2020. In 2010, Google announced it was investing in the Atlantic Wind Connection, an offshore power project that would plug wind turbines into a submerged transmission cable stretching 350 miles from New Jersey to Virginia. This $5 billion project is expected to produce 6,000 MW—sufficient to power two million US homes.

Replacing all US internal combustion cars with electric vehicles powered by wind-generated electricity could cut CO_2 emissions by a third and prevent 15,000 auto-exhaust-related deaths per year. However, this would mean building and installing as many as 144,000 huge 5 MW turbines—just to power the existing number of autos. We would be better off using the electricity to power mass transit systems while we work to develop more compact and energy-efficient cities. Meanwhile, the 158 GW of clean electricity flowing from the world's wind turbines in 2009 prevented the release of 204 million tons of CO_2.[26]

Wind turbines aren't immune from criticism, however. Turbines in the hills of Altamont, California, were implicated in the deaths of endangered raptors, and dozens of YouTube videos document some of the friction-caused fires that, as of June 2010, had destroyed 149 turbines around the world. Nearby buildings have been damaged by ice tossed from the turbine blades and some broken blades have flown as far as four-fifths of a mile. Statistics compiled by the Caithness Windfarm Information Forum, for the period 1995-2011, show 89 fatal wind turbine accidents resulting in 102 deaths (70 wind-energy employees and 32 members of the public.).[27] Nevertheless, while accidents may increase with the expansion of wind farms, the overall dangers remain statistically small. Wind energy is also inconstant, with more wind activity in the winter months than in the summer and the greatest wind-loads found in mountains, on open plains, and at sea. Shipping wind-generated power hundreds of miles over power lines means sustaining significant transmission losses. Still, the average energy returned on energy invested (EROEI) ratio for wind power is 18:1, which is competitive with conventional power technologies (and more efficient than nuclear, at 11:1 EROEI).[28]

Solar Photovoltaics

Land-based photovoltaic potential stands at around 1,700 TW worldwide. California alone boasts 354 MW of installed solar electric power. Two proposed large photovoltaic sites (one in the Mojave Desert and another in Antelope Valley) would add another 783 MW. Meanwhile, the $3 billion California Solar Initiative has helped install more than 50,000 rooftop-mounted generating facilities statewide, accounting for 60 percent of the entire US solar energy market.[29]

While silicon solar cells generally convert 15 to 20 percent of solar radiation to electricity, new designs (like the SolFocus system, which incorporates small, concave mirrors) can reach efficiencies of 30 percent. The sun's light can also be captured, focused, and trained on the surface of small "concentrating photovoltaic" cells. In 2008, a cell developed by Solar Junction and tested by NREL attained a record-setting efficiency of 40.9 percent. Such high concentration photovoltaic (HCP) systems not only double the efficiency of traditional PVpanels, they also do not contain the toxic cadmium telluride found in traditional PV panels. HCPs also appear to maintain their efficiency better than standard PV panels.

The 162,000 solar panels at Spain's 62 MW Olmedilla Photovoltaic Park (the world's largest photovoltaic power station) produce about 87.9 GWh (gigawatt-hours) of electricity per year—enough to power more than 40,000 homes. The Olmedilla plant was constructed in just 15 months at a cost of $350 million.

Concentrated Solar

Concentrated solar installations use large tracts of mirrors to focus the sun's heat on centrally mounted tanks that create steam to spin turbines and produce electricity. Spain expects to be generating 1,789 MW of electricity from concentrated solar power (CSP) plants by the end of 2013, and the NREL expects to see 4,000 MW of CSP installed worldwide by 2015. California has plans for four CSP facilities with a total capacity of 2,660 MW. California's 370 MW Ivanpah Solar Power Facility, under construction in the Mojave Desert, is currently the world's largest solar thermal power project. The 289 MW Mojave Solar Park (also in the Mojave) is set to begin operations in 2014. With Algeria, Australia, Egypt, India, Israel, Libya, Morocco, and South Africa also installing CSP plants, global production is expected to hit 36,859 MW by 2025—the equivalent of 36 nuclear power plants.

A downside: large CSP facilities—like those planned for the Mojave Desert, the Sahara Desert, and elsewhere—have raised environmental concerns over the disruption of animal migration routes, destruction of habitat, and displacement of flora and fauna if the land is graded, sprayed with chemical herbicides, shaded from sunlight, wind, and rain, and subjected to human presence, vehicles, and trash. A far less disruptive approach to solar collection is distributed solar power.

Distributed and Rooftop Solar

Widely distributed wall- and rooftop-mounted photovoltaic and thermal collectors can provide homes and offices with power and heat. Unlike centralized electricity plants, distributed power sources are located right where the energy is used. Homes and cities in the United States occupy about 140 million acres. According to the NREL, placing PV collectors on 7 percent of our rooftops and parking lots could "supply every kilowatt-hour of our nation's current electricity requirements," leaving "a landscape almost indistinguishable from the landscape we know today."[30] An important bonus: using distributed solar would bypass fragile electric transmission grids, greatly improving reliability and resilience.

In March 2012, Pacific Environment, a green think tank based in Oakland, California, published a study, called Bay Area Smart Energy 2020 (BASE 2020), that outlined a plan to add nearly 4,000 MW of rooftop solar power in the Bay Area by converting 25 percent of existing homes and commercial buildings to zero-net-energy buildings by 2020.[31] BASE 2020 argued that placing solar panels on existing urban surfaces would cut electricity-sector greenhouse-gas emissions by 60 percent while avoiding the estimated $15 billion cost of building new transmission lines from distant solar farms—an approach favored by investor-owned utilities. The report argued that the cost of a kilowatt generated by a massive Mojave Desert solar power plant could be 50 percent greater than the cost of drawing the same amount of power from a 500 kW solar system installed on the roof of a large downtown retail outlet.

In recent years, the home-solar market has been given a shot of adrenaline by innovative marketing plans that allow homeowners to "go solar" with no up-front costs. Panels are installed and maintained free of charge and the only cost is a monthly leasing fee paid to the installer. In most cases, the monthly lease proves to be lower than monthly bills previously paid to the Big Grid utilities.

SRS Energy's Solé Power Tiles are fabricated to resemble Spanish tiles, slate, or shake tiles. Each Solé tile contains a flexible solar panel encased in lightweight, recyclable polymer. In some high-heat conditions, these tile-like panels (manufactured by United Solar Ovonic) can outperform flat PV panels.

Solar Thermal Water Tanks

With water heating constituting a home's third largest energy cost, solar water heating provides a simple and affordable means to chop hot water bills in half. A single home equipped with a passive thermal water heater (a common sight on Southern California rooftops during the '30s and still widely used in Greece, Turkey, Israel, China, and India) can prevent the release of more than 50 tons of CO_2 over the course of 20 years.[32] In January 2010, the California Public Utilities Commission launched a $350 million incentive program to install 300,000 solar water heaters on the Golden State's rooftops by 2018. The move is expected to eliminate 100,000 tons of greenhouse gas emissions each year.[33] In February 2010, senator Bernie Sanders (I-Vt.) followed California's lead by introducing the "10 Million Solar Roofs and 10 Million Gallons of Solar Hot Water Act," which called for providing federal rebates to cover up to half the cost of 10 million solar power systems and 200,000 solar water heaters.

Passive Solar

Buildings designed for passive solar heating employ a variety of approaches including south-facing glazed windows, insulation, sunspaces, and Trombe walls to capture the sun's free light and heat. Thick, south-facing Trombe walls and floors built from heat-absorbing materials capture solar heat during the day and slowly release the stored heat to warm the house at night. Day-lighting of north-facing rooms and upper levels is accomplished with a clerestory—a row of windows near the peak of the roof—or solar pipes that collect and redirect sunshine throughout the building.

Thin-Film Solar

Traditional rooftop solar panels may soon become passé, thanks to new thin-film solar coatings. Available on polymer sheets that can be easily applied to windows and walls, these "power sheets" have the potential to provide enough electricity to power entire buildings.[34] An array of thin-film solar panels covering the façade of the Technology Place business center at the British Columbia Institute of Technology, for example, already generates sufficient electricity to provide for 100 percent of the building's lighting.

Thin-film photovoltaics are positioned to become the fastest growing part of the solar module industry because they promise to dramatically reduce production costs for PV applications while generating more power under a wider range of light conditions. Thin-film photovoltaics can be incorporated into solar farms or installed on city rooftops and building facades. Requiring only about 1/200th of the crystalline silicon used in traditional solar cells, thin-film modules promise a quicker energy payback.[35] Nanosolar PowerSheet solar cells may cut the cost of producing solar power from $3 per watt to 30 cents per watt, marking a historic point where solar power becomes cheaper than coal.[36] And, stimulated by new "feed-in" laws that pay homeowners for electricity generated from rooftop panels, Everbrite, a Canadian solar firm, announced plans to build a $500 million manufacturing plant in Ontario capable of producing 150 MW of thin-film solar a year.

Light-Emitting Diodes

Around 20 percent of global electricity is used for lighting. Australia, Brazil, the European Union, and Switzerland have banned incandescent bulbs (which waste 95 percent of their electricity as heat). In the United States, plans for a ban on incandescent bulb were beaten back by Republican lawmakers. Instead of an outright ban, new efficiency standards for incanscent bulbs were mandated to take effect by October 2012. Even though energy-efficient compact fluorescent bulbs use 75 percent less power and last 10 times longer, they contain tiny amounts of toxic mercury and some produce an unsteady ultraviolet light that can trigger epilepsy. The better bet is light-emitting diodes (LEDs), which are 80 percent efficient, can mimic the color of natural daylight, and—with a lifetime of 45,000 hours—will outlast a nuclear fuel rod by two years. Unlike compact fluorescent lights, LEDs do not contain mercury. The DOE estimates that switching the country over to LEDs would save more than $30 billion in energy costs and reduce energy consumption for lighting by one-third—the equivalent of 44 power plants producing 1 MW each.[37]

A lightbulb that produces more light but burns less energy may sound like a good deal—unless you are the owner of a power company that profits from increasing consumption, or you happen to be a Tea Party extremist. Some members of the corporation-backed Tea Party have become absolutely incandescent with rage over the government's attempts to replace old-fashioned, inefficient electric bulbs with LEDs and other new, high-efficiency products. Conservative voices have protested the government's "massive, mandated turnover" from too-hot-to-touch bulbs to cool, efficient, long-lived bulbs, complaining about Washington's "$50 light bulb." But let's do the math. According to an analysis by Treehugger, an online environmental news and information service, if you pay $50 for a bulb that lasts 30 years, the bulb's lifetime electricity consumption would cost around $33. If you insisted on burning—and repeatedly replacing—"traditional" incandescent bulbs, 30 years of lighting would cost around $228.[38]

Super-efficient LEDs

In March 2012, an MIT research lab stunned the engineering world with a report that it had built an LED with an efficiency of 230 percent. In other words, the device produces more energy in the form of light than it consumes in electricity. The super-efficient bulb not only utilizes its own waste heat, it also taps "ambient" heat from the surrounding air to increase its efficiency. The pursuit of technologies that can operate above "unity efficiency"—"over-unity" technologies—has been challenged as a violation of the second law of thermodynamics. The MIT team admits that their experimental LED is small (generating 69 picowatts from 30 picowatts of energy), but it demonstrates that previously unimagined levels of efficiency are no longer simply theoretical. If MIT's LED can be scaled up, future super-efficient LEDs might someday be able to provide both illumination and air-conditioning.[39]

Tidal Energy

Large underwater turbines that can turn the ebb and flow of ocean tides into electricity are being built in Scotland, Ireland, and Australia. France has successfully operated a 240 MW tidal plant at La Rance for more than 40 years.[40] A commercial-size system is already in operation in Nova Scotia, while the Philippines is considering construction of a 2.2 GW tidal fence in the San Bernardino Strait.[41] In the United States, tidal power systems have been proposed for New York's East River, the Mississippi, and under California's Golden Gate Bridge. The caveat is that these technologies remain largely experimental and—like ocean-based wind turbines—are susceptible to formidable corrosion and battering from water and waves.

Wave Energy Converters

Wave energy converters could potentially harness 2 TW of ocean power. Aguçadoura, the world's first wave farm, built off the coast of Portugal, uses three wave energy converters to produce 2.25 MW.[42] Again, the challenge is to perfect converters that can withstand continual pounding by corrosive saltwater.

Hydroelectric

Hydropower represents more than 92 percent of all renewable energy generated worldwide. The world's total hydro potential is nearly 14,370 TWh/year, but at the present time, only about 8,082 TWh/year is economically attainable.[43] In 2008, hydropower delivered 3,288 TWh, 16 percent of world's electricity.[44] Norway generates 98.9 percent of its electricity from dams, followed by Brazil (83.7 percent) and Venezuela (73.9 percent). Between 1980 and 2006, Iceland more than doubled its hydroelectric production. In 2008, Iceland's 1,880 MW of installed hydroelectric capacity covered 75.5 percent of the country's entire electricity needs.[45] The United States is the world's hydropower leader, with 92,000 MW providing 9 percent of the nation's electricity (and 49 percent of the country's renewable energy).

However, with climate change reducing precipitation in the oversubscribed Colorado River watershed, major dams in the Southwest (including Lake Mead and Lake Powell) are nearing "dead pool" status—that is, water reserves are falling to such low levels that they are no longer able to produce power. Another downside is that large dams can become environmental nightmares. In addition to the greenhouse gases produced during construction, forests and vegetation buried beneath a dam's impounded water can decompose to release global-warming CO_2 and methane into the atmosphere. (In 1990, the rate of greenhouse gas emissions per kW produced by the Curuá-Una Dam in Brazil was estimated to be three times greater than that of a comparable oil-burning power plant.)[46]

The massive weight of dammed water has even been known to cause regional earthquakes. Globally, the concentration of vast hydroelectric reserves has had a measurable impact on Earth's rotation. NASA estimates that the 42 billion tons of water rising behind China's Three Gorges Dam will move the planet's axis two centimeters and lengthen the day by 0.06 microseconds.[47]

Small Hydroelectric

Small village-size hydro projects are less destructive to the environment than dams on the scale of the Curuá-Una and Three Gorges. Small and "micro" turbines can be effective in producing mechanical or electric power where river flows are strong and constant year-round. "High head" turbines require water falling from a height greater than 10 feet. "Low head" turbines require a vertical fall of 2 to 10 feet. In the United States, small hydro projects require the approval of the Federal Energy Regulatory Commission and the US Army Corps of Engineers.

Cogeneration

Burning fuels to generate electricity creates heat that can be captured for space or water heating. Cogeneration plants, also known as combined heat and power (CHP) plants, generate electric power and harvest the otherwise lost heat to warm homes and buildings. While conventional coal-fired electric generation is only 33 percent efficient, cogeneration has the ability to perform at 75 to 90 percent efficiencies.[48] Capturing the lost heat from US steel mills and oil refineries could produce 100 GW of electricity while eliminating about 400 million metric tons of CO_2 emissions.[49] The United States currently derives only 6 percent of its electricity from cogeneration or distributed renewables (i.e., small, decentralized power systems like solar PV and wind turbines). In 2003, Denmark met 52 percent of its energy needs with CHP.

Geothermal

In his 2010 book, *Geothermal: Getting Energy from the Earth*, Lester Brown noted that the upper six miles of the planet's crust contains 50,000 times as much energy (captured in the form of heat) as is found in all the world's oil and gas reserves. As of 2010, 70 countries were tapping this vast reservoir of heat beneath our feet. Worldwide geothermal capacity is estimated at nearly 11 GW—equal to the production of a dozen nuclear plants.[50] In addition to warming the subsoil, magma from the planet's molten core also heats subsurface water, producing steam that can be used to power hydrothermal generators. Iceland is the world leader in geothermal and hydrothermal power: Iceland's geothermal resources provide 89 percent of the country's home heating. US hydrothermal resources are estimated to be in the range of 2,400 to 9,600 exajoules.[51] The 15 steam turbines driven by wells at the Geysers hydrothermal site in northern California generate about 725 MW—enough to power a city the size of San Francisco. In Southern California, half a dozen new energy projects are preparing to tap the geothermal potential of Imperial Valley, estimated to approach 2,500 MW.

In addition, low-temperature direct heat can be tapped almost anywhere on Earth using a heat pump with pipes bored only a few meters into the ground. In the colder regions of Europe, Asia, and Canada, heat pump installations have been growing 30 to 40 percent annually.[52] Enhanced geothermal systems (EHS) are a promising alternative that would produce electric power by running turbine fluids through superheated rocks 3 to 10 kilometers underground. While water is a limiting factor, many of the world's poorest regions (in Africa, South America, the Caribbean, and Pacific Islands) are rich in geothermal potential. Another advantage: when sustainably managed, geothermal is the rare renewable resource that provides constant, rather than intermittent, power.

Home/Office Fuel Cells

Small, cost-efficient fuel cells now coming onto the market offer to clip energy bills in half while cutting CO_2 emissions by nearly 40 percent. A suitcase-size 5 kW unit installed outside the home can transform natural gas from existing utility lines into heat and electricity. In addition to being 11 times more efficient than solar panels, stationary fuel cells are a form of distributed power that can produce power directly where it's needed, 24 hours a day, without sustaining the transmission losses endemic to electricity delivered over the power grid. As a combined heat and power system, these home/office fuel cells are predicted to reach efficiencies of 60 to 90 percent.

Thermal-Power Technologies

If there's one thing we can count on in the future, it is that we will be living on a warmer planet. Is there some way to take advantage of some of this added heat? Over the past centuries, human society has burned wood, coal, gas, and atoms to produce heat. Now, new engineering discoveries may make it possible to cool the planet by transforming ambient heat directly into energy. One example is Power Felt, an ingenious new fabric that incorporates nanotubes to give thermoelectric properties to a weave of plastic fibers. Developed by David Carroll, head of the Center for Nanotechnology and Molecular Materials at Wake Forest University, Power Felt can use body heat to recharge a cell phone tucked into a pocket. Added to the insulation inside a home, Power Felt could generate electricity to power household appliances. Installed in a hybrid or all-electric auto, Power Felt could dramatically extend the range of the vehicle. This breakthrough product is also inexpensive. A square foot of Power Felt currently costs no more than $2 to produce.[53]

Biodiesel

There may be no better example of how new energy paths can redefine value than "yellow grease." Up until 2006, this by-product of the restaurant trade was considered a worthless burden that was frequently (and illegally) poured down city storm drains. Now that it is recognized as an excellent feedstock for producing biodiesel fuel, the price of "yellow grease" has jumped to 42 cents per pound and given rise to a new crime—"grease thefts," where poachers break open metal barrels to siphon off the potential fuel. In San Francisco (where the municipal Greasecycle program was begun in 2007), restaurants that once paid to have "yellow grease" hauled away are now reaping a bounty by selling it to refiners who turn it into fuel to run the city's buses.

Wind Hydrogen

Using wind-generated electricity to electrolyze water produces hydrogen that can be banked for later use. This stored hydrogen can be used in an engine or a fuel cell to generate electricity on demand—even when the wind isn't blowing. While hydrogen is traditionally produced by stripping hydrogen atoms from natural gas or fossil fuels (a process that generates greenhouse gases), wind hydro creates no greenhouse gases or other harmful by-products. In a September 2009 wind hydro demonstration, staff from the NREL drove a fuel-cell-powered Mercedes-Benz 110 miles on four pounds of compressed hydrogen. (Of course, the more Earth-friendly option would be to park the Mercedes and hop on a bike.)

For both the near term and the distant future, renewable technologies—wind, wave, geothermal, solar, and so on—offer safer, cheaper, and quicker energy options than risky, costly, cumbersome nuclear power. Renewable technologies, when combined with energy efficiency, can provide half the planet's electricity by 2050, making it possible to cut CO_2 emissions by 50 percent by 2040. The European Renewable Energy Council estimates that investing $11.3 trillion to $14.7 trillion to transition the world's energy sector to renewables (at an annual investment of just 1 percent of global GNP) could cut fuel costs by a quarter and save an estimated $750 billion a year.[54] But despite these impressive figures, it appears unlikely that renewables-plus-efficiencies can sustain industrial society at its present levels of consumption and waste.

Currently, turbine blades and PV panels must be manufactured, delivered, and maintained by drawing on ever-declining reservoirs of cheap oil, dirty coal, and poisonous uranium. While the sun, wind, and tides will still be here 2,000 years from now, renewable energy sources cannot sustain an industrialized economy operating on the current scale. Compared to the 100:1 energy wallop of a barrel of oil, the best EROI for a wind farm is 18:1. The EROI for nuclear power is around 11:1. At some point, we will face the real test: can renewable energy alone build the next generation of solar mirrors, photovoltaic panels, fuel cells, and wind-turbine towers? Further transformations in economic and social priorities are required, as we discuss in the next chapters.

23
Public Policy Reforms

The world has achieved brilliance without wisdom, power without conscience. Ours is a world of nuclear giants and ethical infants. We know more about war than we know about peace, more about killing than we know about living.

—**GENERAL OMAR BRADLEY, US ARMY CHIEF OF STAFF**

THE INDUSTRIAL REVOLUTION of the nineteenth century gave us the Age of Coal. In the twentieth century we entered the Age of Oil. The second half of the twentieth century saw the dawn of the Nuclear Age. The twenty-first century promises to become the Age of Renewable Energy and Conservation, but only if we have sufficient public determination to resist the vested interests that would keep society hooked on the debilitating addiction to costly oil, dirty coal, and deadly atoms—even at the cost of waging resource wars over declining inventories of oil, coal, uranium, and water. Political options must quickly change so that appreciating Earth's limits becomes a constant factor in the way energy policy decisions are made. Aggressive efforts must be made by all levels of government, business, and society to adapt to the realities of a planet being pushed to its limits—and that means applying the conservation imperative first and foremost.

The struggle has become titanic. President Obama's initial 2009 economic stimulus plan included $80 billion for green energy, the Energy Department has promised $150 billion for renewables research, and the Interior Department has established a task force to accelerate development of large-scale, renewable energy projects on federal lands.[1] This is a good beginning, but it's not nearly enough. Large-scale "solutions" may prove to be a luxury that is no longer available in the era we are now entering. In a post-carbon world, we will need to focus more attention on solutions that are local, small-scale, and sustainable.

A Renewables Redirection

Some American businesses, are already undertaking massive energy-efficiency moves. Walmart, the world's largest retailer, has realized that its heavy reliance on transportation makes it vulnerable to increases in oil prices. In response, the company introduced efficiency measures that trimmed consumption of diesel by two-thirds and announced plans "to eliminate 20 million metric tons of greenhouse gas (GHG) emissions from its global supply chain by the end of 2015."[2] The Walt Disney Company—which once promoted the false solution of nuclear power—made good on a promise to reduce its carbon emissions from its theme parks, offices, and cruise ships 50 percent by 2012. The Disney company's ultimate goal is to reduce its carbon emissions to zero by shifting to more efficient clean-energy options.[3] According to energy expert Amory Lovins, learning to use energy "in a way that saves money" will mean that "some big problems like oil dependence, climate change, and the spread of nuclear weapons will go away—not at a cost but at a profit."[4] This may be sound a bit wishful—especially for those who recognize that the idea of a "green" Walmart or "energy-efficient" Disneyworld is incompatible with the realities of a planet that will soon be faced with the need to go on a "power diet."

As the King CONG (coal-oil-nuclear-gas) economy flickers toward its inevitable end, the world's industrialized countries are still providing $60 billion in subsidies to support these outmoded and unsustainable industries. Former British treasury official Nicholas Lord Stern has stated that meeting climate stabilization goals means that "most of the world's electricity production will need to have been decarbonized" by 2050.[5] This would require industrial economies to cut per capita CO_2 emissions by at least 80 percent by midcentury. Developing nations may be asked to cut emissions 20 to 40 percent by 2020. Only carbon-free renewables like solar and wind can be deployed at a speed likely to accomplish this transition.

Many towns, states, and nations are well on the way toward a renewables revolution. Researchers at Johns Hopkins University's Center for Climate Strategies compiled data on climate policies already adopted by 16 US states and, extrapolating to the nation as a whole, estimated that a national commitment to just 23 specific policy approaches would reduce CO_2 emissions 27 percent below 1990 levels by 2020 (4.46 billion metric tons of carbon dioxide equivalent), create 2.5 million "green jobs," and add $159.6 billion to the nation's GDP.[6]

The 2007 Green Jobs Act, sponsored by Representatives John Tierney (D-Mass.) and Hilda Solis (D-Calif.), authorized $125 million to train workers—including low-income youth—to install solar and wind power generators, retrofit old buildings, and construct new "green" buildings. The goal is to create three million green jobs within a decade. The act's Pathways Out of Poverty provision will ensure that the benefits flowing to companies in Silicon Valley also will flow to impoverished jobseekers in East Los Angeles and the Bronx.[7] As the country's new secretary of labor, Hilda Solis is perfectly positioned to promote a green jobs program that simultaneously addresses the two major problems of pollution and poverty.

In April 2010, Texas announced that it had reached its goal of generating 10 GW of renewable energy—15 years ahead of schedule.[8] A bill in the California Senate would put the state on course to generate one-third of its energy from renewable sources by 2020, enough to power every home in the state.[9] Hawaii and Alaska have adopted goals of producing 40 percent and 50 percent, respectively, of their energy from renewable sources. The countries of the European Union expect to derive 100 percent of their power from renewable sources by 2050.[10] Denmark already has the potential to receive as much as one-third of its electricity from wind power.[11] Sweden has announced its intention to become the industrial world's first "oil-free nation" by 2020 and has challenged the rest of the world to transition fully to renewable energy within the next 10 years.[12]

Countries around the world are using public policy to effect these changes and more. Among some of the policy options now being promoted are the following.

Carbon Offsets/Cap-and-Trade

Offset schemes give CO_2 emitters the opportunity to balance their emissions with CO_2 reductions or sequestration. They essentially pay other businesses to counter their carbon impacts. A similar cap-and-trade program, for tradable and gradually decreasing pollution "allowances," was pioneered in the United States to minimize acid rain caused by sulfur dioxide emissions from coal-fired power plants. In 2006, about $5.5 billion worth of carbon offsets were purchased in the compliance market, representing about 1.6 billion metric tons of CO_2 reductions.[13]

Identifying actual reductions is tricky and is subject to fraud, giving rise to some of the same practices that led to the collapse of the US mortgage market. In December 2009, France reported a $209 million "carbon carousel fraud," and in January 2010, EU officials announced carbon fraud losses topping 46.7 billion euros.[14]

Unlike a straightforward carbon tax (see below), cap-and-trade transactions can lead to economically dangerous speculative "bubbles" that can cause broad economic collapse. The immense and often untraceable sums involved in cap-and-trade systems already appear to be contributing to a new global financial bubble. A more effective—and revolutionary—prescription would be to extend carbon offsets to the world's two billion poor who have no access to electricity and should, therefore, rightly represent one of the largest reservoirs of untapped "carbon credits."

Carbon Taxes

The Supreme Court's 2007 ruling that CO_2 was a "pollutant" gave new impetus to advocates of a carbon tax, which economists consider less susceptible to political manipulation and evasion than cap-and-trade systems. Carbon taxes make sense economically and environmentally because they tax carbon directly, using the marketplace to drive change.[15] A carbon tax applied to industries, vehicles, and fuels would quickly make renewables competitive, even with heavily subsidized coal, oil, and nuclear. It would also provide powerful motivation for efficiency improvements in manufacturing and agriculture. A carbon tax would not contribute to the creation of speculative markets in trading "pollution credits"—a practice that cap-and-trade systems are already promoting.

Renewable Offsets

Under the Kyoto Protocol's Clean Development Mechanism (which may be renewed in 2012), industrialized countries can gain "renewable" credits to offset their greenhouse gas emissions if they finance and support renewable energy projects (including reforestation) in developing countries.

Feed-in Tariffs

Feed-in tariffs (FITs) reimburse homeowners for the excess electricity their home-based energy systems contribute to the grid.[16] In this way they favor individuals and small entrepreneurs over large-scale, entrenched corporations. FITs deliver more renewable energy rewards than tax credits or write-offs—and at lower cost. And policy decisions can make feed-in tariffs extremely attracitve. Under the German system, for example, "renewable energy producers are given long-term, fixed-rate contracts, designed to deliver a profit of 7 to 9 percent. This makes green energy a secure bet for both investors and banks."[17]

Germany passed its first feed-in law in 1991 and, thanks to the participation of homeowners, local communities, and small businesses, doubled the country's renewable power production in nine years. Germany had hoped to produce 12 percent of its power from renewables by 2010, but it passed that mark in 2007. Germany now expects to produce half of its electricity cleanly and renewably by midcentury. (And bear in mind that Germany lacks coastal winds and is cloudy much of the time.) Forty countries—including at least 18 of the European Union's 27 members—are following Germany's example. Germany's program has helped launch the country into the green tech market. By 2020, green tech in Germany is expected to account for more than 700,000 jobs, becoming the country's major industry, as new businesses spring up to produce solar panels and wind turbines for buyers as far away as New York and Texas.

In Canada, Ontario's provincial power authority has introduced a FIT program that pays Native Canadian villages and homeowners up to 80.2 cents (79 cents US) for every excess kWh that flows from their rooftop solar panels into the local power grid. Large multi-megawatt producers also get paid, but at a lower rate of 44 cents per kWh. Community-based wind and solar projects should prosper under this plan.[18] (In May 2012, pro-nuclear forces attempted to blame renewable technologies for rising electricity rates in Ontario, but an investigation by the Ontario Energy Board established that 45 percent of the increases were linked to nuclear energy, while renewables only accounted for 6 percent of the increase.)

Gainesville, Florida, the first US city to enact feed-in tariffs, requires utilities to buy power generated from local homeowners and businesses at a rate slightly higher than the cost of production. FIT policies have generated more than 7 MW of renewable power for Gainesville's 125,000 residents (outpacing California, China, France, and Japan for installed solar per capita). In late 2011, Gainsville homeowners were being paid 32 cents per kWh for electricity created by their rooftop panels.[19] California, Hawaii, Maine, and eight other states are moving to enact similar laws.

Feebates

A new policy tool to encourage smart market choices is the "feebate." Described as "a cross between a fee and a rebate," it provides rewards for people who choose to purchase more efficient homes and vehicles. People who invest in larger, less-fuel-efficient vehicles or bigger, less-energy-efficient buildings pay an additional "fee" that provides the funds to "rebate" purchasers of more sustainable options. In France, automobile feebates were so successful that the program has been extended to cover a wide range of consumer products. Feebates also work to shift the market toward more sustainable production and have proven more popular than fuel taxes and efficiency standards.[20]

Decoupling and Shared Savings

In 2008, two US states introduced a novel incentive program that allowed electric and gas utilities to keep any profits they realized from saving their customers money (usually through improved efficiencies). The experiment proved so popular that, by 2009, 25 states had either adopted or were planning to adopt "decoupling and shared savings" programs. According to energy expert Amory Lovins, "Decoupling utilities' profits from how much energy they sell" is one of the best tools policy planners could use to promote the wise use of energy. The policy of rewarding utilities for reducing consumption was actually adopted by the country's state utility commissioners—in 1988.[21]

A Smart Grid

The US electricity grid is a vast patchwork governed by mechanical switches that date from the analog era. This aging network is susceptible to unpredictable blackouts and is inherently inefficient since transmission lines lose power for every mile their electrons must travel. Routine outages cost the economy $150 billion a year. Proponents argue that a smart grid, resembling the Internet in complex interactivity, could juggle power needs with digital speed—maximizing reliability and minimizing losses, while increasing transmission efficiency by 50 percent. Improving distribution efficiency is critical since electricity often costs more to distribute than to generate.

But there are drawbacks. Building a national grid could cost as much as $50 billion, and—like the Internet—the advantages might not be shared equally by poorer members of society.[22] The "mandatory" installation of smart meters in California and Texas outraged customers who discovered that errant meters were producing false hikes in their energy bills. Others have complained that EMF radiation generated by the wireless signals used to broadcast smart meter readings have triggered health problems including insomnia, headaches, nausea, heart palpitations, tinnitus, memory loss, dizziness, and disruptions of the immune, nervous, and hormonal systems, as well as posing long-term cancer risks and potential damage to DNA.

Smart meters also raise privacy concerns, since detailed monitoring of appliance use can be analyzed to profile residents' behavior (a tempting tool for commercial data mining). Smart meter transmissions also can be read remotely by anyone with the proper equipment, and the meters are vulnerable to hackers. (One simulation found that an attack on just one smart meter could progressively "infect" 15,000 meters in a single day.) Even a smart grid is vulnerable to cyber attack. Russian and Chinese operatives have successfully installed software in the US grid[23] and, in 2002, 70 percent of US energy providers reported instances of "serious cyberattack."[24]

Smart meters give commercial utilities the power to control a homeowner's electricity remotely. While this could be useful in rationing consumption during peak use to avoid brownouts, some consumers object to the unprecedented corporate intrusion into their private lives.

Instead of a balkanized grid—bankrolled by private industry and guarded by proprietary smart meters—a nationalized grid that is government-run as a public utility might better serve the country.

Microgrids

The Santa Rita Jail in Dublin, east of San Francisco, now generates all the electricity it needs to care for its 3,000 inmates and support staff from its own $11 million "microgrid." Now that the jail is off the grid—fully powered by 1.2 MW of solar power panels, five 2.5 kw wind turbines, and a fuel cell—the facility's power bills have plummeted. Whenever it becomes necessary to recharge the on-site batteries, the jail can tap electrons from the commercial grid late at night, when electricity prices are lowest.[25]

Zero-Energy Buildings

A new generation of self-sustaining buildings that use only the energy they generate is springing up around the world, from America to Zimbabwe. The National Renewable Energy Lab's new 222,000-square-foot office uses passive heating and cooling along with wind and photovoltaic power to service the office needs of 800 employees. The Center for Environmental Studies at Oberlin College relies on rooftop and passive solar, natural ventilation, and geothermal heat pumps to achieve its zero-energy status. Premier Gardens, a "solar subdivision" in Rancho Cordova, California, is a "first-of-its-kind" cluster of 100 zero-energy homes that combine solar panels with improved energy efficiency. The Eastgate office complex in Harare, Zimbabwe, uses 90 percent less energy than similar-size buildings at a savings of more than $35 million.[26] The savings were accomplished through the application of "biomimicry." The architects incorporated techniques that African termites use to naturally heat and air-condition their mounds.

Energy-Efficient Eco-Cities

The old model of cities powered by the mining of coal, oil, or uranium is giving way to a new urban option—large cities where homes and offices are heated, cooled, and illuminated by free energy that flows from the sky or blows across the land. In Aspen, Colorado, 75 percent of the city's municipal electricity comes from renewable sources, mainly wind, and the city expects to be carbon neutral by 2015. Austin, Texas (a city with 206 parks, 12 preserves, 26 greenbelts, and 50 miles of hiking and biking trails), expects to be carbon neutral by 2020. San Jose, California, plans to be generating 100 percent of its electricity from renewable sources by 2022.[27]

Making the transition to renewables makes all the differences. Case in point: Driving an all-electric vehicle in California will reduce CO_2 emissions, but driving that same car in Florida, where electricity comes from coal-burning power plants, will actually increase greenhouse gas emissions. The eco-urban transformation has to go beyond introducing better energy choices: the city itself must be transformed. While plug-in hybrids are superior to oil-burning autos, the largest energy savings (and greatest greenhouse mitigations) will come from redesigning our cities, replacing urban sprawl with compact communities designed for "access by proximity" that replaces single-driver automobiles with efficient, affordable mass transit. One goal of the New Urbanism movement is to create communities where you can reach everything you need by walking no more than five minutes.

One way to reduce time- and energy-intensive trips to shop and work is to build more compact and diverse urban centers by narrowing the distances with "infill building." Instead of building new housing and offices on land far from the urban core, 81 percent of developers who responded to a 2007 poll now favor redeveloping existing neighborhoods with attractive amenities that include integrated shopping, recreation, work, schools, and parks.[28] The California Infill Builders Association (CIBA) anticipates that intermingling housing, small commerce, and workplaces could cut California's driving miles by one-third and save $4.3 billion in infrastructure costs. Mixed-use neighborhoods could cut commuting by 3.7 trillion miles—a climate-friendly impact that would be the equivalent of removing every car from the state's roads for 12 years. CIBA estimates that urban consolidation would save the average household $6,500 in auto and utility expenses.[29]

The 2,000-Watt Society

The Swiss Council of the Federal Institute of Technology has proposed that industrialized countries should begin a transition to a world in which every citizen can live comfortably on a continuous requirement of 2,000 watts (17,520 kWh per year). The average Swiss citizen consumes around 5,000 watts; the average American consumes 12,000 watts. The technologies—energy efficiencies, zero-energy buildings, heat pumps, and renewables—already exist. And living the 2,000-watt lifestyle is possible, even in the United States and even at an altitude of 8,000 feet. Since 1983, energy expert Amory Lovins has run his Colorado mountain home and office on about 1,200 watts—10 percent of the electricity used by the average US household—while using only 1 percent of the "normal" energy consumed for space and water heating.[30] In 2009, Lovins installed additional solar panels, boosting his solar electric capacity to 9.7 kilowatts. His home-office now produces more power than it consumes.[31]

Community Aggregation

Many cities have formed municipal entities to purchase renewable power at discounted rates. Other cities already operate their own municipal power plants. At the local level, the nonprofit One Block at a Time encourages aggregation of entire city blocks to install small-scale power systems, a process that brings neighbors together for discounted group purchases, thus bringing down the individual costs of solarization.[32] Locally controlled and locally owned power systems work to keep investments and revenue in the local economy. These decentralized systems are also more easily maintained, more resilient to weather or quake damage, and structurally immune to massive blackouts. Traditional commercial utilities naturally see aggregation as a threat. In California, the Pacific Gas & Electric Company (PG&E) placed an initiative on the June 2010 ballot to block cities from establishing their own municipal power systems. Despite spending $45 million on the campaign, PG&E's power play was voted down by the people.

Transition Towns

Transition Towns, which got their start in the United Kingdom, are now sprouting up in Europe, Asia, and the Americas. Discouraged with governments' failure to act on the climate change threat or to begin meaningful preparations for an "energy descent action plan,"[33] citizens in hundreds of these Transition Towns have committed to ending their dependency on fossil fuels in 10 to 20 years. Calling the movement "a social experiment on a massive scale," the Transition Network provides a 12-step guide to a low-carbon economy. "We truly don't know if this will work," the organizers admit. "What we do know is this: if we wait for governments, it'll be too little, too late; if we act as individuals, it'll be too little; but if we act as communities, it might just be enough, just in time."[34]

Meanwhile, other "localizing" campaigns are sprouting up around the world. The Post Carbon Institute has been providing blueprints for "a more resilient, equitable, and sustainable world" since 2003.[35] In San Francisco, Bay Localize has created a Community Resilience Toolkit that has been adopted by hundreds of groups around the world.[36]

PRIVATIZING SUNBEAMS: THE RISE OF BIG SOL

With solar power on the rise, the US Department of Energy (DOE) announced a $1.2 billion loan guarantee to help the company SunPower build a 250 MW centralized solar plant in Southern California; another $2.1 billion loan guarantee went to Solar Trust of America (STA) for the first of two 242 MW solar-thermal electricity plants. Another large cluster of plants, the $2.2 billion Ivanpah Solar Electric Generating System, is being built in the Mojave Desert under the auspices of US power giant NRG Energy and Google. As Solar Trust's CEO noted, federal money is helping to kick-start "a new era of utility-scale solar development."[37] (But even with federal backing, new businesses still run the risk of failure. Following the collapse of its German-based partner, Solar Trust declared bankruptcy in April 2012. In June, Oakland, California-based BrightSource won a bid to take over one of STA's partially built plants.)

On the surface, this sounds like good news. Go deeper, though, and you can detect a familiar corporate agenda: these taxpayer-backed plants are being built to serve large commercial interests. Like the sums on the federal checks from Washington, the message behind them is "writ large." In addition to the old bogeymen of Big Oil and Big Coal, we are now witnessing the birth of Big Sol.[38]

It's not enough to simply invent new ways of producing electricity if you don't also fashion better ways of delivering it. The traditional model of centralized ownership that dominates the US economy—with its Darwinian celebration of "red tooth and claw" capitalism—is now vying for control of emerging energy alternatives. Across the country, local communities are engaged in struggles to demand the right to choose—and sometimes to operate—their own renewable energy systems. The pitched battle over a future of centralized vs. distributed power has become nothing less than a struggle for the democratization of energy.

BP was one of the first petroleum conglomerates to position itself for a takeover of the emerging renewables market. BP's branding teams moved to "green up" the company's image by announcing that the corporate initials no longer stood for "British Petroleum"; instead, they now stood for "Beyond Petroleum." (As it turns out, "BP" instead has become shorthand for "Botched Procedures," a fitting epitaph for the company that caused a massive oil spill that devastated the Gulf of Mexico.) Shell, ExxonMobil, and other oil-soaked multinationals are also maneuvering for position on the inside track of the renewables racetrack.

Who "Wins" the Race to Renewables?

Energy secretary Steven Chu has defended the DOE's Big Sol loans with the pro-American argument that the United States "can either sit on the sidelines and watch the competition pass us by, or we can get in the race and play to win." But Solar Trust is a joint venture of two German firms—Solar Millennium AG and Ferrostaal Inc.—while the French oil giant Total controls SunPower. Foreign ownership is one problem; corporate control is another. Solar Trust has partnered with Chevron, while SunPower will supply electricity to PG&E (California's largest private utility). The Ivanpah facility is locked in to supply electricity to PG&E and Southern California Edison (California's second-largest private utility) for at least 20 years.[39]

As David Myers, director of California's Wildlands Conservancy, observed, "There is a corporate lobbying push to focus federal stimulus funds on projects that keep renewable energy behind the meter. In other words, to stop the democratization of energy. Think about it. If, in ten years, we are driving electric cars, owning our own rooftop solar systems with battery backups would be like owning our own gas stations. Business models would be threatened."[40] The technological means to short-circuit the corporate control of sunshine is already at hand. Instead of imposing vast, mirrored complexes on untrammeled wilderness, solar farms can be planted atop urban brownfields, parking structures, and shopping malls.

A FIT Solution

Ultimately, the antidote to corporate centralization is decentralized microgeneration—a solar array for every home. In California (which leads the nation in the installation of solar panels), residential and commercial rooftops are projected to generate more electricity than the state consumes on the hottest summer day. In November 2011, the state marked a milestone with 100,000-plus rooftop panels producing more than 1 gigawatt of photovoltaic power. The state is on track to generate 3 GW by 2016. If California were a country, it would be the world's sixth largest producer of solar electricity.[41] But

Big Sol has some tricks for reducing the economic competitiveness of sunbeams harvested right off your rooftop.

Under "net metering" programs, utilities offer "credits" for excess power grabbed by their central grid whenever a homeowner's electric meters "spin backward." The utility profits by reselling this free excess power at a premium. As energy analyst Robert Freehling notes, net metering can put photovoltaics "in conflict with efficiency and conservation efforts" because it favors "energy hogs that use three or more times the 'baseline' amount of electricity." As Freehling goes on to explain: "Frugal customers that only pay 12 cents/kWh for all their electricity, will get little benefit from installing solar projects on their homes, while energy hogs that use three or more times the average amount of electricity will get the most benefit because the solar will be valued at nearly 30 to 40 cents/kWh."[42]

A better approach is the feed-in tariff (FIT), a system that pays businesses and homeowners directly for excess solar production at an agreed-upon price. Thanks to FITs, homeowners in Ontario, Canada, receive a major payback for their excess rooftop power—more than commercial generators receive. In Germany, where FITs rule, citizens pay 40 percent less to go solar than net-metered Americans. FIT expert Paul Gipe argues that adopting Germany's feed-in tariffs would save US ratepayers money "and result in windfall profits."[43] Not surprisingly, Big Sol prefers net metering.

Policies that concentrate photons in the hands of an energy elite prioritize stockholders' gain at the cost of biosphere's pain. While off-the-grid independence is the ideal, short of that, we can still cast a shadow over Big Sol's power grab by demanding a FIT price for farming photons from atop our own solar homesteads.

Meanwhile, the off-the-grid movement is making surprising strides, not only in rural America and the villages of Europe but also in some of the world's most impoverished regions. For the first time in their lives, struggling rice farmers in Bangalore, India, are enjoying clean, solar-powered indoor lighting thanks to an organization called Simpa Networks. Simpa places photovoltaic panels on the rooftops of village shacks and allows people to light their homes for around $1 a month. Not only is the cost significantly less than the black market price of kerosene, but the light is brighter and does not cloud homes with unhealthy fumes and soot.

The Democratization of Energy

As the Bloomberg news service reported on April 12, 2012, the world now stands on "the crest of an electricity revolution that's sweeping through power markets and threatening traditional utilities' dominance of the world's supply. From the poorest parts of Africa and Asia to the most-developed regions of the United States and Europe . . . small-scale wind and biomass generators promise to extend access to power to more people than ever before."[44]

In forgotten regions of the globe, where no corporation would ever bother to build a coal, gas, or nuclear power plant, small start-ups have assembled renewable power resources into microgrids to power local homes and small businesses. Jeremy Rifkin, a prolific author and visionary, has celebrated this historic transition in the title of his latest book, *The Third Industrial Revolution*. Rifkin sees the evolution of a new form of "lateral power" that weds renewable technology with smart grids empowered by the connectivity of the Internet. Rifkin believes that this development promises a social and economic revolution that will rival the invention of the steam engine and the creation of the printing press. If the production and distribution of energy is freed from corporate control and given to the commons, the 1960s cry of "Power to the People" could finally be realized as industrial society shape-shifts toward sustainability.[45]

John Farrell is another prophet of the energy democratization movement. His 2011 report *Democratizing the Electricity System* identifies some bright roadmarks on the path to a new energy future. For example, 16 US states now boast energy policy portfolios that endorse "distributed" solar PV expansion. California alone expects to produce 12,000 MW from distributed power by 2020.[46]

Applying his celebrated blend of visionary and actuarial brilliance, author and energy expert Amory Lovins has spent the better part of four decades providing "soft path" energy blueprints to promote efficiency and renewable energy technologies as the best replacements for nuclear, oil, and coal. If the United States were merely to adopt the efficiency and alternative energy measures already being employed in many US states, Lovins believes this move would save $5 trillion, grow the economy by 158 percent, and cut carbon emissions 82–86 percent below 2000 levels.[47]

"Within a decade," the *Bloomberg Market Magazine* predicts, "installing photovoltaic panels may be cheaper for many families than buying power from national grids in much of the world, including the United States, Japan, Brazil, and the United Kingdom. . . . The ultimate losers in this shifting balance of power may be established utilities. They've invested billions of dollars in centralized networks that are slowly being edged out of markets they've dominated."[48]

24

Powering Down: The Conservation Imperative

The most important environmental issue is one that is rarely mentioned, and that is the lack of a conservation ethic in our culture.

–GAYLORD NELSON, FORMER US SENATOR AND PRINCIPAL FOUNDER OF EARTH DAY

TO MINIMIZE cascading climate calamities and impending wars over shrinking resources, industrialized economies need to formally recognize the limits of the earth's carrying capacities, put the brakes on growth-oriented economics, and begin a wide-ranging process of "powering down"—learning to live better with less, to unplug from the grid, to simplify our lives, to become more localized and self-reliant. The watchwords for a Brave New Power-Down World are already entering our vocabulary: "less and local," "carbon footprint," "slow food," "locavore." All have come to mean a change in our values and habits—a rejection of economies based on consumption and growth in favor of solutions that emphasize localization and less use of energy and materials in all economic activity.

While the looming inevitability of downsizing poses a challenge for the industrialized world, the move to conservation economies may be more easily negotiated by the "underdeveloped" world where marginal survival—rather than overconsumption—is the norm. Unfortunately, the world's richest nations have offered the false promise of universal prosperity to the entire world. As a result, many in the world's poorest nations feel entitled to experience the questionable pleasures of private automobiles, 24-hour electricity, microwave convenience, and processed food.

The harsh reality is that, instead of seeing the world's poor majority joining the world's wealthy minority, the damage our overconsumption has inflicted on the planet's forests, oceans, and mineral resources means there is now less for everyone to share. Simple justice requires that it be the world's wealthier residents who adjust their lifestyles. Attaining a just and sustainable future will also benefit greatly by efforts to reduce the size of the human population. We are all in the same boat, and that boat doesn't expand with the growing number of passengers—it just gets more crowded and comes closer to sinking.

Smaller Is Better

In a powered-down world, new homes would be smaller and, hence, cheaper to heat and cool. All new construction would follow green building principles to conserve resources and reduce energy consumption. Insulation requirements would be tripled to match Germany's R-70 standard. (The R-value is a measure of thermal resistance used by the building and construction industries. In February 2010, the US Department of Energy ordered insulation standards increased from R-38 to R-60 in the nation's coldest regions.)[1]

Food would be grown locally (Beijing's vegetables are raised within 60 miles of the city). Water-intensive lawns would be replaced by gardens, and vacant city lots would be transformed into productive urban farms. Solar-box cookers and solar window-box heaters would cut cooking and heating costs. Instead of replacing aging sewer systems, composting dry toilets could be installed in homes to save billions of gallons of precious water now used to flush wastes. Rainwater harvesting and reuse of "gray water" from household sinks and bathtubs would save even more.

Achieving a powered-down, reduced-scale, steady-state economy will require a new economic system—one that is not based on speculation and ever-expanding debt. As the Center for the Advancement of the Steady State Economy argues, "perpetual economic growth is neither possible nor desirable. Growth, especially in wealthy nations, is already causing more problems than it solves."[2] Because the banking industry will not voluntarily reform the system, transformation must begin from the bottom up, with the empowering of local economies. Citizens can localize credit by moving funds to locally owned and controlled cooperatives, credit unions, and small banks. Local currencies backed by tangible assets—labor, craft skills, home care, artistic talents—can help sustain essential jobs that support food, housing, and health care services.[3]

The Uppsala Protocol outlines a global power-down plan featuring equitable sharing of the planet's remaining oil-based energy resources based on a world depletion rate, which requires that current and future consumption of oil be limited to the amount of oil extracted in a given year.[4] Colin Campbell, a geologist with the Uppsala Hydrocarbon Depletion Study Group at Sweden's Uppsala University, proposed the protocol in hopes of establishing an international accord to prevent profiteering from the global oil shortage, discourage wasteful consumption, and stimulate the development of alternative energy options.

The world has already seen some examples of successful power-down economies. Kerala, an unindustrialized state in southwest India, never "powered up." Kerala has a zero-growth economy and one of the country's lowest per-capita incomes. Nonetheless, Kerala's citizens enjoy higher levels of health, education, social mobility, economic stability, and gender equality than any other Indian state. Similarly, after the collapse of the Soviet Union, Cuba was forced to power down. Without imported petroleum-based fuel, pesticides, and fertilizers, Cuban agriculture went organic, and community gardens now cover one-third of Havana's urban landscape. Cuban transport largely relies on bicycles and mass transit. The country now enjoys one of the world's highest levels of literacy, health, and overall quality of life.[5]

On personal, community, national, and global levels, our choices in energy policy will determine whether we achieve a stable, durable, peaceful society. Maximum energy and goods consumption cannot guarantee a secure and joyful future.

Ingredients for a New Sustainable Economy

In his 2009 report, *Searching for a Miracle*, author and futurist Richard Heinberg outlined some of the steps required to craft an emerging economy of sufficiency, equity, sustainability, and peace.[6] Here is a brief overview of some essential goals:

Adopt the Oil Depletion Protocol

Adopting the Oil Depletion Protocol[7] (a.k.a. the Uppsala Protocol) will enable a rapid transition from carbon-based energy to renewable energy. We must reject dangerous and unsustainable "alternatives" such as "clean coal," "new-generation nuclear," "industrial-scale biofuels," and "waste-to-fuel incineration." We need to place global limits on the expenditure of declining fossil fuel reserves and place sufficiency standards on manufactured goods to safeguard sustainability goals while assuring the equitable reallocation of dwindling resources.

Adopt Small-Scale Energy

Swift adoption of small-scale, locally owned, ecologically sustainable renewable energy systems will accelerate the process of "powering down" by dramatically increasing conservation and efficiency. Additionally, adoption of the "precautionary principle" (which requires initial evidence that a technology, product, or practice is safe—not that it is "not known to be unsafe") will protect the public against the introduction of goods, services, or practices that could harm humans, societies, or the environment.

Reallocate Resources

Much of the West's inordinate wealth stems from an ofttimes brutal colonial history. Nations have grown rich and powerful by occupying other nations, extracting their resources for profit, and leaving them impoverished and in debt. In a world of diminishing resources, this historic misappropriation of wealth needs to be addressed. Avoiding a grim future of local and global "resource wars" will require moving from a world of "haves and have-nots" to a world of "share and want not." A reallocation of global resources will be needed to achieve an equitable balance between nations.

Reject Economic Globalization

The agenda of economic globalization serves to sustain the growth of global corporations at the expense of the planet's environmental carrying capacity and to the detriment of local communities and businesses. To survive, we must reject the main driving forces of economic globalization, which include hyper-growth; export-oriented food, energy, and commodity production; deregulation that enables corporate abuses; profit-based privatizing of the shared public commons; privatization of public services; and the promotion of global economic goals over local needs.

Minimize Consumption

A new generation must be taught to reject the market's invitation to indulge in excessive consumption. Spending less time working for cash to acquire nonessentials will buy more time to spend with family and friends and to engage in social, recreational, cultural, and spiritual pursuits.

Rely on Local Production

The inevitable passing of the Age of Cheap Oil will require the creation of "survival economies" that rely on less long-distance trade and less movement of capital across time zones. Society will need to become more self-sufficient by "relocalizing"—relearning the skils required to create essential goods and services from local resources to serve local needs (a process that could be called "import substitution"). Government, industry, agriculture, and commerce all will benefit from new, sustainable, and democratic investment rules that favor local ownership and mandate community participation on all corporate boards.

We must redesign existing urban and suburban living environments into compact, self-sufficient, pedestrian-friendly eco-cities[8] built to accommodate—and transcend—the limitations of a post-carbon world. We must seek regional and home-based production of economic necessities, such as food, clothing, transportation, and shelter. The production of essential goods and services must be determined by democratic consensus, not by the imperatives of global capitalism. Wall Street must yield to Main Street with prohibitions on economic speculation and debt-based finance. Growth-based national economies must be replaced by no-growth, steady-state local economies that function within natural limits.[9]

Deindustrialize and Reinvigorate Agriculture

Land removed from traditional agriculture must be converted back to croplands and cultivated under local community ownership. Post-carbon agriculture will rely on renewable and natural inputs rather than fossil fuel by-products and manufactured chemicals. The depletion of cheap oil will require the deindustrialization of agriculture and the adoption of organic, climate-wise, and locally sustainable agroecology and micro-farming. The strategies of permaculture[10] will reconnect agriculture with natural ecologies to create productive, stable, and resilient systems. While petroleum-dependent industrial agriculture pours greenhouse gases into the atmosphere, organic farming practices actually capture excess carbon from the air and return it safely—and productively—to the soil.

Make Polluters Pay

We must adopt of the principle of "polluter pays." At all levels, the full ecological and social costs of production must be identified and born by the producer.

Use New Parameters to Measure Well-Being

The old models used to measure economic well-being—gross domestic product (GDP) and gross national product (GNP)—must be abandoned. New measurements must emphasize the satisfaction of basic human needs (rather than economic wealth) and securing "economic sufficiency" for all people—including shelter, food, education, health, and access to healthy environments. Rather than worshipping exponential growth, corporate profit, and the accumulation of personal wealth, societies will need to focus on economic and environmental sustainability, the preservation of "natural capital," and the collective well-being of all members of the community. As a means to achieve this goal, green economists have proposed a genuine progress indicator (GPI) to evaluate the degree to which a country's economic performance improves the overall welfare of its people and the environment.[11] Factors tracked by the GPI include: the value of housework and volunteer work; ecosystem services freely provided by wetlands, farmland, and forests; and the negative costs of crime, commuting, underemployment, pollution, and the loss of finite natural resources.

The annual per capita energy required for human well-being is believed to be 50 to 70 gigajoules (Gj). Consumption of energy beyond 100 Gj produces no appreciable improvement in one's personal sense of well-being, nor does it show any objective gain in measures of well-being. In North America, current annual per capita energy consumption stands at around 325 Gj.[12]

Heinberg lists five fundamentals that can serve to gauge a successful survival society: "ecological sustainability; degree of 'net energy gain' or loss; degree of social equity, well-being and 'sufficiency' (rather than surplus consumption and wealth); democratic decision-making processes; and nonviolent conflict resolution."

"A deliberate embrace of limits does not amount to the end of the world," Heinberg writes, "but merely a return to a more normal pattern of human existence. We must begin to appreciate that the 20th century's highly indulgent, over-consumptive economic patterns were a one-time-only proposition and cannot be maintained."[13]

Conclusion

THE STORY of nuclear power is not only a tale of science and engineering, it is also a study of hubris and denial. Nuclear advocates believed they had built a power source they could control, one that would prove immune to the ravages of age, human error, or natural calamity. When the technology began to fail and assumptions about the power of nature exceeded "design parameters," the evidence was diminished, dismissed, or ignored.

Fukushima forced the world's nuclear nations to confront some unpleasant issues. What happens in the aftermath of a nuclear disaster? Can government provide an adequate response? Will industry be held accountable for the damage to lives and property? The Fukushima disaster provided a textbook example of what can happen when a nuclear accident occurs in a heavily populated country: faced with a disaster that neither TEPCO nor Tokyo was able to fully manage, the government came close to collapse.

A postmortem conducted by the nonprofit Rebuild Japan Initiative Foundation (RJIF) concluded that both TEPCO and the government "were thoroughly unprepared on almost every level for the cascading nuclear disaster. This lack of preparation was caused, in part, by a public myth of 'absolute safety' that nuclear power proponents had nurtured over decades."[1] The RJIF's Yoichi Funabashi and Kay Kitazawa explained that, since the 1970s, "disaster risk has been deliberately downplayed by what has been called Japan's nuclear *mura*," a cabal of nuclear advocates in industry, government, and academia that feared "if the risks related to nuclear energy were publicly acknowledged, citizens would demand that plants be shut down."[2]

Tokyo has been forced to acknowledge the danger of entrusting nuclear safety to the same ministry that was created to promote nuclear power. With this in mind, Japan is considering a new safety agency that would operate independent of the Nuclear Safety Commission and Nuclear and Industrial Safety Agency, possibly serving as a branch of the Environment Ministry. If Washington took this lesson to heart, the NRC would no longer be entrusted with oversight of nuclear safety. Instead, this critical responsibility might be better served by the Environmental Protection Agency.

Japan now faces a daunting challenge. In order to deal with winter's chill and the heat of summer (when peak electricity demand can surpass supplies by 15 percent), the government has introduced radical energy conservation measures. Office heating and air-conditioning use has been cut back, along with the use of energy-hogging copiers and printers. With elevators turned off, office workers have been enjoying the cardiovascular benefits of climbing stairs—sometimes assisted by flashlights, since nonessential lighting has been dimmed or disconnected. Companies have chipped in by moving some work schedules to weekends and early mornings.

On August 6, 2011, the sixty-sixth anniversary of President Harry Truman's nuclear attack on the city of Hiroshima, then prime minister Kan declared, "We will deeply reflect over the myth that nuclear energy is safe. . . . [T]o secure safety, we'll implement fundamental measures while also decreasing the degree of dependence on nuclear power generation, to aim for a society that does not rely on nuclear power."[3]

When President Jimmy Carter called energy conservation "the moral equivalent of war" and challenged citizens to "put up with inconveniences and to make sacrifices," the energy-powers-that-be pounced, memorably reframing Carter's wise admonitions as a prescription for "shivering in the dark." In response, "conservation without sacrifice" became a widespread corporate mantra. Advertisements assure us that, by resorting to "greensumption," we can shop our way to an environmental nirvana where we can continue to enjoy "warm showers and cold beers"—our "sacrifices" limited to the choice between driving a Prius or an Escalade.

A more realistic view of our energy future—and the lifestyles it will support—requires a more complex analysis. Traditional energy planning has been predicated on the assumption that populations will continue to grow and that humans have a right to make continually greater demands on the world's resources. But since we live on a finite planet, designing solutions to sustain *growth* are not *ecologically* sustainable and hence fail as solutions. They are merely mechanisms for sustaining the problem.

Rather than allowing food shortages, wars, and social breakdown to control population, today's leaders should be looking at ways to restrain population growth. And they should help us reframe human happiness as a goal that is not dependent on the ever-growing consumption of goods.

What we need is a society of resilience, a society of sharing and mutual support. Disasters will come. Even a smart grid will remain vulnerable to massive outages caused by everything from winter storms to solar flares to computer hackers disrupting the grid via the Internet. Such vulnerability can be minimized through the decentralization that comes with widely distributed rooftop and backyard generation. Once millions of off-the-grid homes and businesses are equipped with stand-alone power-generating systems, massive blackouts will no longer be a threat. Community-based generation will provide depth and redundancy that can improve power reliability and reinforce national, as well as local, security.

Switching energy sources and "going green" is only part of the solution. After all, a pistol redesigned to fire biodegradable bullets is still a weapon. If the Age of Industrial Solar produces a legacy of "green tailings"—choking landfills with cast-off batteries and PV panels leaching heavy metals, acidic fluids, and caustic chemicals into our groundwater—we will have failed to address the fundamental issues of sustainable survival. Our goal must be total closed-loop recycling, where every industry's waste becomes another industry's raw material, imitating nature's wise and economical use of resources.

Many of the sicknesses of contemporary America and the rest of the "developed world" stem from excess: too many calories, too many miles driven, too many ostentatious living spaces, too much throwaway stuff. Instead of globalization's promise of "More from Everywhere," a better slogan for the future will be "Less and Local." Our long-term survival in the twenty-first century depends not on consumption but on social solidarity, cooperation, sharing, resourcefulness, knowledge, and health. The future belongs to those whose basic needs and well-being can be sustained locally, in low-impact, self-supporting communities.

We need decentralized solutions and community-based alternatives. Instead of bailing out speculators, we should be investing to create small-scale, decentralized, sustainable, and democratic communities supported by millions of small, local, organic farms. We will need to respond to massive economic, social, and climate changes and we had better be flexible and agile. In the New Post-Oil World, old ideas about energy will disappear. We might need to understand the economic advantages of owning a mule rather than a tractor. Someone who knows how to make and repair a boot could become more valued than someone who knows how to boot up a computer. Many capital-intensive, extractive centralized industries will collapse, leaving it to citizens and communities to care for an endangered commons beset by polluted air, damaged land, tainted waterways, chemically poisoned croplands, and a broken oceanic food chain.

Turning away from the false solution of nuclear power is not only good in itself but will help us focus on the real necessities for a better future. We will need to redesign our technologies to be smarter and more efficient and our cities to be more compact, convenient, and congenial. We will need to learn to live within our means—doing without extravagance, just as our thrifty forebears did, and living better with less, more in tune with each other and our shared planet. If we can escape the easy temptations of the consumerist mentality, we will be able to address the inequality that threatens the fundamental stability of our nation and our world. The old combustion economy promoted massive inequities in the distribution of wealth and power. What the world needs now is a sustainable compassion economy that will usher in a new era of stewardship and sharing.

Acknowledgments

WE WOULD LIKE to express our great appreciation to several people who were very important to the creation and completion of this project. Claire Greensfelder initiated the project while she was communications director of the International Forum on Globalization (IFG) and provided the initial research and outlines. Victor Menotti, Kate Damasco, and Eileen Hazel supported the project through several iterations. Laura Friedenbach saved the day more than once with her technical and communications skills. Daniela Sklan handled the design and production of the original IFG publication with speed, elegance, and alacrity. Thanks especially to Jack Santa Barbara and the Santa Barbara Family Foundation, as well as Wade Greene and one anonymous donor for a generous gift of financial support. The author thanks his partner, Cynthia Mahabir, for generous collegial support, even in the face of competing deadlines. And, finally, thanks to the friendly and supportive crew at Chelsea Green: senior editor Brianne Goodspeed, production coordinator Hillary Gregory, and peerless copy editor Nancy Ringer.

Notes

INTRODUCTION

1. Timothy Mousseau, "Abundance of Birds in Fukushima as Judged from Chernobyl," *Environmental Pollution*, May 2012, Vol. 164, pp. 36-39, http://sc.academia.edu/TimothyMousseau/Papers/1421710/Abundance_of_birds_in_Fukushima_as_judged_from_Chernobyl.
2. Greenpeace International, "Lessons from Fukushima," February 2012, p. 9, http://www.greenpeace.org/international/Global/international/publications/nuclear/2012/Fukushima/Lessons-from-Fukushima.pdf.

INTRODUCTION

1. US Energy Information Administration, *Annual Energy Review 2010* (Washington, D.C.: US Government Printing Office, October 2011).
2. "Energy Department Must Disclose More Information About Loan Guarantee Program, Groups Say," Public Citizen, March 17, 2010, http://www.citizen.org/pressroom/pressroomredirect.cfm?ID=3066.
3. Darius Dixon, "Jaczko Dissents as NRC Approves Vogtle Permits," Politico, February 9, 2012, http://www.politico.com/news/stories/0212/72680.html.
4. Thomas Fanning, "Southern Company Subsidiary Receives Historic License Approval for New Vogtle Units, Full construction Set to Begin," Southern Company press release, February 9, 2012, www.southerncompany.com/nuclearenergy/presskit/docs/COL_press_release.pdf.
5. "NRC Concludes Hearings on Vogtle New Reactors," US Nuclear Regulatory Commission news release, no. 12-013, February 9, 2012, http://pbadupws.nrc.gov/docs/ML1204/ML120410133.pdf.
6. Hon. Edward Markey, "Markey to NRC: Vogtle Reactor Vote is an Abdication of Duty," press release, February 9, 2012, http://markey.house.gov/press-release/markey-nrc-vogtle-reactor-vote-abdication-duty.
7. Allison Fisher, "With Japanese Nuclear Crisis Still Unfolding, New Nuclear Reactor Approval Is Move in Wrong Direction," statement from the press office of Public Citizen, February 9, 2012, http://www.citizen.org/pressroom/pressroomredirect.cfm?ID=3523.
8. "NRC Commissioners Approve 2 New AP1000s at Vogtle by 4 to 1 Vote," Beyond Nuclear, February 10, 2012, http://www.beyondnuclear.org/new-reactors/2012/2/10/nrc-commissioners-approve-2-new-ap1000s-at-vogtle-by-4-to-1.html.
9. Karl Grossman, "Nuclear Power in the US: A Rigged System," Counterpunch, January 23, 2012, http://www.counterpunch.org/2012/01/23/nuclear-power-in-the-us-a-rigged-system/.
10. Kristi E. Swartz, "New Reactors? We'll Hear Today," *Atlanta Journal-Constitution*, February 9, 2012.
11. "As Fukushima Worsens, US Approves New Nukes," Common Dreams, March 30, 2012. http://www.commondreams.org/headline/2012/03/30-7.
12. Jeff Johnson, "Going Nuclear Again," *Chemical & Engineering News*, February 10, 2012, http://cen.acs.org/articles/90/web/2012/02/Nuclear-Again.html.

13. "DOE Denies FOIA Request Over Plant Vogtle Loan," SACE press release, October 13, 2011. http://www.cleanenergy.org/index.php?/Press-Update.html?form_id=8&item_id=247]
14. Amy Goodman, "The Bipartisan Nuclear Bailout," Democracy Now!, March 8, 2012, http://www.democracynow.org/blog/2012/3/8/the_bipartisan_nuclear_bailout_by_amy_goodman.
15. Ayesha Rascoe, "US Approves First New Nuclear Plant in a Generation," Reuters, February 9, 2012, http://www.reuters.com/article/2012/02/09/us-usa-nuclear-nrc-idUSTRE81820920120209.
16. Eric Wesoff, "Global Nuclear Generation Capacity Falls to 366.5 GW," Green Tech Media, December 7, 2011, http://www.greentechmedia.com/articles/read/global-nuclear-generation-capacity-falls-to-366.5-gw/.
17. Patricia Reaney, "Global Support for Nuclear Energy Drops after Fukushima," Reuters, June 22, 2011, http://www.reuters.com/article/2011/06/22/us-energy-nuclear-poll-idUSTRE75L4BY20110622.
18. Worldwatch Institute, "Global Nuclear Generation Capacity Falls," highlights from the *Vital Signs Online* Worldwatch report for 2011, http://www.worldwatch.org/global-nuclear-generation-capacity-falls-0.
19. "Obama Administration Calls for No Expansion to Nuclear Loan Guarantee Program in FY2013 Budget!" Beyond Nuclear, February 15, 2012, http://www.beyondnuclear.org/nuclear-loan-guarantees/2012/2/15/obama-administration-calls-for-no-expansion-to-nuclear-loan.html.
20. Eric Talmadge (Associated Press), "Japan Shuts Down Its Next-to-Last Nuclear Reactor," Yahoo! News, March 26, 2012, http://news.yahoo.com/japan-shuts-down-next-last-nuclear-reactor-165907044.html.
21. "France's 'Red Green' Alliance Advocates Near 50% Nuclear Shutdown by 2050," Beyond Nuclear, November 18, 2011, http://www.beyondnuclear.org/france-whats-new/2011/11/18/french-red-green-alliance-advocates-near-50-nuclear-shutdown.html.
22. Eric Wesoff, "Global Nuclear Generation Capacity."
23. "Fix 'em or Shut 'em French Nuclear Safety Agency Tells EDF," Beyond Nuclear, January 4, 2012, http://www.beyondnuclear.org/the-nuclear-retreat/2012/1/4/fix-em-or-shut-em-french-nuclear-safety-agency-tells-edf.html.
24. Peggy Hollinger, "France Aims to Rebalance Its Energy Mix," *The Financial Times*, July 10, 2011, http://www.ft.com/intl/cms/s/0/e490bc14-ab04-11e0-b4d8-00144feabdc0.html.
25. Fiona Harvey and Terry Macalister, "Wind Power Cheaper than Nuclear, Says EU Climate Chief," *Guardian*, March 17, 2011.
26. Zoë Casey, "French Nuclear Set to Become More Expensive than Wind Power," European Wind Energy Association February 3, 2012, http://blog.ewea.org/2012/02/french-nuclear-set-to-become-more-expensive-than-wind-power/.
27. "France Faces 79-bn-euro Charge for Nuclear Power," Agence France-Presse, January 31, 2012, http://www.nuclearpowerdaily.com/reports/France_faces_79-bn-euro_charge_for_nuclear_power_auditor_999.html.
28. "No Nukes for Siemens," Beyond Nuclear, September 19, 2011, http://www.beyondnuclear.org/germany/2011/9/19/no-nukes-for-siemens.html.
29. Paul Gipe, "Renewables Helped France Avoid Freezing in the Dark," RenewableEnergyWorld.com, February 10, 2012, http://www.renewableenergyworld.com/rea/news/article/2012/02/renewables-helped-france-avoid-freezing-in-the-dark.
30. Meteor Blades, "Germany's Solar Installations Generate 10% of the Nation's Electricity in May This is a Failure?," Daily Kos, June 12, 2012, http://www.dailykos.com/story

/2012/06/12/1099511/-Germany-s-solar-installations-generate-10-of-the-nation -s-electricity-in-May-This-is-a-Failure.

31. World Nuclear Association, "Nuclear Power in France," June 2012, http://www.world -nuclear.org/info/inf40.html.
32. Evan Jones, "Hollande May Not Like It, but French Nuclear Is Full Steam Ahead," The Conversation, May 18, 2012, http://theconversation.edu.au/hollande-may-not-like-it -but-french-nuclear-is-full-steam-ahead-7073.
33. Ibid.

CHAPTER 1

1. Steve Connor, "Warning: Oil Supplies Are Running Out Fast," *Independent*, August 3, 2009.
2. Greenpeace, *Nuclear Power: Undermining Action on Climate Change*, Greenpeace International Case Study and Alternatives briefing, reference no. GN089 (Amsterdam: Greenpeace International, December 2007), http://www.greenpeace.org/usa/en/media -center/reports/nuclear-power-undermining-ac.pdf.
3. Stephen Pacala and Robert Socolow, "Stabilization Wedges: Solving the Climate Problem for the Next 50 Years With Current Technologies," *Science*, August 13, 2004, 305 (5686), pp. 968-972, http://www.princeton.edu/mae/people/faculty/socolow/Science -2004-SW-1100103-PAPER-AND-SOM.pdf.
4. Energy Information Agency, "Nuclear Power: 12 Percent of America's Generating Capacity, 20 Percent of the Electricity," http://www.physics.ohio-state.edu/~wilkins /energy/Resources/Lectures/nuclearpower.html.
5. Tom Doggett, "NRC Expects Requests for 7 New Nuclear Reactors," Reuters, March 18, 2009, http://uk.reuters.com/article/2009/03/18/us-usa-nuclear-reactors -idUKTRE52H4BP20090318.
6. Greenpeace, *Nuclear Power.*
7. Arnie Gundersen, "Report for Southern Alliance on Clean Energy on TVA Bellefonte Plant," Fairewinds Energy Education video transcript, August 9, 2011, http://www .fairewinds.org/content/fairewinds-report-southern-alliance-clean-energy-tva -bellefonte-plant.
8. Christopher Paine, "Will Climate Change Revive Nuclear Power?" (presentation, American Enterprise Institute conference, Washington, D.C., October 6, 2006).
9. Arnulf Grubler, "An Assessment of the Costs of the French Nuclear PWR Program 1970–2000," International Institute for Applied Systems Analysis Interim Report IR-09-036, October 6, 2009, http://www.nirs.org/neconomics/frenchnukecosts1970_2000.pdf.
10. Kate Sheppard, "Obama's Nuclear Blind Spot," *Mother Jones*, March 8, 2010, http://www.motherjones.com/politics/2010/03/obama-nuclear-loan-guarantee.
11. "NRC Informs Westinghouse of Safety Issues with AP1000 Shield Building," NRC press release No. 09-173, October 15, 2009, http://pbadupws.nrc.gov/docs/ML0928 /ML092880421.pdf.
12. "Warnings about a New Nuclear Reactor," an episode for Public Radio International's *Living on Earth* program, April 15, 2011, http://www.loe.org/shows/segments.html ?programID=11-P13-00015&segmentID=2.
13. "US NRC Slams Westinghouse AP1000's Flawed Design," *Nuclear Monitor* no. 697 (November 6, 2009), http://www.nirs.org/mononline/nm697.pdf.
14. "Groups Urge Feds to Suspend Nuclear Licensing; Westinghouse Reactor Defect Was Missed by Regulators," Nuclear Information and Resource Service press release, April 21, 2010, http://www.nirs.org/press/04-21-2010/1.

15. Arnold Gundersen, *Post Accident AP1000 Containment Leakage: An Unreviewed Safety Issue* (Fairewinds Associates, April 21, 2010), http://fairewinds.org/content/post-accident-ap1000-containment-leakage.
16. Matthew L. Wald, "Is a New Reactor Rust-Prone?" *New York Times*, June 28, 2010.
17. US Nuclear Regulatory Commission, "100.11 Determination of Exclusion Area, Low Population Zone, and Population Center Distance" (part of the NRC regulations), accessed March 7, 2012.
18. Gloria Tatum, "AP1000 Design for New Vogtle Reactors a Problem for Georgia Power," *Atlanta Progressive News*, November 25, 2011, http://www.atlantaprogressivenews.com/interspire/news/2011/11/25/ap1000-design-for-new-vogtle-reactors-a-problem-for-georgia-power.html.
19. Ibid.
20. Ibid.
21. Harvey Wasserman, "Will You Pay as New Reactors Jump $900 Million in Three Months?" *The Free Press*, May 15, 2012, http://www.freepress.org/columns/display/7/2012/1932.
22. "Safety Procedures in Disarray at Finland's Olkiluoto 3 Nuclear Construction Site," Greenpeace International, August 13, 2008, http://www.greenpeace.org/international/en/news/Blogs/nuclear-reaction/breaking-news-safety-procedures-in-disarray-a/blog/10682/.
23. "EDF Delays Flamanville 3 EPR Project," Nuclear Engineering International, July 20, 2011, http://www.neimagazine.com/story.asp?storyCode=2060192.
24. Cathy Garger, "France, Stay Home! Your 74% Safety Record Is Not Okay Here!" Baltimore Independent Media Center, April 8, 2009, http://baltimore.indymedia.org/newswire/display/18720/index.php.
25. "Plans for New Reactors Worldwide," World Nuclear Association, February 2012, http://www.world-nuclear.org/info/inf17.html.
26. Eileen O'Grady, "US Utilities, Regulator Disagree on Generation," Reuters, May 6, 2009, http://uk.reuters.com/article/idUKN0551402820090505.
27. Brian Wingfield, "GE Sees Solar Cheaper than Fossil Power in Five Years," Bloomberg News, May 26, 2011.

CHAPTER 2

1. "Nuclear Can Be Safe Or It Can Be Cheap . . . But It Can't Be Both," WashingtonsBlog, December 23, 2011, http://www.washingtonsblog.com/2011/12/nuclear-can-be-safe-or-it-can-be-cheap-but-it-cant-be-both.html.
2. Katie Fehrenbacher, "Nuclear Costs to Soar Post Japan Disaster," Reuters, March 25, 2011, http://www.reuters.com/article/2011/03/25/idUS423443138820110325.
3. Komanoff Energy Associates (on behalf of Greenpeace), *Fiscal Fission: The Economic Failure of Nuclear Power* (New York: Komanoff Energy Associates, December 1992).
4. "Nuclear Power in the USA," World Nuclear Association (updated June 6, 2012), http://www.world-nuclear.org/info/inf41.html.
5. Charles Komanoff, "The Real Cost of Nuclear Power," *Multinational Monitor* 7, no. 9 (May 1986).
6. Christopher Paine, "Will Climate Change Revive Nuclear Power?" (presentation, American Enterprise Institute conference, Washington, D.C., October 6, 2006).
7. Jim Riccio, "US Should Bring an End to the Nuclear Era," US News, February 3, 2012, http://www.usnews.com/debate-club/should-nuclear-power-be-expanded/us-should-bring-an-end-to-the-nuclear-era.

8. Nirsnet [pseud.], "The Nuclear 'Renaissance' Stalls with Pending Collapse of Calvert Cliffs," Daily Kos (blog), August 5, 2010, http://www.dailykos.com/story/2010/08/05/889695/-The-nuclear-renaissance-stalls-with-pending-collapse-of-Calvert-Cliffs.
9. Ibid.
10. Judy Pasternak, "Nuclear Energy Lobby Working Hard to Win Support," Investigative Reporting Workshop at the American University School of Communication, January 24, 2010.
11. Harvey Wasserman, "Another Astonishing Holiday no new Nukes Victory," The Free Press, December 23, 2010, http://freepress.org/columns/display/7/2010/1866.
12. "Nuclear Follies," *Forbes Magazine*, February 11, 1985.
13. Physicians for Social Responsibility, *The American Power Act Includes Big Nuclear Giveaways That Harm Taxpayers, Ratepayers, and Public Safety*, May 12, 2010, http://www.psr.org/resources/apa-nuclear-subtitle-summary.pdf.
14. United Nations Development Programme (UNDP), *World Energy Assessment: Energy and the Challenge of Sustainability* (New York: UNDP, 2000).
15. Citizen Sarah, "$18.2 Billion! Amores Nucleares: San Antonio Nuclear Expansion Soap Opera Update," Public Citizen in Texas, December 23, 2009, http://texasvox.org/2009/12/23/18-2-billion-amores-nucleares-san-antonio-nuclear-expansion-soap-opera-update/.
16. "Turkey Point, Florida," Public Citizen, 2012, http://www.citizen.org/Page.aspx?pid=659.
17. John Muraski, "Florida's Crystal River Nuclear Plant Vexes Progress Energy," *Raleigh News & Observer*, March 1, 2012.
18. Nuclear Information Research Service (NIRS), "Don't Let the Nuclear Power Industry Hijack the Climate Bill!" (an alert to subscribers from NIRS), July 24, 2009, http://org2.democracyinaction.org/o/5502/t/5846/campaign.jsp?campaign_KEY=1864.
19. Nuclear Information Research Service (NIRS), "Stop the 'Clean Energy' Bank" (an alert to subscribers from NIRS), May 14, 2009, http://www.nirs.org/alerts/05-14-2009/1.
20. Nuclear Information Research Service (NIRS), "Urgent: Take Action Now to Stop Dirty Energy Bank" (an alert to subscribers from NIRS), January 27, 2010, http://www.nirs.org/alerts/01-27-2010/1.
21. Citizen Carol [pseud.], "Ramming Nuclear Power Subsidies into War Appropriations Bill Is Inappropriate and Should Be Blocked," Texas VOX, May 24, 2010, http://texasvox.org/2010/05/24/ramming-nuclear-power-subsidies-into-war-appropriations-bill-is-inappropriate-and-should-be-blocked/.
22. Anco Blazev, "US Energy Recovery Funding Goes to Foreign Companies," Examiner.com, September 10, 2010, http://www.examiner.com/article/us-energy-recovery-funding-goes-to-foreign-companies.
23. "Groups: Foreign Companies, Workers Are Big Early Winners under Federal Loan Guarantees for Nuclear Reactors," Nuclear Information Research Service press release, July 1, 2010, http://www.nirs.org/press/07-01-2010/2.
24. David Schlissel, Michael Mullett, and Robert Alvarez, *Nuclear Loan Guarantees: Another Taxpayer Bailout Ahead?* (Cambridge, Mass.: Union of Concerned Scientists, March 2009), http://www.ucsusa.org/assets/documents/nuclear_power/nuclear-loan-guarantees.pdf.
25. Benjamin K. Sovacool, "How Much Will New Nuclear Power Plants Cost?" Scitizen, November 2, 2008, http://scitizen.com/future-energies/how-much-will-new-nuclear-power-plants-cost-_a-14-2287.html.
26. Peter Bradford and David Schlissel, *Why a Future for the Nuclear Industry Is Risky* (Friends of the Earth et al., January 2007), http://www.nirs.org/factsheets/fctsht.htm/whynewnukesareriskyfcts.pdf.
27. Ibid.

28. "The Case against Nuclear Power," Greenpeace briefing paper, January 8, 2008, http://www.greenpeace.org.uk/files/pdfs/nuclear/nuclear-power-briefing.pdf.
29. Joseph Romm, "The Staggering Cost of New Nuclear Power," Center for American Progress, January 5, 2000, http://www.americanprogress.org/issues/2009/01/nuclear_power.html.
30. Lorie Lam, "Shoreham Nuclear Power Plant," Encyclopedia of the Earth, August 25, 2009, http://www.eoearth.org/article/Shoreham_Nuclear_Power_Plant.
31. Safe Energy Communication Council (SECC), "Is the Price-Anderson Act a Subsidy?" SECC fact sheet, http://www.nirs.org/reactorwatch/paa/priceandersonsubsidyseccfctsht.pdf.
32. Pacific Northwest National Laboratory, *An Analysis of Federal Incentives Used to Stimulate Energy Production*, (Springfield, Va.: National Technical Information Service, December 1978), http://www.osti.gov/bridge/servlets/purl/6489241-ITh4rn/6489241.pdf.
33. V. L. Sailor et al., *Severe Accidents in Spent Fuel Pools in Support of Generic Safety, Issue 82*, report no. NUREG/CR-4982 (Uphaven, N.Y.: Brookhaven National Laboratory, 1987), http://www.osti.gov/bridge/servlets/purl/6135335-5voofL/6135335.pdf.
34. ICF International, "The Impacts on Nuclear Power," January 1, 2005, http://www.icfi.com/insights/white-papers/2005/impacts-on-nuclear-energy.
35. "Is the Price-Anderson Act a Subsidy? Government Studies Determined It Is," Safe Energy Communication Council Factsheet, 2002, p. 2, www.nirs.org/reactorwatch/paa/priceandersonsubsidyseccfctsht.pdf.
36. Harvey Wasserman, "Stewart Brand Is Wrong about Nukes—and Is Losing," Huffington Post (blog), July 26, 2010.
37. "Advanced Fuel Cycle Initiative," DOE (accessed June 27, 2012), http://www.ne.doe.gov/AFCI/neAFCI.html.
38. Craig Severance, "Business Risks and Costs of New Nuclear Power: The Staggering Cost of New Nuclear Power," Energy Economy Online, http://energyeconomyonline.com/Nuclear_Costs.html.
39. Greenpeace, *Nuclear Power: Undermining Action on Climate Change*, Greenpeace International Case Study and Alternatives briefing, reference no. GN089 (Amsterdam: Greenpeace International, December 2007), http://www.greenpeace.org/international/Global/international/planet-2/report/2008/3/nuclear-power-undermining-ac.pdf.
40. Michael Grunwald, "Nuclear's Comeback: Still No Energy Panacea," *Time* magazine, December 31, 2008, 38.
41. "The Continuing Struggle for a Nuclear-Free Philippines: The Bataan Nuclear Power Plant," WISE News Communiqué, October 16, 1998, http://www10.antenna.nl/wise/index.html?http://www10.antenna.nl/wise/499-500/4935.html.
42. Harvey Wasserman, "The Reactor Relapse Takes 3 Hits to the Head," Free Press, November 12, 2009, http://freepress.org/columns/display/7/2009/1783.
43. Oliver Wright, "Power Politics: French Threat to UK Energy," *The Independent*, May 16, 2012, http://www.independent.co.uk/environment/green-living/power-politics-french-threat-to-uk-energy-7754470.html.
44. Jonathon Porritt, "The Eye-Watering Expense of Nuclear Power," *Guardian* Environment (blog), May 4, 2012, http://www.guardian.co.uk/environment/blog/2012/may/04/expense-nuclear-power-energy-coalition.
45. Fiona Harvey and Terry Macalister, "Ministers Planning 'Hidden Subsidies' for Nuclear Power," *Guardian*, April 20, 2012.
46. "Chernobyl Still a Menace 20 Years after Meltdown," Environment New Service, April 26, 2006, http://www.ens-newswire.com/ens/apr2006/2006-04-26-04.html.

47. Pabitra L. De, "Costs of Decommissioning Nuclear Power Plants," *International Atomic Energy Agency Bulletin*, March 1990, 39-42, http://www.iaea.org/Publications/Magazines/Bulletin/Bull323/32304783942.pdf.
48. Daniel Barlow, "Experts Say Vt. Yankee's Nuke Waste Is Here to Stay," *Vermont Times-Argus*, January 30, 2009.
49. Amory Lovins, Imran Sheikh, and Alex Markevich, "Forget Nuclear," *RMI Solutions* (newsletter of the Rocky Mountain Institute), April 2008.
50. Globalization 101, a project of the State University of New York Levin Institute, "Solar Power," http://www.globalization101.org/solar-power/.
51. Jurriaan Kamp, "There Is Plenty Renewable Energy to Fight Global Warming," Huffington Post (blog), December 30, 2009, http://www.huffingtonpost.com/jurriaan-kamp/there-is-plenty-renewable_b_397237.html.
52. Nuclear Information Research Service (NIRS), "Got Solar!" NIRS fact sheet, http://www.nirs.org/factsheets/gotsolar.pdf.
53. Craig A. Severance, *Business Risks and Costs of New Nuclear Power* (Climate Progress, January 2, 2009), http://energyeconomyonline.com/uploads/Business_Risks_and_Costs_of_New_Nuclear_Power_Reprint_-_Jan_2__2009_Craig_A._Severance.pdf.
54. Joseph Romm, *The Self-Limiting Future of Nuclear Power* (Center for American Progress Action Fund, June 2008), http://www.americanprogressaction.org/issues/2008/pdf/nuclear_report.pdf.
55. Amory Lovins, Imran Sheikh, and Alex Markevich, "Forget Nuclear," *RMI Solutions* (newsletter of the Rocky Mountain Institute), April 2008.
56. Scott Learn, "Despite Billions Spent on Cleanup, Hanford Won't Be Clean for Thousands of Years," *Oregonian*, February 10, 2010.
57. Peter Eisler, "Hanford: America's Nuclear Nightmare," *USA Today*, January 25, 2012.
58. Ibid.
59. Emily Yehle, "New Reports Raise More Safety Concerns at Hanford," *E&E News*, January 13, 2012.
60. Eisler, "Hanford."

CHAPTER 3

1. North American Electric Reliability Corporation, "Generating Availability Reports," 2012, at http, http://www.nerc.com/page.php?cid=4|43|47.
2. "Maintenance of a Nuclear Power Plant," Nuclear Tourist, http://www.nucleartourist.com/operation/mtce1.htm.
3. "Fuel/Refueling Outages," Nuclear Energy Institute, http://www.nei.org/resourcesandstats/nuclear_statistics/fuelrefuelingoutages.
4. "Nuclear Power's Green Promise Dulled by Rising Temps," *Christian Science Monitor*, August 10, 2006.
5. Matthew Wald, "Nuclear Reactor's Life Is Prolonged in New Jersey," *New York Times*, April 2, 2009, http://www.nytimes.com/2009/04/02/science/earth/02nuclear.html.
6. Karl Grossman, "Inviting Atomic Catastrophe," Huffington Post, June 3, 2012, http://www.huffingtonpost.com/karl-grossman/nuclear-regulatory-commission_b_1565916.html.
7. Energy Information Administration (EIA), *Annual Energy Outlook 2008* (Washington, D.C.: EIA, September 2008), www.eia.gov/oiaf/archive/aeo08/index.html.
8. "US Reactor Uprates Boost Capacity," World Nuclear News, September 8, 2008, http://www.world-nuclear-news.org/NN_US_reactor_uprates_boost_capacity_0809082.html.
9. Daniel Fineren, "Factbox: European Nuclear Plant Life Extensions," Planet Ark, June 17, 2009, http://planetark.org/enviro-news/item/53381.

10. World Nuclear Association Reactor Database (accessed March 17, 2012), http://world-nuclear.org/NuclearDatabase/rdresults.aspx?id=27569&ExampleId=130.
11. Mark Henderson, "New Generation of Nuclear Reactors Promises 'Greener and Safer' Energy," *Times* (London), January 11, 2008.
12. Terry Macalister, "New Generation of Nuclear Power Stations 'Risk Terrorist Anarchy,'" *Guardian*, March 16, 2009, http://www.guardian.co.uk/environment/2009/mar/16/nuclearpower-nuclear-waste.
13. "Uranium Mining," World Nuclear Association (updated October 2011), http://www.world-nuclear.org/education/mining.htm.
14. *Reserven, Ressourcen und Verfügbarkeit von Energierohstoffen 2005* [Reserves, Resources, and Availability of Energy Resources 2005] (Hannover, Germany: Bundesanstalt für Geowissenschaften und Rohstoffe [Federal Institute for Geosciences and Natural Resources], 2006).
15. Karl Burkart, "Uranium: The New 'Foreign Oil,'" Mother Nature Network, August 31, 2009, http://www.mnn.com/green-tech/research-innovations/blogs/uranium-the-new-foreign-oil.
16. Organisation for Economic Co-operation and Development (OECD) and the International Atomic Energy Agency, *Uranium 2007: Resources, Production and Demand* (Paris: OECD Publishing, June 17, 2008).
17. Toni Johnson, "Global Uranium Supply and Demand," Council on Foreign Relations, January 14, 2010, http://www.cfr.org/energy/global-uranium-supply-demand/p14705.
18. "Supply of Uranium," World Nuclear Association (updated September 2011), http://www.world-nuclear.org/info/inf75.html.
19. *Nuclear Power: Myth and Reality* (Johannesburg: Heinrich Boll Foundation, 2006).
20. *Nuclear Energy: No Solution to Climate Change* (Netherlands: Women in Europe for a Common Future, November 2007), http://www.umweltinstitut.org/down-load/engl_klimaretter_atomkraft.pdf.
21. Thørring H, Ytre-Eide MA, Liland A Konsekvenser, "Consequences in Norway after a Hypothetical Accident at Sellafield," Norwegian Radiation Protection Authority, December 20, 2010, http://www.regjeringen.no/upload/MD/2011/vedlegg/rapporter/sellafieldrapport_straalevernet_250111.pdf
22. Mark Diesendorf and Peter Christoff, "CO2 Emissions from the Nuclear Fuel Cycle," Energy Science fact sheet 2, November 2006, www.docstoc.com/docs/42535055/CO2-Emissions-from-the-Nuclear-Fuel_Cycle.
23. Amory Lovins, interview by Amy Goodman, *Democracy Now!*, "Amory Lovins: Expanding Nuclear Power Makes Climate Change Worse," July 16, 2008, http://www.democracynow.org/2008/7/16/amory_lovins_expanding_nuclear_power_makes.
24. C. Johnson, "Climate Effects of Electric Power Plants," University of Chicago (updated May 24, 2012), http://mb-soft.com/ public2/powerplt.html.
25. Matthew Wald, "NRG Abandons Project for 2 Reactors in Texas," *New York Times*, April 19, 2011.
26. Arnie Gundersen, "New Containment Flaw Identified in the BWR Mark 1," Fairewinds Associates, February 9, 2012, http://www.fairewinds.org/content/new-containment-flaw-identified-bwr-mark-1.
27. Paul Gunter, "Hazards of Boiling Water Reactors in the United States," Nuclear Information and Resource Service factsheet, Updated March 2011, http://www.nirs.org/factsheets/bwrfact.htm
28. Karl Grossman, "After Fukushima, Media Still Buying Nuclear Spin," *Extra!*, May 2011, http://www.fair.org/index.php?page=4335.

CHAPTER 4

1. Testimony of Peter A. Bradford, "Three Mile Island: Thirty Years of Lessons Learned," Senate Committee on Environment and Public Works Subcommittee on Clean Air and Nuclear Safety, March 24, 2009, http://www.nirs.org/reactorwatch/accidents/bradfordtestimony32409tmi.pdf.
2. "Davis-Besse Nuclear Plant Comes Close to Disaster as Lax Regulator Places Company Interests ahead of Public Safety," Nuclear Information & Resource Service press release, March 13, 2002, http://www.nirs.org/press/03-13-2002/1.
3. Testimony of Peter A. Bradford, "Three Mile Island."
4. Harvey Wasserman and Norman Soloman, *Killing Our Own: The Disaster of America's Experience with Atomic Radiation* (New York: Dell Publishing, 1982), 224–25, 238–39.
5. Harvey Wasserman, "Cracking the Corporate Media's Iron Curtain around Death at Three Mile Island," Free Press, April 1, 2009, http://freepress.org/columns/display/7/2009/1736.
6. A. Yablokov, V. Nesterenko, and A. Nesternko, "Chernobyl: Consequences of the Catastrophe for People and the Environment," *Annals of the New York Academy of Sciences* 1181 (December 2009).
7. Harvey Wasserman, "Who Will Pay for America's Chernobyl Roulette?" Free Press, April 26, 2009, http://freepress.org/columns/display/7/2009/1742.
8. Project for Excellence in Journalism, *The State of the News Media: An Annual Report on American Journalism* (Pew Research Center Project for Excellence in Journalism, 2004).
9. Pam Sohn, "Nuclear Plant's 'Red' Rating to Cost TVA Millions," *Chattanooga Times Free Press*, August 31, 2011.
10. Benjamin J. Sovacool, "The Dirt on Nuclear Power," Project Syndicate, March 16, 2011, http://www.project-syndicate.org/commentary/the-dirt-on-nuclear-power.
11. Greenpeace International and the European Renewable Energy Council (EREC), *Energy [R]evolution: A Sustainable World Energy Outlook* (Greenpeace International and EREC, October 2008), http://www.greenpeace.org/international/Global/international/publications/climate/2010/fullreport.pdf.
12. Harvey Wasserman, "Yet Another $50 Billion for Rust-Bucket Nukes?" Free Press, April 10, 2009, http://nukefree.org/news/yetanother50billionforrustbucketnukes.
13. Karl Grossman, "A Radioactive Extension for Aging Nuclear Plants," *CounterPunch*, April 13, 2009, http://www.counterpunch.org/2009/04/13/a-radioactive-extension-for-aging-nuclear-plants/.
14. Sandia National Laboratories, *Calculation of Reactor Accident Consequences* (CRAC-2), November 1, 1982, http://www.beyondnuclear.org/storage/CRAC%202%20chart.pdf.
15. Office of Nuclear Regulatory Research, "State-of-the-Art Reactor Consequence Analyses (SOARCA) Report," Nuclear Regulatory Commission, January 2012, http://pbadupws.nrc.gov/docs/ML1209/ML12090A794.pdf.
16. Cindy Folkers, "The Demonic Reality of Fukushima and the Absurdity of the NRC," Counterpunch, March 2–4, 2012, http://www.counterpunch.org/2012/03/02/the-demonic-reality-of-fukushima/.
17. Joel Davis, "Nuclear Reactions," *Omni Magazine*, May 1988, 46.
18. William J. Broad, "Military Crew Said to Be Exposed to Radiation, but Officials Call Risk in US Slight," *New York Times*, March 13, 2011.
19. Reiji Yoshida, "Pump Failure Nearly Brings No. 5 to a Boil," *Japan Times* Online, May 30, 2011, http://www.japantimes.co.jp/text/nn20110530a1.html.
20. "Japanese Government: No Plans to Re-Start Fukushima Daini," NEI Nuclear Notes (blog), February 17, 2012, http://neinuclearnotes.blogspot.com/2012/02/japanese-government-no-plans-to-re.html.

21. "Accident at Second Japanese Nuclear Complex," Washington's Blog (blog), January 12, 2012, http://www.washingtonsblog.com/2012/01/accident-at-second-japanese-nuclear-complex.html.
22. Martin Fackler, "Japan Weighed Evacuating Tokyo in Nuclear Crisis," *The New York Times*, February 27, 2012, http://www.nytimes.com/2012/02/28/world/asia/japan-considered-tokyo-evacuation-during-the-nuclear-crisis-report-says.html.
23. Kenji Hall and Julie Makinen, "Workers Suffer Hardships to Stabilize Fukushima Plant," *Los Angeles Times*, March 29, 2011.
24. Martin Fackler, "Japan Weighed Evacuating Tokyo in Nuclear Crisis," *New York Times*, February 27, 2012.
25. World Nuclear News, "Japan 'Unprepared' for Fukushima Accident," June 8, 2011, http://www.world-nuclear-news.org/RS-Japan_unprepared_for_Fukushima_accident-0806114.html.
26. Akio Matsumura, "The Fourth Reactor and the Destiny of Japan," Akomatsumura.com, September 29, 2011, http://akiomatsumura.com/2011/09/the-fourth-reactor-and-the-destiny-of-japan.html.
27. Akio Matsumura, "Fukushima Daiichi Site: Cesium-137 is 85 Times Greater than at Chernobyl Accident," Akiomatsumura.com, April 3, 2012, http://akiomatsumura.com/2012/04/682.html
28. Senator Ron Wyden, Letter to His Excellency Ichiro Fujisaki, April 16, 2012, http://www.wyden.senate.gov/download/letter-to-japanese-ambassador-on-fukushima-recovery-efforts.
29. "Nuclear Crisis in Japan," Nuclear Information Research Service, December 7, 2011, http://www.nirs.org/fukushima/crisis.htm.
30. "Interim Report: Tepco Prepared to Airdrop Ice into Reactor No. 1 Spent Fuel Pool—3.5 Ice Brought to Fukushima," ENENews, March 17, 2012, http://enenews.com/interim-report-tepco-prepared-airdrop-ice-reactor-1-spent-fuel-pool-35-tons-brought-fukushima-dai-ni, quoting "Accident Response at TEPCO's Fukushima Dai-ichi NPS," Investigation Committee on the Accidents at the Fukushima Nuclear Power Station, February 23, 2012, http://icanps.go.jp/eng/interim-report.html, chap. 4.
31. "Fukushima No. 2 Reactor Radiation Level up to 73 Sieverts per Hour," *Mainichi Daily News*, March 28, 2012.
32. "Experts: Radiation at Fukushima Plant Far Worse than Thought," Common Dreams, March 28, 2012, https://www.commondreams.org/headline/2012/03/28-1.
33. Michael Marriote, "Fukushima Anniversary To Be Marked By Protests, Rallies, Flash Mobs Across US and Entire World," NIRS, March 9, 2012, www.nirs.org/fukushima/fukushimaanniversaryactions3912.pdf.
34. Daniel Tovrov, "Japan Earthquake 2012: Tokyo at Risk for Major Disaster in Near Future," *International Business Times*, March 15, 2012, http://www.ibtimes.com/articles/314733/20120315/japan-earthquake-2012-tokyo.htm. See also: "Prediction Offered for Big Tokyo Quake," UPI, January 23, 2012, http://www.upi.com/Science_News/2012/01/23/Prediction-offered-for-big-Tokyo-quake/UPI-59791327359282/

CHAPTER 5

1. Jeff Donn (Associated Press), "Safety Rules Loosened for Aging Nuclear Reactors," MSNBC.com, June 20, 2011, http://www.msnbc.msn.com/id/43455859/ns/us_news-environment/t/safety-rules-loosened-aging-nuclear-reactors.
2. Ibid.
3. "Update on Defending Great Lakes against Risky Atomic Reactors," Beyond Nuclear, March 7, 2012, http://www.beyondnuclear.org/home/2012/3/7/update-on-defending-great-lakes-against-risky-atomic-reactor.html.

4. Ibid.
5. Tina Lam, "Michigan's Palisades Nuclear Plant May Be Named One of the Nation's 5 Worst," *Detroit Free Press*, January 15, 2012.
6. Beyond Nuclear, "Entergy Nuclear, Infamous for 'Buying Reactors Cheap, then Running Them into the Ground,'" February 19, 2012, http://www.beyondnuclear.org/home/2012/2/19/entergy-nuclear-infamous-for-buying-reactors-cheap-then-runn.html.
7. Ibid.
8. Beyond Nuclear, "Palisades: 'It's an Accident Waiting to Happen,'" January 25, 2012, http://www.beyondnuclear.org/home/2012/1/25/palisades-its-an-accident-waiting-to-happen.html.

CHAPTER 6

1. John Gofman, *Radiation-Induced Cancer from Low-Dose Exposure: An Independent Analysis* (San Francisco: Committee for Nuclear Responsibility, 1990).
2. *Hidden Radioactive Releases from Nuclear Power Plants in the United States* (Washington, D.C.: Nuclear Information Research Service, November 2005), http://www.nirs.org/factsheets/drey_usa_pamphlet.pdf.
3. Ramaswami Ashok Kumar, "Indian Infant Mortality and Nuclear Power," cited by Leuren Moret in "Population Exposed to Environmental Uranium: Part 2," *Namaste Magazine* 11, no. 1 (January 11, 2007): 35; See also: R. A. Kumar, "Indian Infant Mortality and Nuclear Tests," Stop Nuclear Energy Programmes (blog), December 28, 2007, http://plutoniumaradiumabillionpeoplehitdna.blogspot.com/2007/12/health-effects-of-worldwide-nuclear.html, and R. A. Kumar, http://indiainfantmortalityandnuclearpower.blogspot.com/.
4. Beyond Nuclear, "Shifting the Radiation Exposure Paradigm from Protecting the Industry to Protecting the Public," comments from Beyond Nuclear to the Nuclear Radiation and Studies Board of the National Academy of Sciences, September 23, 2011, http://www.beyondnuclear.org/storage/NAScommentsFINALSept28.pdf.
5. Ibid.
6. Diane D'Arrigo, Letter to the NRC, February 10, 2011, http://pbadupws.nrc.gov/docs/ML1104/ML110460033.pdf.
7. Beyond Nuclear, "Comments from Beyond Nuclear," September 23, 2011, http://www.beyondnuclear.org/storage/NAScommentsFINALSept28.pdf.
8. Eartha Jane Melzer, "Cancer Questions Grow around Fermi Nuclear Plant," *Michigan Messenger*, February 17, 2009.
9. "Curiosity Grows as Long-Delayed Anti-radiation Pills Are Ready Near Michigan Nuclear Plants," *Grand Rapids Press*, September 14, 2009.
10. John Vidal, "Chernobyl Nuclear Accident: Figures for Deaths and Cancers Still in Dispute," *The Guardian*, January 10, 2010.
11. Körblein A, Küchenhoff H, "Perinatal Mortality in Germany Following the Chernobyl Accident," *Radiation and Environmental Biophysics*, 1997 June, 36 (2):137.
12. P.J. Baker, D.G. Hoel, "Meta-Analysis of Standardized Incidence and Incidence an Mortality Rates of Childhood Leukaemia in Proximity of Nuclear Facilities," *European Journal of Cancer Care*, 2007, 16:355-363.
13. Victoria Gill, "Chernobyl Zone Shows Decline in Biodiversity," BBC News, July 30, 2010, http://www.bbc.co.uk/news/science-environment-10819027.
14. Charles Hawley, "Radioactive Boar on the Rise in Germany," *Der Spiegel* Online, July 30, 2010, http://www.spiegel.de/international/zeitgeist/0,1518,709345,00.html.
15. Frank Joseph Smecker, "The Nuclear Goliath: Confronting Industrial Energy," Toward Freedom, March 30, 2009, http://towardfreedom.com/home/content/view/1552/1/.

16. M. C. Hatch, et al. "Cancer Near the Three Mile Island Nuclear Plant: Radiation Emissions," *American Journal of Epidemiology* 132, no. 3 (1990): 397–412.
17. Jay M. Gould, *The Enemy Within: The High Cost of Living Near Nuclear Reactors* (New York: Four Walls Press, 1996), p. 187.
18. "Child Leukemia Rates Increase Near US Nuclear Plants," Salem-News.com, May 18, 2009, http://www.salem-news.com/articles/may182009/kids_leukemia_5-18-09.php.
19. Claire Sermage-Faure, et al., "Childhood Leukemia around French Nuclear Power Plants: The Geocap Study, 2002–2007," *International Journal of Cancer* 131, no. 1 (January 2012): 1–12.
20. Peter Kaatsch, et al., "Childhood Leukemia in the Vicinity of Nuclear Power Plants in Germany," *Deutsches Ärsteblatt* 105 (October 2008): 725–32.
21. "Radiation Exposure Limits Weakened in Departing Bush Move," Public Employees for Environmental Responsibility press release, January 21, 2009, http://www.commondreams.org/newswire/2009/01/21-0.
22. Harvey Wasserman and Norman Solomon, "Uranium Milling and the Church Rock Disaster," *Killing Our Own: The Disaster of America's Experience with Atomic Radiation* (orig. pub. New York: Dell Publishing, 1982); available online at the website of Black Mesa Indigenous Support, http://blackmesais.org/2001/10/uranium-milling-and-the-church-rock-disaster/, chap. 9.
23. MaryLynn Quartaroli, "'Leetso,' the Yellow Monster: Uranium Mining on the Colorado Plateau," Land Use History of North America—Colorado Plateau, 2002, http://www.cpluhna.nau.edu/Change/uranium.htm.
24. Keith Schneider, "Dying Nuclear Plants Give Birth to New Problems," *The New York Times*, October 31, 1988, http://www.nytimes.com/1988/10/31/us/dying-nuclear-plants-give-birth-to-new-problems.html.
25. Dr. Arjun Makhijani, "A Nuclear Incident 'Worse Than Three Mile Island,'" interview by Steve Curwood for Public Radio International's *Living on Earth* program, January 20, 2006, http://www.loe.org/shows/segments.html?programID=06-P13-00003&segmentID=1.
26. Ibid.
27. Ibid.
28. *Radiological Background Study Report: Santa Susana Field Laboratory* (Washington, D.C.: US Environmental Protection Agency, July 2011).
29. Michael Collins, "Radiation Readings Soar at Rocketdyne," EnviroReporter.com, March 6, 2012, http://www.enviroreporter.com/2012/03/radiation-readings-soar-at-rocketdyne/.

CHAPTER 7

1. "World Uranium Mining," World Nuclear Association (updated May 2012), http://www.world-nuclear.org/info/inf23.html.
2. Winona LaDuke, "Navajos Ban Uranium Mining," *Earth Island Journal*, Autumn 2005, http://www.earthisland.org/journal/index.php/eij/article/navajos_ban_uranium_mining/.
3. Chris Shuey, *Uranium Exposure and Public Health in New Mexico and the Navajo Nation: A Literature Summary* (Albuquerque, N.Mex.: Southwest Research and Information Center, October 14, 2008), 1.
4. Frank Joseph Smecker, "The Nuclear Goliath," http://towardfreedom.com/home/content/view/1552/1/.
5. "Nuclear Waste Dumps—Where They Plan to Put Them," Prairie Island Coalition, http://www.no-nukes.org/prairieisland/dumps.html.
6. David Thorpe, "Extracting a Disaster," *Guardian*, December 5, 2008, http://www.guardian.co.uk/commentisfree/2008/dec/05/nuclear-greenpolitics.

7. Stefan Simanowitz, "Hostage Negotiation in Mali and Niger: The Silent Trade," *CounterPunch*, January 28, 2009, http://www.counterpunch.org/2009/01/28/the-silent-trade/.
8. Jim Green, "Water, Uranium and Nuclear Power," Friends of the Earth Australia, May 13, 2012, www.foe.org.au/anti-nuclear/issues/oz/water-nuclear/long.
9. "Uranium Mine Called "World's Worst Practice,'" *Flinders News* (Port Pirie, South Africa), August 21, 2009, http://www.theflindersnews.com.au/news/local/news/business/uranium-mine-called-worlds-worst-practice/1602768.aspx?storypage=0.
10. Grey RM, Tsingine R, Yazzie MH. Navajo AML Reclamation Program, Presentation to Navajo Abandoned Uranium Mines Collaboration Annual Meeting, (Albuquerque, N.M.), May 1, 2003.
11. Environment America Research and Policy Center et al., *Grand Canyon at Risk: Uranium Mining Doesn't Belong Near Our National Treasures* (Environment America Research and Policy Center, Summer 2011), 16.

CHAPTER 8

1. Tom J. Chalko, "Earthquake Energy Rise on Earth," *NU Journal of Discovery*, May 2008, 1–2, http://nujournal.net/EarthquakeEnergyRise.pdf.
2. Dina Cappiello and Jeff Donn (Associated Press), "Quake Risk to Reactors Greater than Thought," Yahoo! News, September 2, 2011, http://news.yahoo.com/quake-risk-reactors-greater-thought-071301249.html.
3. Central and Eastern United States Seismic Source Characterization for Nuclear Facilities (CEUS), a joint project of the Electric Power Research Institute, the US Department of Energy, and the US Nuclear Regulatory Commission, January 31, 2012, http://www.ceus-ssc.com.
4. NRC, "New Seismic Model Will Refine Hazard Analysis at U.S. Nuclear Plants," Press release, January 31, 2012, http://pbadupws.nrc.gov/docs/ML1203/ML120330098.pdf.
5. Amanda Peterson Beadle, "Report: 27 US Nuclear Reactors Need Upgrades to Avoid Severe Damage from Earthquakes," Climate Progress, September 2, 2011, http://thinkprogress.org/climate/2011/09/02/310816/report-27-u-s-nuclear-reactors-need-upgrades-to-avoid-severe-damage-from-earthquakes/.
6. Rebecca Smith and Mark Maremont, "Earthquake Risks Probed at US Nuclear Plants," *Wall Street Journal*, July 19, 2011.
7. Cappiello and Donn, "Quake Risk to Reactors Greater than Thought."
8. Jim Sciutto, Jack Cloherty, and Lee Ferran, "East Coast Quake: Nuclear Reactors Taken Offline," ABC News, August 23, 2011, http://abcnews.go.com/US/earthquake/east-coast-quake-nuclear-reactors-offline/story?id=14365268.
9. Brian Todd, "Officials: Virginia Quake Shifted Nuclear Plant's Storage Casks," CNN, September 1, 2011, http://www.cnn.com/2011/US/09/01/virginia.quake.nuclear/index.html.
10. "Fracking Could Have Caused East Coast Earthquake," *Russia Today*, August 24, 2011, http://rt.com/usa/news/fracking-earthquake-virginia-dc-817-061/.
11. John Daly, "US Government Confirms Link Between Earthquakes and Hydraulic Fracturing," OilPrice.com, November 8, 2011, http://oilprice.com/Energy/Natural-Gas/US-Government-Confirms-Link-Between-Earthquakes-And-Hydraulic-Fracturing.html.
12. Rebecca Smith and Mark Maremont, "Earthquake Risks Probed."
13. "Nuclear Regulator Commission Advises US Nuclear Reactor Safety Upgrade," *PBS NewsHour*, July 19, 2011, http://www.pbs.org/newshour/bb/environment/july-dec11/nuclearsafety1_07-19.html.

14. Rebecca Smith and Mark Maremont, "Earthquake Risks Probed."
15. Bruce Barcott, "Totally Psyched for the Full-Rip Nine," *Outside* online, August 25, 2011, http://www.outsideonline.com/outdoor-adventure/nature/Totally-Psyched-for-the-Full-Rip-Nine.html.
16. Simon Winchester, "The Scariest Earthquake Is Yet to Come," *Newsweek*, March 21, 2011.
17. David Perlman, "'Monster' Earthquake May Hit off California Coast within 50 Years," *San Francisco Chronicle*, October 10, 2011.
18. Greenpeace International, "Lessons from Fukushima," February 2012, p.11, http://www.greenpeace.org/slovenia/Global/ slovenia/Dokumenti/Lessons-from-Fukushima.pdf.
19. NRC, "Severe Accident Risks: An Assessment for Five US Nuclear Power Plants (*NUREG-1150)*," December 1990, http://www.nrc.gov/reading-rm/doc-collections/nuregs/staff/sr1150/.
20. Anonymous, "From Hiroshima to Fukushima: The Political Background to the Nuclear Disaster in Japan," World Socialist Web Site, June 23 and 24, 2011, http://wsws.org/articles/2011/jun2011/fuku-j23.shtml.
21. Yoichi Funabashi and Kay Kitazawa, "Fukushima in Review: A Complex Disaster, a Disastrous Response," *Bulletin of the Atomic Scientists* 68, no. 2 (March/April 2012), 9–21, http://bos.sagepub.com/content/68/2/9.

CHAPTER 9

1. *Climate Chaos and Nuclear Power*, Beyond Nuclear, February 19, 2008, http://www.beyondnuclear.org/storage/Climate%20Chaos%20and%20Nuclear%20Power%20Fast%20Facts%202%2019%202008.pdf.
2. Mari Yamaguchi (Associated Press), "More Leaks Found at Crippled Japan Nuclear Plant," *Star Advertiser* (Hawai'i), February 3, 2012, http://www.staradvertiser.com/news/breaking/138708919.html?id=138708919.
3. "Far from 'Solving Global Warming,' Atomic Power Too Risky to Operate in a Destabilized Climate," Beyond Nuclear, July 4, 2011, http://www.beyondnuclear.org/climate-change-whats-new/2011/7/4/far-from-solving-global-warming-atomic-energy-is-too-risky-t.html.
4. David Kraft, "Nuclear Power Won't Work in Global Warming World," Nuclear Enery Information Service, August 7, 2009, http://www.neis.org/Campaigns/YCNGW/its_the_water_stupid.shtml.
5. Eric Fleischauer, "Browns Ferry Cuts Production by 50%," DecaturDaily.com (Tennessee), July 26, 2010, http://www.decaturdaily.com/detail/65259.html.
6. David Lochbaum, "Nuclear Heat," cited in *Climate Chaos and Nuclear Power*, Beyond Nuclear, p. 6.
7. Sue Sturgis, "Gulf Coast Nuclear Plants at Risk from BP Oil Spill," *Texas Observer*, June 30, 2010, http://www.texasobserver.org/cover-story/gulf-coast-nuclear-plants-at-risk-from-bp-oil-spill.
8. Arnie Gundersen, "Gundersen Discusses the Situation at the Flooded Ft. Calhoun and Cooper Nuclear Power Plants,'" CNN, June 28, 2011, http://fairewinds.org/content/gundersen-discusses-situation-flooded-ft-calhoun-and-cooper-nuclear-power-plants.
9. Beyond Nuclear, "Far from Solving Global Warming, Atomic Power Too Risky to Operate in a Destabilized Climate," July 4, 2011.
10. Tony Phillips, "Getting Ready for the Next Big Solar Storm," NASA, June 21, 2011, http://science.nasa.gov/science-news/science-at-nasa/2011/22jun_swef2011/.
11. Richard A. Lovett, "What If the Biggest Solar Storm on Record Happened Today?" National Geographic Daily News, March 2, 2011, http://news.nationalgeographic.com/news/2011/03/110302-solar-flares-sun-storms-earth-danger-carrington-event-science/.

12. National Research Council, *Severe Space Weather Events—Understanding Societal and Economic Impacts: A Workshop Report* (Washington, D.C.: The National Academies Press, 2008).
13. Ibid.
14. Bill Hemmer, "NASA Warns of Solar Superstorm 2012," Fox News, http://www.youtube.com/watch?v=0aeqSEIXH-8 (accessed on June 17, 2012).
15. International Business Times, "Massive Solar Storm Could Cause Catastrophic Nuclear Threat in US," August 6, 2011,
16. For more information on the Solar Shield Bill, visit the website of EMPact America (www.empactamerica.org) and www.ShieldAct.com.

CHAPTER 10

1. Climate One, "Nuclear Revival?" Commonwealth Club, potcast audio, June 14, 2012, http://climate-one.org/blog/nuclear-revival.
2. Karin Wurzbacher (Munich Environmental Institute), "Fact Sheet 5: Safe Energy," Women in Europe for a Common Future, December 2007, http://umweltinstitut.org/download/5_nuclearwaste_en.pdf.
3. "US Department of Energy Releases Revised Total System Life Cycle Cost Estimate and Fee Adequacy Report for Yucca Mountain Project," DOE news release, August 5, 2008, http://energy.gov/articles/us-department-energy-releases-revised-total-system-life-cycle-cost-estimate-and-fee.
4. The 1982 Nuclear Waste Policy Act as Amended, Office of the General Counsel, Nuclear Regulatory Commission, cited in "Nuclear Regulatory Legislation (NUREG-0980)," 111th Congress, 2nd Session, January 2011, Vol. 1, No. 9, pp. 411-487, http://www.nrc.gov/reading-rm/doc-collections/nuregs/staff/sr0980/v1/sr0980v1.pdf.
5. Daniel Barlow, "Experts Say Vt. Yankee's Nuke Waste Is Here to Stay," *Times-Argus* (Vermont), January 30, 2009.
6. Greenpeace International and the European Renewable Energy Council (EREC), *Energy [R]evolution: A Sustainable World Energy Outlook.*
7. Marjorie Mazel Hecht, "The Beauty of Completing the Nuclear Fuel Cycle," *21st-Century Science and Technology* 18, no. 4 (Winter 2005–2006), http://www.larouchepub.com/eiw/public/2006/2006_10-19/2006-18/pdf/56-60_618_technukes.pdf.
8. Arnie Gundersen, "Why Fukushima Can Happen Here: What the NRC and Nuclear Industry Don't Want You to Know," Fairewind Associates, July 10, 2011, http://www.fairewinds.org/content/why-fukushima-can-happen-here-what-nrc-and-nuclear-industry-dont-want-you-know.
9. National Research Council, "Spent Fuel Pool Storage," *Safety and Security of Commercial Spent Nuclear Fuel Storage: Public Report* (Washington, D.C.: National Academies Press, 2006), chap. 3.
10. Arnie Gundersen, "This Could Become Chernobyl on Steroids," Democracy Now!, March 15, 2011, http://www.democracynow.org/2011/3/15/this_could_become_chernobyl_on_steroids.
11. "Cabinet Kept Alarming Nuke Report Secret," *Japan Times*, January 22, 2012.
12. "Experts: US Has Agreed to Store Enough Nuclear Reactor Waste to Fill Two Yucca Mountains . . . or Face Billions of Dollars in New Penalties," Institute for Energy and Environmental Research news release, March 24, 2010, http://ieer.org/wp/wp-content/uploads/2010/03/032410-IEER-Yucca-2-news-release-FINAL4.pdf.
13. Mary Olson, "Reprocessing is Not the Solution to the Nuclear Waste Problem," Nuclear Information Research Service fact sheet, http://www.nirs.org/factsheets/reprocessisnotsolution.pdf.

14. Mycle Schneider and Yves Merignac, *Nuclear Spent Fuel Reprocessing in France*, research report no. 4 (Princeton, N.J.: International Panel on Fissile Materials, April 2008), http://www.fissilematerials.org/ipfm/site_down/rr04.pdf.
15. Public Health "Consequences of MOX Fuel: NRC Reactor Licensing Issues," Nuclear Control Institute background paper, January 21, 1999, http://www.nci.org/i/ib12199.htm.
16. "Europe's Radioactive Secret," Greenpeace International briefing paper, November 18, 2005, http://www.greenpeace.org/international/Global/international/planet-2/report/2006/6/european-rad-secret.pdf.
17. Susan M. Jablonski, "Radioactive Waste," Pollution Issues, April 19, 2012, http://www.pollutionissues.com/Pl-Re/Radioactive-Waste.html.
18. US Government Accounting Office (GAO), *Radiation Standards: Scientific Basis Inconclusive, and EPA and NRC Disagreement Continues*, report no. GAO/RCED-00-152 (Washington, D.C.: GAO, June 2000); "Public Health and Environmental Radiation Protection Standards for Yucca Mountain, NV: Final Rule," EPA press release, June 13, 2001, http://nepis.epa.gov/Exe/ZyPURL.cgi?Dockey=00000E2C.txt.
19. "Final Commission Report," Blue Ribbon Commission on America's Nuclear Future press release, January 26, 2012, http://brc.gov/sites/default/files/brc_final_report_-_press_release_012612.pdf. Full report and updates are available from the BRC website at http://brc.gov/.
20. Blue Ribbon Commission on America's Nuclear Future, *Disposal Subcommittee Report to the Full Commission* (January 2012), http://brc.gov/sites/default/files/documents/disposal_report_updated_final.pdf.

CHAPTER 11

1. "Fact Sheet on Decommissioning Nuclear Power Plants," Nuclear Regulatory Commission, updated March 29, 2012, www.nrc.gov/reading-rm/doc-collections/fact-sheets/decommissioning.html.
2. Ibid.
3. Ibid.
4. Ibid.
5. "Ex-FBI Agent Charges Feds with Radioactive Cover-up at Rocky Flats," *Grist*, January 22, 2005, http://grist.org/politics/little-rockyflats/.
6. Laura Snider, "Study: Rocky Flats Area Still as Contaminated with Plutonium as 40 Years Ago," *The Daily Camera*, February 18, 2012, http://www.dailycamera.com/boulder-county-news/ci_19995436.
7. "Fact Sheet on Decommissioning Nuclear Power Plants."
8. Office of Nuclear Reactor Regulation, "Summary Findings Resulting from the Staff Review of the 2010 Decommissioning Funding Status Reports for Operating Power Reactor Licensees," Nuclear Regulatory Council report SECY-11-0149, October 26, 2011, http://www.nrc.gov/reading-rm/doc-collections/commission/secys/2011/2011-0149scy.pdf.
9. Jay Hancock, "Customers Protest Fund Switch by Constellation Suitor Exelon," *Baltimore Sun*, August 6, 2011, http://articles.baltimoresun.com/2011-08-06/business/bs-bz-hancock-exelon-nuclear-decommis20110806_1_exelon-spokeswoman-krista-lopykinski-trust-fund-comed.
10. Robert Alvarez, "Fixing America's Nuclear Waste Storage Problem," *Nation*, June 20, 2011, http://www.thenation.com/article/161500/fixing-americas-nuclear-waste-storage-problem. For a more detailed account, see the Institute for Policy Studies report *Spent Nuclear Fuel Pools in the US: Reducing the Deadly Risks of Storage* by Robert Alvarez (Washington, D.C.: IDP, May 2011), http://www.ips-dc.org/reports/spent_nuclear_fuel_pools_in_the_us_reducing_the_deadly_risks_of_storage.

11. "Spent Nuclear Fuel Storage Fact Sheet," Union of Concerned Scientists, December 2011, http://www.ucsusa.org/assets/documents/nuclear_power/fact-sheet-spent-fuel-storage.pdf.
12. Robert Alvarez, "Fixing America's Nuclear Waste Storage Problem," Institute for Policy Studies, June 20, 2011, http://www.ips-dc.org/articles/fixing_americas_nuclear_waste_storage_problem.
13. Joaquin Sapien, "While Nuclear Waste Piles Up in US, Billions in Fund to Handle It Sit Unused," ProPublica, March 30, 2011, http://www.propublica.org/article/while-nuclear-waste-piles-up-in-U.S.-billions-in-fund-to-handle-it-sits-unu.

CHAPTER 12

1. Mark Evanoff, "Civilian Nuclear Myth Explodes," *Not Man Apart*, October 1982, cited by *The Ecologist* online at http://exacteditions.theecologist.org/read/ecologist/digest-(vol-12-no-6-1982)-6454/4/3.
2. Jeff Lindemyer, "The Global Nuclear Energy Partnership: Proliferation Concerns and Implications," *Nonproliferation Review* 16, no. 1 (March 2009): 79.
3. Sharon Squassoni, "Iran's Nuclear Program: Recent Developments," Congressional Research Service, September 6, 2006, http://www.fas.org/sgp/crs/nuke/RS21592.pdf.
4. Dafna Linzer, "Past Arguments Don't Square with Current Iran Policies," *Washington Post*, March 27, 2005.
5. Helen Caldicott, "After Fukushima: Enough Is Enough!" *International Herald Tribune*, December 2, 2011.
6. "AREVA, Northrop Grumman Break Ground on AREVA Newport News Facility, Marking Concrete Step in US Nuclear Energy Revival," AREVA news release, July 22, http://us.areva.com/EN/home-1300/areva-inc-press-release.html.
7. Mark Z. Jacobson, "Review of Solutions to Global Warming, Air Pollution and Energy Security," *Energy & Environmental Science* 2 (2009): 148–73.
8. V. Gilinsky, M. Miller, and H. Hubbard, *A Fresh Examination of the Proliferation Dangers of Light Water Reactors* (Washington, D.C.: Nonproliferation Policy Education Center, October 22, 2004), http://www.npolicy.org/files/20041022-GilinskyEtAl-LWR.pdf.
9. Tetsuo Arima, *Genpatsu. Shoriki. CIA.* [Nuclear Power Plants. Shoriki. CIA.], (Tokyo: Shinchosha Publishers, 2008). The April 20, 2011 edition of Japan's daily newspaper *Mainichi Shimbun* also reported on Tetsuo Arima's disclosures regarding Shoriki's career and *Wall Street Journal* reporter Eleanor Warnock provided additional coverage in a June 1, 2012 blog entitled "Japan's Nuclear Industry: The CIA Link," http://blogs.wsj.com/japanrealtime/2012/06/01/japans-nuclear-industry-the-cia-link/.
10. "Problems of the Japan Atomic Energy," Association of Democratic Scientists, Department of Physics, 1953, cited in "From Hiroshima to Fukushima: The Political Background to the Nuclear Disaster in Japan," World Socialist Web Site, June 20, 2011, http://www.wsws.org/articles/2011/jun2011/fuku-j23.shtml.
11. "Nuclear Weapons Program," Federation of American Scientists, http://www.fas.org/nuke/guide/japan/nuke/.
12. David Snell, "Japan Developed Atom Bomb; Russia Grabbed Scientists," *Atlanta Constitution*, October 2, 1946, http://www.reformation.org/atlanta-constitution.html.
13. Joseph Trento, "United States Circumvented Laws to Help Japan Accumulate Tons of Plutonium," National Security News Service, D.C. Bureau, April 9, 2012, http://www.dcbureau.org/201204097128/national-security-news-service/united-states-circumvented-laws-to-help-japan-accumulate-tons-of-plutonium.html.
14. Ibid.

15. David McNeill, "Japan Must Develop Nuclear Weapons, Warns Tokyo Governor," *Independent*, March 8, 2011, http://www.independent.co.uk/news/world/asia/japan-must-develop-nuclear-weapons-warns-tokyo-governor-2235186.html#.
16. Taro Maki, "Japan's Leaders Must Face Country's 'Latent' Possession of Nuclear Weapons," *Mainichi Daily News*, October 28, 2011.
17. Gruppe Okologie, AntiAtom International, and Okologie Institut, *35 Years Promotion of Nuclear Energy: The International Atomic Energy Agency: A Critical Documentation of the Agency's Policy*, 3rd ed. (Hannover, Germany: Gruppe Okologie, 1993)
18. "Nuclear Power in the World Today," World Nuclear Association (updated April 2012), http://www.world-nuclear.org/info/inf01.html.
19. Mark Hibbs, "Moving Forward on the US-India Nuclear Deal," Carnegie Endowment for International Peace, April 5, 2010, http://www.carnegieendowment.org/2010/04/05/moving-forward-on-u.s.-india-nuclear-deal/bl3b.
20. *Nuclear Power: Myth and Reality* (Johannesburg: Heinrich Boll Foundation, 2006).
21. Summer Said, "Saudi Arabia, China Sign Nuclear Cooperation Pact," Wall Street Journal, January 16, 2012, http://online.wsj.com/article/SB10001424052970204468004577164742025285500.html.
22. "International Framework for Nuclear Energy Cooperation (Global Nuclear Energy Partnership)," World Nuclear Association (updated April 2012), http://www.world-nuclear.org/info/inf117_gnep.html.
23. "Nuclear Reprocessing: Dangerous, Dirty, and Expensive," updated May 4, 2011, Union of Concerned Scientists, http://www.ucsusa.org/nuclear_power/nuclear_power_risk/nuclear_proliferation_and_terrorism/nuclear-reprocessing.html.
24. "The Global Nuclear Energy Partnership."
25. Office of Nuclear Energy, US Department of Energy, "Notice of Cancellation of the Global Nuclear Energy Partnership (GNEP) Programmatic Environmental Impact Statement (PEIS)," Notices, *Federal Register* 74, no. 123 (June 29, 2009): 31017–18.
26. US Department of Defense, *Nuclear Posture Review* (2002).

CHAPTER 13

1. Brian Martin, "The Global Health Effects of Nuclear War," *Current Affairs Bulletin*, Vol. 59, No. 7, December 1982, pp.14-26, http://www.bmartin.cc/pubs/82cab/index.html.
2. Bennet Ramberg, "Military Sabotage of Nuclear Facilities: The Implications," *Annual Review of Energy*, November 1985, Vol. 10, pp. 495-514.
3. Jim Riccio, "Nuclear Plants Are Vulnerable to Earthquakes, Hurricanes, and Attacks—Are You at Risk?" Greenpeace blog, August 26, 2011, http://www.greenpeace.org/usa/en/news-and-blogs/campaign-blog/nuclear-plants-are-vulnerable-to-earthquakes-/blog/36524/.
4. Bill Dedman, "Nuclear Neighbors: Population Rises Near US Reactors," MSNBC, April 14, 2011, http://www.msnbc.msn.com/id/42555888/ns/us_news-life/t/nuclear-neighbors-population-rises-near-us-reactors/.
5. Conrad V. Chester and Rowena O. Chester, "Civil Defense Implications of the US Nuclear Power Industry During a Large Nuclear War in the Year 2000," *Nuclear Technology*, Vol. 31, December 1976, pp. 326–338.
6. Samuel Glasstone, Ed., "Effects of Nuclear Weapons," Revised edition, Atomic Energy Commission, April 1962, pp. 195, 200, http://www.atomicarchive.com/Docs/Effects/index.shtml.
7. US Congressional Office of Technology Assessment, "The Effects of Nuclear War," May 1979, pp. 64-67, http://www.fas.org/nuke/intro/nuke/7906/.
8. Gar Smith, "Nuclear Reactors and Nuclear War," *Not Man Apart*, June 1983, pp. 12–13.

9. Sir Brian Flowers, Concerns about the 'Plutonium Economy' cited from "Nuclear Power and the Environment," the Sixth Report of the UK Royal Commission on the Environment, 1976, http://www.ccnr.org/Flowers_plute.html.
10. Pol Heanna D'Huyvetter, Mayors for Peace, "Mayors for Peace: Background," Nuclear Free Virginia, September 2, 2011, http://nuclearfreeva.blogspot.com/2011/09/mayors-for-peace.html.
11. Steve Fetter and Kosta Tsipis, *Catastrophic Nuclear Radiation Releases*, report no. 5 (Cambridge, Mass.: Department of Physics, Massachusetts Institute of Technology, September 1980), and "Catastrophic Releases of Radioactivity," *Scientific American* 244, no. 4 (April 1981): 41–47.
12. David M. Bearden and Anthony Andrews, "Radioactive Tank Waste from the Past Production of Nuclear Weapons: Background and Issues for Congress," Congressional Research Service, January 3, 2007, http://www.dtic.mil/cgi-bin/GetTRDoc?AD=ADA467150.
13. Robert Scheer, *With Enough Shovels: Reagan, Bush & Nuclear War* (New York: Random House, 1982), pp. 20, 21.
14. J. R. Buchanan, "Siting of Nuclear Facilities," NRC Office of Regulatory Research, July 1976, p. 9, http://www.osti.gov/bridge/servlets/purl/7357370/7357370.pdf.
15. Buchanan, pp. 99–100.
16. Benjamin Sovacool, "The Dirt on Nuclear Power," Project Syndicate, March 16, 2011, http://www.project-syndicate.org/commentary/the-dirt-on-nuclear-power.
17. Mycle Schneider, "The Threat of Nuclear Terrorism: from Analysis to Precautionary Measures, Appendix 2, Examples of Attacks on Nuclear Interests or Involving Nuclear Materials: 1977-1990," WISE-Paris, December 2001, pp. 14-16, www.wise-paris.org/english/reports/conferences/011210Terrorisme.pdf. See also: Superphénix Rocket Attack," Wikipedia, http://en.wikipedia.org/wiki/Superph%C3%A9nix#Rocket_attack.

CHAPTER 14

1. "Generation IV Nuclear Energy Systems: Program Overview," US Department of Energy, http://www.ne.doe.gov/geniv/neGenIV1.html; "Generation IV Nuclear Reactors," World Nuclear Association (updated December 2010), http://www.world-nuclear.org/info/inf77.html.
2. Frederic Varaine, Jean-Paul Grouiller, and M. Delpech, "Results on Transient Scenarios towards GEN IV Systems," French Atomic Energy Commission,http://www.oecd-nea.org/pt/iempt8/abstracts/Abstracts/Abstract_Varaine.doc.
3. Dan Yurman, "Pebble-Bed Reactor Loses Funding," Energy Collective, February 25, 2010, http://theenergycollective.com/djysrv/30611/pebble-bed-reactor-loses-funding.
4. TerraPower: The Need for Innovation in Energy," Intellectual Ventures, http://www.intellectualventures.com/OurInventions/TerraPower.aspx.
5. Ibid.
6. Camille Ricketts, "Gates, Khosla-Backed TerraPower Lands $35 Million to Launch a New Era of Nuclear," VentureBeat/Green, June 14, 2010.
7. Steven Chu, "America's New Nuclear Option," *Wall Street Journal*, March 23, 2011.
8. Tonyjack [pseud.], "TVA, SCANA, US Army and DoE Discuss SMR Deployment," Nuclear Energy Insider, January 14, 2011, http://analysis.nuclearenergyinsider.com/small-modular-reactors/tva-scana-us-army-and-doe-discuss-smr-deployment.
9. Savannah River Site Citizen's Advisory Board, "Annual Work Plan," June 2, 2000, p. 10.
10. David H. Freedman, "On the Road with a Supersalesman," *Inc.* magazine, April 1, 2010, p. 91, http://www.inc.com/magazine/20100401/on-the-road-with-a-supersalesman.html.

CHAPTER 15

1. "Nuclear Agency Head Requested Cancellation of Safety Study in 2006," *Kyodo News*, March 16, 2012.
2. "Radiation Fears Haunt Food Shoppers in Japan," *Japan Today*, March 4, 2012.
3. "Radiation Found in Japan Whales," News24, June 15, 2011, http://www.news24.com/SciTech/News/Radiation-found-in-Japan-whales-20110614.
4. "Excessive Cesium Levels again Found in Iwate Cows," *Japan Times*, September 9, 2011, http://www.japantimes.co.jp/text/nn20110909a8.html.
5. House of Japan, "Radiation Monitoring System Launched at Schools," February 22, 2012, http://www.houseofjapan.com/local/radiation-monitoring-system-launched-at-schools.
6. Winifred Bird, "As Fukushima Cleanup Begins, Long-term Impacts are Weighed," *Environment 360*, January 9, 2012, http://e360.yale.edu/feature/as_fukushima_cleanup_begins_long-term_impacts_are_weighed/2482/.
7. "Cesium in Sea May Return in 20 to 30 Years," *Japan Times*, September 14, 2011, http://www.japantimes.co.jp/text/nn20110915a4.html.
8. Anne Ambrose, "Cal State Long Beach Faculty Measure Radioactive Fallout in California Kelp Beds from Damaged Japanese Nuclear Reactor," California State University at Long Beach, April 16, 2012, http://urd.csulb.edu/news-events/story.cfm?hackid=1777.
9. Kate Clifford, "Radiation Cloud 'Not Harmful,'" *Sunshine Coast Daily* (Australia), January 14, 2012, http://www.sunshinecoastdaily.com.au/story/2012/01/14/radiation-cloud-not-harmful-sunshine-coast/.
10. "San Francisco Bay Area Milk Sample Has Highest Amount of Cesium-137 Since Last June—Almost Double EPA's Maximum Contaminant Level," *ENENews*, April 10, 2012, http://enenews.com/april-milk-sample-highest.
11. "Fukushima Evacuees Still in Limbo as TEPCO Drags Its Feet," *Japan Today*, March 7, 2012 (expired on original site; article available at ONTD Political, http://ontd-political.livejournal.com/9411809.html, May 22, 2012).
12. "Fukushima Residents Report Various Illnesses," Al Jazeera, March 9, 2012, http://www.aljazeera.com/video/asia-pacific/2012/03/20123914421232874.html.
13. Dahr Jamail, "Fukushima Radiation Alarms Doctors," Al Jazeera, August 18, 2011, http://www.aljazeera.com/indepth/features/2011/08/201181665921711896.html.
14. "Noda Changing Kan's N-plant Stance," *Yomiuri Shimbun*, September 25, 2011, http://www.yomiuri.co.jp/dy/national/T110924003182.htm.
15. Michael McAteer, "Japan in Uproar over Censorship of Emperor's Anti-Nuclear Speech," *Atlantic* online, March 2012, http://www.theatlantic.com/international/archive/2012/03/japan-in-uproar-over-censorship-of-emperors-anti-nuclear-speech/255025/.
16. Akiko Fujita, "Japan Restarts Nuclear Reactors," ABC News, June 16, 2012, http://abcnews.go.com/International/japan-restarts-nuclear-reactors/story?id=16585084#.T-EF547483c.
17. Christine McCann, "Fukushima Nuclear Crisis Update for May 11th–May 14th, 2012," Greenpeace International blogpost, May 15, 2012, http://www.greenpeace.org/international/en/news/Blogs/nuclear-reaction/fukushima-nuclear-crisis-update-for-may-11th-/blog/40442/.

CHAPTER 16

1. Michio Kaku, "Fukushima 'Still a Ticking Timebomb,'" CNN, June 21, 2011, http://youtu.be/r4k77vjeEL0.
2. "Yokohama Finds High Strontium-90 Levels," *Japan Times*, October 13, 2011, http://www.japantimes.co.jp/text/nn20111013a3.html.

3. John D. Boice, Jr., ScD, "Congressional Hearing on Nuclear Energy Risk Management," May 13, 2011, http://hps.org/documents/John_Boice_Testimony_13_May_2011.pdf.
4. American Nuclear Society, "Fukushima Radiation Q&A," April 12, 2011, www.ans.org /misc/FukushimaRadiationQ%26A_LS.pdf.
5. Joseph Mangano, "The Dangerous Myths of Fukushima," Third World Resurgence, March 2012, No. 259, pp. 29-30, http://www.twnside.org.sg/title2/resurgence/2012/259 /special1.htm. See also: "Experts Say No Health Effects Expected from Fukushima Daiichi Accident," Nuclear Energy Institute, March 8, 2012, http://safetyfirst.nei.org /public-health/experts-say-no-health-effects-expected-from-fukushima-daiichi-accident/.
6. Rob Edwards, "Revealed: British Government's Plan to Play Down Fukushima," *Guardian*, June 30, 2011, http://www.guardian.co.uk/environment/2011/jun/30 /british-government-plan-play-down-fukushima.
7. David Jackson, "Obama: Experts Say Radiation Will Not Reach US," *USA Today*, March 15, 2011, http://content.usatoday.com/communities/theoval/post/2011/03/obama-not -worried-that-japan-radiation-will-reach-us/1#.UCF_VkQmbWk.
8. Arnie Gundersen, "One-third of Fukushima Kids Have Lumps in Thyroids," Collapse Network radio interview, February 28, 2012, http://collapsenet.com/free-resources/ collapsenet-public-access/must-see-videos/item/6671-1/3-of-fukushima-kids-have -lumps-in-thyroids-arnold-gunderson.
9. "Hot Particles Found at 2 out of 3 US Monitoring Stations during April, including Boston," ENENews, November 1, 2011, http://enenews.com/nuke-expert-hot-particles -found-at-2-out-of-3-us-monitoring-stations-in-april-there-will-be-an-increase-in -cancers-especially-on-the-west-coast-video.
10. "'High Concentrations' of Radiation Hit US and Canada," ENENews, October 28, 2011, http://enenews.com/high-concentrations-cesium-137-hit-canada.
11. Joseph J. Mangano and Janette D. Sherman, "An Unexpected Mortality Increase in the United States Follows Arrival of the Radioactive Plume from Fukushima: Is There a Correlation?" *International Journal of Health Services* 42, no. 1 (December 2011): 47–64, http://janettesherman.com/wp-content/uploads/2011/12/122011_IJHS_Article_42-1F.pdf.
12. Joseph J. Mangano, "Excess US Deaths after Fukushima Rise from 14,000 to 22,000 after Analysis Updated," Radiation and Public Health Project, February 23, 2012, http://www.radiation.org/reports/JapanUpdateTo22000.pdf.
13. Sandi Doughton, "Universities Come Through in Monitoring for Radiation," *Seattle Times*, April 5, 2011, http://seattletimes.nwsource.com/html/localnews/2014693490 _nukemonitors06m.html.
14. Ibid.
15. Alexander Higgins, "Confirmed: EPA Rigged RADNET Japan Nuclear Radiation Monitoring Equipment to Report Lower Levels of Fukushima Fallout," Alexander Higgins Blog, June 21, 2011, http://blog.alexanderhiggins.com/2011/05/19/confirmed -epa-rigged-radnet-japan-nuclear-radiation-monitoring-equipment-report-levels-nuclear -fallout-22823/.
16. Steven Mufson, "Messages Show Conflict within NRC after Japan's Earthquake and Tsunami," *The Washington Post*, February 7, 2012, http://www.washingtonpost.com /business/economy/messages-show-conflict-within-nrc-after-japan-earthquake-and -tsunami/2012/01/09/gIQA2ll6uQ_story.html.
17. "'Very High Concentrations' of Hot Particles in Pacific NW during April, May," ENNews, June 29, 2011, http://enenews.com/air-seattle-loaded-hot-particles-april-very -high-concentrations-pacific-nw-different-tokyo-audio.
18. EPA, "Frequently Asked Questions," http://www.epa.gov/japan011/japan-faqs.html.

19 Sandi Doughton, "Universities Come Through in Monitoring for Radiation."
20. Jeff McMahon, "Why Does the FDA Tolerate More Radiation than the EPA?" *Forbes*, April 14, 2011.
21. "UCB Food Chain Sampling Results," University of California at Berkeley College of Engineering Air Monitoring Station, April 8, 2011, http://www.nuc.berkeley.edu/UCBAirSampling/FoodChain.
22. Jeff McMahon, "Radiation Detected in Drinking Water in 13 More US Cities, Cesium-137 in Vermont Milk," *Forbes*, April 9, 2011.
23. "Suit to Air Internal EPA Protests on Radiation Exposure Plan," Public Employes for Environmental Responsibility (PEER) press release, October 28, 2009, http://www.peer.org/news/news_id.php?row_id=1273 and "EPA Halts Heightened Monitoring of Fukushima Fallout," PEER press release, May 9, 2011, http://www.peer.org/news/news_id.php?row_id=1480
24. Helen Caldicott, "Unsafe at Any Dose," *New York Times*, April 30, 2011.
25. Tony Muga, "Plumegate: Tales from the Script: Inside the NRC FOIA Documents," IntelHub, March 17, 2012, http://philosophers-stone.co.uk/wordpress/2012/03/plumegate-tales-from-the-script-inside-the-nrc-foia-documents-part-1/.
26. Greenpeace, *Lessons from Fukushima* (Netherlands: Greenpeace International, February 28, 2012).

CHAPTER 17

1. Matthew L. Wald, "New Plan on Nuclear Licenses Is Barred," *New York Times*, November 3, 1990.
2. Karl Grossman, "A Radioactive Extension for Aging Nuclear Plants," *CounterPunch*, April 13, 2009, http://www.counterpunch.org/2009/04/13/a-radioactive-extension-for-aging-nuclear-plants/.
3. Ralph Vartabedian, "NRC Issues Nuclear Safety Regulations," *Los Angeles Times*, July 13, 2011.
4. Jeff Donn (Associated Press), "Safety Rules Loosened for Aging Nuclear Reactors," MSNBC.com, June 20, 2011, http://www.msnbc.msn.com/id/43455859/ns/us_news-environment/t/safety-rules-loosened-aging-nuclear-reactors.
5. Joe Rubin and Serene Fang (Center for Investigative Reporting), "Danger Zone: Aging Nuclear Reactors," episode on *People & Power*, Al Jazeera, February 23, 2012, http://www.aljazeera.com/programmes/peopleandpower/2012/02/2012222134934495461.html.
6. Jeff Donn, "Safety Rules Loosened."
7. Ibid.
8. Ibid.
9. Ibid.
10. Ibid.
11. Serene Fang, "Danger Zone: Aging Nuclear Reactors," Center for Investigative Reporting, March 6, 2012, http://cironline.org/reports/danger-zone-aging-nuclear-reactors.
12. Kevin Kamps (Beyond Nuclear), "Davis-Besse Atomic Reactor: 20 MORE Years of Radioactive Russian Roulette on the Great Lakes Shore?!" November 19, 2010, http://www.beyondnuclear.org/storage/Davis_Besse_Backgrounder.pdf.
13. Ibid.
14. NRC Office of the Inspector General, "Audit of NRC's License Renewal Program," report OIG-07-A-15, September 6, 2007, http://pbadupws.nrc.gov/docs/ML0724/ML072490486.pdf.
15. Donn, "Safety Rules Loosened."

16. Jeff Donn (Associated Press), "Radioactive Tritium Leaks Found at 48 US Nuke Sites," MSNBC.com, June 21, 2011, http://www.msnbc.msn.com/id/43475479/ns/us_news-environment/t/radioactive-tritium-leaks-found-us-nuke-sites/.
17. Ibid.
18. Jeff Donn, "Safety Rules Loosened."
19. Ibid.
20. Reuters, "Update 1–NRC Proposes First Post-Fukushima Safety Orders," February 22, 2012, http://www.reuters.com/article/2012/02/23/utilities-nrc-idUSL2E8DMD2B20120223.
21. David Lochbaum and Edwin Lyman, *US Nuclear Power Safety One Year after Fukushima* (Cambridge, Mass.: Union of Concerned Scientists, March 2011), http://www.ucsusa.org/assets/documents/nuclear_power/fukushima-anniversary-report-3-5-12.pdf.
22. Ibid.
23. Ibid.
24. Ibid.
25. "Spent Nuclear Fuel Storage Fact Sheet," Union of Concerned Scientists, December 2011, http://www.ucsusa.org/assets/documents/nuclear_power/fact-sheet-spent-fuel-storage.pdf.
26. "Nuclear Power Safety and Security Recommendations," UCS, July 14, 2011, http://www.ucsusa.org/nuclear_power/nuclear_power_risk/safety/ucs-nuclear-safety-recommendations.html.
27. John M. Broder and Matthew L. Wald, "Chairman of NRC to Resign Under Fire," *The New York Times*, May 21, 2012, http://www.nytimes.com/2012/05/22/us/gregory-jaczko-to-resign-as-nrc-chairman-after-stormy-tenure.html.
28. Bernie Sanders, "Statement on Nuclear Regulatory Commission Chairman," May 21, 2012, http://www.sanders.senate.gov/newsroom/news/?id=5e39fe6f-e750-464b-bf57-b90dc04249b5.
29. Alison Sr. John, "Nuclear Regulatory Commission Chairman Jaczko Resigns," KPBS, May 21, 2012, http://www.kpbs.org/news/2012/may/21/nrc-chairman-jaczko-resigns/.
30. Ryan Grim, "Gregory Jaczko Resigns: Nuclear Regulatory Commission Chairman Steps Down," *The Huffington Post*, June 22, 2012, http://www.huffingtonpost.com/2012/05/21/gregory-jaczko-resigns-nrc-nuclear-regulatory-commission_n_1531805.html.
31. Jeff Donn, "Feds Quietly Cut Safety Drills at Nuclear Power Plants," Associated Press, May 17, 2012.

CHAPTER 18

1. Kevin Kamps, "Davis-Besse Atomic Reactor."
2. Ibid.
3. *Regulatory Implications of the Davis-Besse Mishap: Hearing before the Subcommittee on Energy Conservation and Power of the Committee on Energy and Commerce*, House of Representatives, Ninety-ninth Congress, first session, October 2, 1985, http://books.google.com/books/about/Regulatory_implications_of_the_Davis_Bes.html.
4. "NRC Needs to More Aggressively and Comprehensively Resolve Issues Related to the Davis-Besse Nuclear Power Plant's Shutdown," General Accounting Office (GAO-04-415), May 17, 2004, http://www.gao.gov/products/GAO-04-415.
5. "Statement of Congressman Dennis J. Kucinich on the GAO Report on the Davis-Besse Nuclear Power Plant," press release from the office of congressman Dennis J. Kucinich, May 18, 2004, http://kucinich.house.gov/News/DocumentSingle.aspx?DocumentID=26006.
6. Kevin Kamps, "Davis-Besse Atomic Reactor."
7. Ibid.

8. "FirstEnergy Report Raises Further Questions Regarding the Safety of Nuclear Plant," press release from the office of congressman Dennis J. Kucinich, March 3, 2012, http://kucinich.house.gov/news/email/show.aspx?ID=OLH3TP7XI3J3BFKTB27GKZVYD4.
9. Kevin Kamps, "Davis-Besse Challenged Again Based on Kucinich Revelations," Beyond Nuclear news release, February 27, 2012, http://www.beyondnuclear.org/storage/Feb%2027%202012%20D%20B%20media%20release.pdf.
10. David Sneed, "Diablo Has Fail-Safe in Case of Emergency," *Tribune* (San Luis Obispo), October 12, 2011, http://www.sanluisobispo.com/2011/10/02/1780388/diablo-has-fail-safe-nuclear.html.
11. David R. Baker, "Diablo Canyon Nuclear Plant 'Near Miss' in Report," *San Francisco Chronicle*, March 18, 2011.
12. Susan Rust, "Safety Concerns for State, World Nuclear Plants," *San Francisco Chronicle*, June 1, 2011.
13. Senators Barbara Boxer and Dianne Feinstein, "Boxer and Feinstein Urge NRC To Do Comprehensive Inspection of San Onofre and Diablo Canyon Nuclear Plants in California," Press release, March 16, 2011, http://boxer.senate.gov/en/press/releases/031611c.cfm.
14. Susanne Rust, "Scientists Concerned about Fault System near Diablo Canyon Plant," California Watch, July 15, 2011, http://californiawatch.org/environment/scientists-concerned-about-fault-system-near-diablo-canyon-plant-11460.
15. Joe Rubin and Serene Fang (Center for Investigative Reporting), "Danger Zone: Aging Nuclear Reactors," episode on *People & Power*, Al Jazeera, February 23, 2012, http://www.aljazeera.com/programmes/peopleandpower/2012/02/2012222134934495461.html.
16. Susanne Rust (California Watch), "PG&E, USGS Disagree on Diablo Canyon Fault Danger," *San Francisco Chronicle*, July 17, 2011, : http://www.sfgate.com/news/article/PG-E-USGS-disagree-on-Diablo-Canyon-fault-danger-2354326.php.
17. Ibid.
18. Joe Rubin and Serene Fang, "Danger Zone."
19. Norman Solomon, "It's Time to Close California's Nuclear Power Plants," Huffington Post (blog), April 28, 2011, http://www.huffingtonpost.com/norman-solomon/its-time-to-close-califor_b_854978.html.
20. Scott Hendrick, "State Restrictions on New Nuclear Power Facility Construction," National Conference of State Legislatures, December 2010, http://www.ncsl.org/issues-research/env-res/states-restrictions-on-new-nuclear-power-facility.aspx.
21. Joseph Mangano, "20,000 Excess Cancer Cases in 15 Years Near Indian Point: New Report Suggests Radiation Exposure May Be One Cause," Radiation and Public Health Project, November 18, 2010, http://www.radiation.org/reading/pubs/101118_IndianPointreport.pdf.
22. "Vision and Mission," Entergy, http://www.entergy-nuclear.com/about_us/vision_and_mission.aspx.
23. "Indian Point 2 Nuclear Reactor Back in Service," Huffington Post, January 19, 2012, http://www.huffingtonpost.com/2012/01/19/indian-point-2-nuclear-reactor_n_1216562.html. "Officials: Leak Contained after Indian Point Reactor Taken Offline," News 12 TV video, January 10, 2012, http://www.news12.com/articleDetail.jsp?articleid=303668&position=1&news_type=news. Viewed online, January 15, 2012. Video has since been removed.
24. Richard Brodsky, "Indian Point, the NRC and 2012," Huffington Post (blog), January 6, 2012, http://www.huffingtonpost.com/richard-brodsky/indian-point-the-nrc-and-_b_1187154.html.
25. "Siding With A.G. Schneiderman, Feds Deny Indian Point's Requests For Over 100 Fire Safety Exemptions," *The Yonkers Tribune*, February 1, 2012, http://yonkerstribune

.typepad.com/yonkers_tribune/2012/02/siding-with-ag-schneiderman-feds-deny-indian-points-requests-for-over-100-fire-safety-exemptions.html.

26. Greg Clary, "NRC Wants Fast Upgrades at Indian Point, Other Nuclear Plants after Japan Study," LoHud.com (*Journal News* of the Lower Hudson Valley, N.Y.), October 20, 2011, http://www.lohud.com/article/20111020/NEWS02/110200373/NRC-wants-fast-upgrades-Indian-Point-other-nuclear-plants-after-Japan-study.
27. Gregory Jaczko, "Chairman Jaczko's Comments on SECY-11-0137: Prioritization of Recommended Actions to be Taken in Response to Fukushima Lessons Learned," NRC Notation Vote, 2011, http://www.nrc.gov/reading-rm/doc-collections/commission/cvr/2011/2011-0137vtr-gbj.pdf.
28. Marvin Fertel, "The High Cost of Closing Indian Point," *The New York Post*, February 17, 2012, http://www.nypost.com/p/news/opinion/opedcolumnists/the_high_cost_of_closing_indian_gPKjUSO1fRSoEsmM3a6ebO.
29. Joe Rubin and Serene Fang, "Danger Zone."
30. Joe Rubin and Serene Fang, "Danger Zone."
31. "Cuomo Orders Safety Review of Indian Point Nuclear Plant," CBS New York, March 16, 2011, http://newyork.cbslocal.com/2011/03/16/cuomo-orders-safety-review-of-indian-point-nuclear-plant/.
32. Greg Clary, "Critics Want Indian Point Replaced, Cite Risks to Region," LoHud.com (*Journal News* of the Lower Hudson Valley), October 18, 2011, http://www.lohud.com/article/20111018/NEWS01/110180327/Critics-want-Indian-Point-replaced-cite-risks-region.
33. Kevin A. Cahill, "Assembly Committees' Preliminary Findings Show Indian Point Can Be Shut Down," New York Assembly District 101, February 1, 2012, http://assembly.state.ny.us/mem/Kevin-A-Cahill/story/46160/.
34. "SONGS to Upgrade Key Components of Unit 2 During Planned Outage," Southern California Edison, January 19, 2012, http://www.songscommunity.com/news.asp.
35. Michael R. Blood (Associated Press), "Small Radiation Amount 'Could Have' Escaped Plant," Yahoo! News, February 1, 2012, http://news.yahoo.com/small-radiation-amount-could-escaped-plant-192144016.html.
36. "Unprecedented Damage Found in over 800 Tubes in San Onofre Reactor 2 Alone," Enformable, February 2, 2012, http://enformable.com/2012/02/unprecedented-damage-found-in-over-800-tubes-in-san-onofre-reactor-2-alone/.
37. Ibid.
38. "NRC—San Onofre Damage Seen in Multiple Nuclear Reactors across United States," Enformable, February 4, 2012, http://enformable.com/2012/02/nrc-san-onofre-corrosion-damge-seen-in-multiple-nuclear-reactors-across-united-states/.
39. Michael R. Blood (Associated Press), "NRC Chief Promises Tough Probe at Cal Nuke Plant," Yahoo! News, April 6, 2012, http://news.yahoo.com/nrc-chief-promises-tough-probe-cal-nuke-plant-232200325.html.
40. "New Report Reveals Scale of Edison Steam Generator Failures at San Onofre Nuclear Plant," Friends of the Earth, May 15, 2012, http://www.foe.org/news/news-releases/2012-05-new-report-reveals-scale-of-edison-steam-generator; "San Onofre's Steam Generator Failures Could Have Been Prevented," Friends of the Earth, May 14, 2012, http://libcloud.s3.amazonaws.com/93/01/3/1442/SO_Steam_Generator_Analysis_May.pdf.
41. Paul Sisson, "Worker at Nuclear Plant Fell into Reactor Pool Last Week," *North County Times*, February 3, 2012, http://www.nctimes.com/news/local/oceanside/san-onofre-worker-at-nuclear-plant-fell-into-reactor-pool/article_519d459c-19f4-55ed-87c6-8131872e6760.html.

42. Tom Reopelle, "NRC: San Onofre Worker Returned To Work After Fall into Nuclear Reactor Pool," KNX 1070, February 3, 2012, http://losangeles.cbslocal.com/2012/02/03/san-onofre-worker-returned-to-work-after-falling-into-nuclear-reactor-pool/.
43. Lucas W. Hixon, "San Onofre Reactor 2 Operated for Several Months with Degraded Nuclear Fuel Cladding," Enformable, February 17, 2012, http://enformable.com/2012/02/san-onofre-reactor-2-operated-for-several-months-with-degraded-nuclear-fuel-cladding/.
44. "An Assessment of California's Nuclear Power Plants: AB 1632 Report," California Energy Commission, November 2008, http://www.energy.ca.gov/2008publications/CEC-200-2008-002/CEC-200-2008-002-CMF.pdf.
45. Senator Barbara Boxer and Diane Feinstein, "Boxer and Feinstein Urge NRC To Do Comprehensive Inspection."
46. Barbara Diamond, "Laguna Takes Stand on San Onofre," *Coastline Pilot*, February 16, 2012, http://articles.coastlinepilot.com/2012-02-16/news/tn-cpt-0217-sanonofre-20120210_1_nuclear-plants-low-safety-significance-nuclear-power.
47. Statement of Senator Barbara Boxer at a joint hearing of the US Senate Committee on Environment & Public Works and the Subcommittee on Clean Air and Nuclear Safety, "Review of the NRC's Near-Term Task Force Recommendations for Enhancing Reactor Safety in the 21st Century," December 15, 2011, press release, http://epw.senate.gov/public/index.cfm?FuseAction=Majority.PressReleases&ContentRecord_id=426c2a48-802a-23ad-432a-c3b0a1258452.
48. "NRC Relicenses US Reactor that Is Same Design as Fukushima," Beyond Nuclear, March 22, 2011, http://www.beyondnuclear.org/home/2011/3/21/nrc-relicenses-us-reactor-that-is-same-design-as-fukushima.html.
49. "Safety of Vermont Nuclear Plant Cables Questioned," *Burlington Free Press*, April 4, 2011.
50. Vermont Attorney General's Criminal Investigation Report, July 6, 2011, http://www.atg.state.vt.us/assets/files/Office_of_the_Attorney_Generals_Criminal_Investigation_Report_on_Vermont_Yankee.pdf.
51. "Vermont Appeals Court Ruling on Yankee Nuclear Plant," *Burlington Free Press*, February 18, 2012.
52. "Midlife Crisis: Approaching 40, Vermont Yankee's Ability to Keep Operating Debated," *Burlington Free Press*, March 9, 2012.
53. "Vermont Appeals Court Ruling on Yankee Nuclear Plant," *Burlington Free Press*, February 18, 2012.
54. "Vermont Yankee Says No to More Tests in Well for Radioactive Tritium," *Burlington Free Press*, January 27, 2012.
55. Alan Panebaker, "Entergy to PSB: Don't Shut Us Down," Vermont Digger, March 9, 2012, http://vtdigger.org/2012/03/09/entergy-to-psb-dont-shut-us-down/.
56. Joe Rubin and Serene Fang, "Danger Zone."

CHAPTER 19

1. John Murawski, "Unqualified Progress Energy Workers Caused Fluke Mishap at Nuclear Plant," NewsObserver.com (of the *News & Observer*, Raleigh, N.C.), January 26, 2012, http://www.newsobserver.com/2012/01/26/1809319/unqualified-progress-energy-workers.html.
2. David Lochbaum, *The NRC and Nuclear Power Plant Safety in 2011: Living on Borrowed Time* (Cambridge, Mass.: Union of Concerned Scientists, March 2012), http://www.ucsusa.org/assets/documents/nuclear_power/nrc-nuclear-safety-2011-full-report.pdf.

3. Jils J. Diaz, "The Very Best-Laid Plans (the NRC's Defense-in-Depth Philosophy)," NRC, June 3, 2004, http://www.nrc.gov/reading-rm/doc-collections/commission/speeches/2004/s-04-009.html.
4. "Defense-in-Depth," NRC, Updated March 29, 2012, http://www.nrc.gov/reading-rm/basic-ref/glossary/defense-in-depth.html.
5. David Lochbaum, "The NRC and Nuclear Power Plant Safety."
6. John Sullivan and Ariel Wittenberg, "Nuclear Plants and Disasters: NRC Inspection Results," ProPublica, June 29, 2011, http://www.propublica.org/article/nuclear-plants-and-disasters-nrc-inspection-results.
7. "Mitigating Strategies," FireDirect, May 12, 2011, http://www.firedirect.net/index.php/2011/05/mitigating-strategies/.

PART 3 INTRODUCTION

1. Christian Schwägerl, "Germany's Unlikely Champion of a Radical Green Energy Path," *Yale Environment 360*, May 9, 2011, http://e360.yale.edu/feature/germanys_unlikely_champion_of_a_radical_green_energy_path/2401/.
2. Stephen Lacey, "German Solar Output Increases by 60% in 2011," Climate Progress (blog), January 1, 2012, http://thinkprogress.org/climate/2012/01/01/395922/german-solar-output-increases-2011/.
3. Paul Gipe, "51 Percent of German Renewables Now Owned by Citizens," Wind-Works, January 5, 2012, http://www.wind-works.org/coopwind/CitizenPowerConference tobeheldinHistoricChamber.html.
4. Paul Gipe, personal email to author, June 23, 2012.
5. "German Village Produces 321% More Energy Than It Needs!" ZeitNews, August 21, 2011, http://www.zeitnews.org/energy/german-village-produces-321-more-energy-than-it-needs.html.
6. Paul Gipe, "51 Percent of German Renewables Now Owned by Citizens."
7. Christian Schwägerl, "Germany's Unlikey Champion of a Radical Green Energy Path."
8. Paul Gipe, "Saudi Arabia Launches Massive Renewable Program with Hybrid FITs," Wind-Works, May 14, 2012, http://www.wind-works.org/FeedLaws/Saudi%20Arabia/SaudiArabiaLaunchesMassiveRenewableProgramwithHybridFITs.html.

CHAPTER 20

1. Millennium Ecosystem Assessment, "Overview of the Millennium Ecosystem Assessment" (http://www.maweb.org/en/About.aspx) and "Responses Assessment" (http://www.maweb.org/en/Responses.aspx).
2. Richard Heinberg, *Searching for a Miracle: "Net Energy" Limits & the Fate of Industrial Society*, False Solutions Series no. 4 (International Forum on Globalization and the Post Carbon Institute, September 2009), http://www.postcarbon.org/new-site-files/Reports/Searching_for_a_Miracle_web10nov09.pdf.
3. "The End of Nuclear," The Worldwatch Institute, December 2011, http://www.worldwatch.org/node/8069. For the full report, see: Mycle Schneider, Antony Froggatt, Steve Thomas, *Nuclear Power in a Post-Fukushima World: The World Nuclear Industry Status Report 2010-2011*, The Worldwatch Institute, 2011, http://www.worldwatch.org/system/files/pdf/WorldNuclearIndustryStatuesReport2011_FINAL.pdf.
4. David Brower, *Let the Mountains Talk, Let the Rivers Run* (San Francisco: HarperCollinsWest, 1995), introduction, 2.
5. United Nations Development Programme, "How Many Planets?" table 1.2 of the *Human Development Report 2007/2008* (New York: Palgrave Macmillan, 2007), http://hdr.undp.org/en/statistics/data/climatechange/planets/.

CHAPTER 21

1. Rex Weyler, "Net Energy: The Cost of Living," blog entry, March 18, 2010, http://rexweyler.com/2010/03/18/net-energy-the-cost-of-living.
2. Nick Hodge, "Profiting from the Fifth Fuel," Green Chip Stocks, January 19, 2010. http://www.greenchipstocks.com/articles/energy-efficiency-companies/723.
3. John Vidal, "Nuclear Plants Bloom," *Guardian*, August 14, 2004.
4. Christopher Shea, "Picturing a Ton of Carbon," *Boston Globe*, December 20, 2009.
5. "Energy Conservation," *National Geographic*, March 2009, 80–81.
6. Michael Grunwald, "America's Untapped Energy Resource: Boosting Efficiency," *Time* magazine, December 31, 2008, http://www.time.com/time/magazine/article/0,9171,1869224,00.html.
7. Susan Kraemer, "Nearly 200 Gigawatts of US Energy Is Wasted," Clean Technica, August 9, 2010, http://s.tt/12uSN.
8. David Chandler, "Manufacturing Inefficiency," MIT News Office, March 17, 2009, http://web.mit.edu/newsoffice/2009/energy-manufacturing-0317.html. See also, Matthew T. Gutowski, Jeffrey B. Dahmus, Alissa J. Jones, Alexandre Thiriez and Dusan Sekulic, "Thermodynamic Analysis of Resources Used in Manufacturing Processes," *Environmental Science & Technology* January 29, 2009, 43, pp. 1584-90. http:// meche.mit.edu/documents/gutowski_CV.pdf
9. International Institute for Sustainable Development, "Sustainable Business Practices: IISD's Checklist," 2010, http://www.iisd.org/business/tools/principles_sbp.aspx.
10. "Ban Urges Leaders at Davos to Forge 'Green New Deal' to Fight World Recession," UN News Centre, January 29, 2009, http://www.un.org/apps/news/story.asp?NewsID=29712&Cr=Ban&Cr1=Climate+change
11. Amory Lovins, "More Profit with Less Carbon," *Scientific American*, September 2005, 74–82.
12. "The Cheapest, Cleanest Energy There Is," *New Energy News*, September 25, 2009, http://newenergynews.blogspot.com/2009/09/cheapest-cleanest-energy-there-is.html.
13. Harald N. Rostvik, *The Sunshine Revolution*, (Staavanger, Norway: SUN-LAB Publishers, 1992).
14. Steven Nadel, "Yes, US Oil and Gas Production Is Increasing, but Energy Efficiency Is Still the Number One Resource," American Council for an Energy-Efficient Economy, June 19, 2012, http://www.aceee.org/blog/2012/06/yes-us-oil-and-gas-production-increas.
15. "Energy Use Flows Downhill," Lawrence Livermore National Laboratory Report, August 30-September 3, 2010, https://www.llnl.gov/news/lab_report/2010/LL_Report_9-3-10.pdf.
16. "US Carbon Dioxide Emissions in 2009: A Retrospective Review," US Energy Information Administration, May 5, 2010. http://www.eia.doe.gov/oiaf/environment/emissions/carbon/index.html.
17. "Recession, Cleaner Fuels Drive Down US CO2 Emissions," ClimateBiz, May 7, 2010, http://www.greenbiz.com/news/2010/05/07/recession-cleaner-fuels-drive-down-co2-emissions-us.
18. "The Problem of Leaking Electricity-Energy Vampires," California Energy Center, California Energy Commission, http://www.consumerenergycenter.org/home/appliances/small_appl.html.
19. Ibid.
20. Amory Lovins, "The Negawatt Revolution," *Across the Board* 27, no. 9 (September 9, 1990): 18–23.
21. "Energy Facts," Cornell University Cooperative Extension factsheet citing DOE and Maryland Energy Administration, n.d.

22. Arjun Makhijani, "Carbon-Free and Nuclear-Free: A Roadmap for US Energy Policy," *Science for Democracy*, Vol. 15, No. 1, p. 1, August 2007, http://www.helencaldicott.com/roadmap_summary.pdf.
23. Gar Smith, *A Harvest of Heat: Agribusiness and Climate Change* (Agribusiness Action Initiatives North America, spring 2010).
24. Greenpeace International and the European Renewable Energy Council (EREC), *Energy [R]evolution*.
25. Amory Lovins, interview by Amy Goodman of *Democracy Now!*
26. Genevieve Head-Gordon, "City Adopts Final Version of Climate Action Plan," *Daily Californian*, June 3, 2009, http://archive.dailycal.org/article/105828/city_adopts_final_version_of_climate_action_plan.

CHAPTER 22

1. Amory Lovins, interview by Amy Goodman of *Democracy Now!*
2. Amory Lovins, "Stewart Brand's Nuclear Enthusiasm Falls Short on Facts and Logic," review of *Whole Earth Discipline*, by Stewart Brand, *Grist*, October 14, 2009, http://grist.org/article/2009-10-13-stewart-brands-nuclear-enthusiasm-falls-short-on-facts-and-logic/.
3. IPCC, *IPCC Special Report on Renewable Energy Sources and Climate Change Mitigation*, prepared by Working Group III of the Intergovernmental Panel on Climate Change (Cambridge, U.K., and New York: Cambridge University Press, 2011).
4. "New Renewables to Power 40 Percent of Global Electricity Demand by 2050," *Science Daily*, March 11, 2009, http://www.sciencedaily.com/releases/2009/03/090311103609.htm. See the full report at: Katherine Richardson et al., *Synthesis Report*, from the International Alliance of Research Universities International Scientific Congress "Climate Change: Global Risks, Challenges & Decisions," University of Copenhagen, March 10–12, 2009, http://climatecongress.ku.dk/pdf/synthesisreport/.
5. Mark Z. Jacobsen and Mark A. Delucchi, "A Path to Sustainable Energy by 2030," *Scientific American*, November 2009, 58–65.
6. Trevor Curwin, "Momentum for Clean Coal Conversion Burning Out," CNBC, February 8, 2010, http://www.cnbc.com/id/35295212/Momentum_For_Clean_Coal_Conversion_Burning_Out.
7. M. A. Palmer, et al., "Mountaintop Mining Consequences," *Science*, January 8, 2010, pp. 148-149, http://www.sciencemag.org/content/327/5962/148.summary.
8. Joe Romm, "Is 450 ppm Possible? Part 5: Old Coal's Out, Can't Wait for New Nukes, So What Do We Do Now?" Climate Progress (blog), May 8, 2008, http://climateprogress.org/2008/05/08/is-450-ppm-possible-part-5-old-coals-out-cant-wait-for-new-nukes-so-what-do-we-do-now/.
9. John O. Blackburn and Sam Cunningham, "Solar and Nuclear Costs—The Historic Crossover: Solar Energy Is Now the Better Buy," NC WARN, July 2010, http://www.ncwarn.org/wp-content/uploads/2010/07/NCW-SolarReport_final1.pdf.
10. Ibid.
11. Ken Bossong, "Renewable Energy Provided 11% of Domestic Energy Production in 2010," *Renewable Energy World*, April 6, 2011, http://www.renewableenergyworld.com/rea/news/article/2011/04/renewable-energy-provided-11-of-domestic-energy-production-in-2010.
12. "Cars Produced in the World," Worldometers, http://www.worldometers.info/cars/.
13. Mark Z. Jacobsen and Mark A. Delucchi, "A Path to Sustainable Energy by 2030."
14. Lovins, interview, "Expanding Nuclear Power."

15. "GWEC: Wind Energy Powers Ahead Despite Economic Turmoil," German Wind Energy Association, February 10, 2012, http://www.wind-energie.de/en/infocenter/articles/gwec-wind-energy-powers-ahead-despite-economic-turmoi.
16. "Large-Scale Offshore Wind Power in the United States," NREL, October 7, 2010, http://www.nrel.gov/wind/pdfs/40745.pdf.
17. "Global Energy Statistics 2011," Global Wind Energy Council, February 2012, http://www.gwec.net/fileadmin/images/News/Press/GWEC_-_Global_Wind_Statistics_2011.pdf.
18. Energy Information Administration (EIA), *Electric Power Annual 2008* (Washington, D.C.: EIA, January 2010), http://www.eia.gov/electricity/annual/archive/03482008.pdf.
19. Dallas Kachan, "Wind Beating Down Nuclear and Coal in Europe," Clean Tech News, February 4, 2010. Posted by Nuclear Free Virginia, February 5, 2010, http://nuclearfreeva.blogspot.com/2010/02/wind-beating-down-nuclear-and-coal-in.html.
20. "Revenue for Solar, Wind and Biofuels Reached US$139bn in 2009," *Renewable Energy Focus*, March 17, 2010, http://www.renewableenergyfocus.com/view/8081/revenue-for-solar-wind-and-biofuels-reached-us139bn-in-2009.
21. Zachary Shahan, "2009 Global Wind Power Report (by Country)," CleanTechnica (blog), February 5, 2010, http://www.cleantechnica.com/2010/02/05/2009-global-wind-power-report-by-country-wind-up-despite-economy-and-china-us-up-big/
22. Duane Sharp, "US Wind Installations Drop in Second Quarter of 2010," Suite101, August 7, 2010, http://suite101.com/article/us-wind-installations-drop-in-second-quarter-of-2010-a270608.
23. James Burgess, "Global Wind Power Capacity Up 21% in 2011," Oil Price.com, February 9, 2012, http://oilprice.com/Alternative-Energy/Wind-Power/Global-Wind-Power-Capacity-up-21-in-2011.html.
24. Christian Kjaer, "Blowing Away Nuclear Power," March 26, 2012, http://www.europeanvoice.com/page/3323.aspx?LG=1&ArtID=73977&SecName=Energy&SectionID=15.
25. "AWEA First Quarter 2012 Market Report," American Wind Energy Association, http://www.awea.org/learnabout/industry_stats/index.cfm.
26. Zachary Shahan, "2009 Global Wind Power Report."
27. "Summary of Wind Turbine Accident Data to 31 March 2012," Caithness Windfarm, http://www.caithnesswindfarms.co.uk/accidents.pdf.
28. Cutler Cleveland and Ida Kubisewski, "Energy Return on Investment (EROI) for Wind Energy," Oil Drum (blog), http://www.theoildrum.com/node/1863.
29. David Hochchild, "Protecting Net Metering: The Civil Rights Movement for Solar Energy," RenewableEnergyWorld.com, April 27, 2009, http://www.renewableenergyworld.com/rea/news/article/2009/04/protecting-net-metering-the-civil-rights-movement-for-solar-energy.
30. National Renewable Energy Laboratory, *PV FAQs*, no. DOE/GO-102004-1835 (Washington, D.C.: US Department of Energy Office of Energy Efficiency and Renewable Energy, 2004), http://www.nrel.gov/docs/fy04osti/35097.pdf.
31. Bill Powers, *Bay Area Smart Energy 2020* (Pacific Environment, March 2012), http://pacificenvironment.org/downloads/BASE2020_Full_Report.pdf.
32. "Solar Water Heaters, Photovoltaic Panels, Geothermal Heat Pumps," Sustainable Choices, a website of the Stanford University School of Earth Sciences, http://sustainablechoices.stanford.edu/actions/in_the_home/solar.html.
33. Renee Cho, "Solar Water Heaters Sprouting on Rooftops Worldwide," Inside Climate News, March 14, 2010, http://insideclimatenews.org/news/20100314/solar-water-heaters-sprouting-rooftops-worldwide.

34. Ambrose Evans-Pritchard, "Monday View: Cheap Solar Power Poised to Undercut Oil and Gas by Half," *Telegraph*, February 19, 2007, http://www.telegraph.co.uk/finance/comment/ambroseevans_pritchard/2804574/Monday-view-Cheap-solar-power-poised-to-undercut-oil-and-gas-by-half.html.
35. "DuPont Apollo Ltd. Opens Thin-Film Photovoltaic Production Facility," DuPont news release, November 17, 2009, http://www2.dupont.com/Photovoltaics/emus/news_events/article20091117.html.
36. J. Hunt, "Nanosolar's Breakthrough: Solar Now Cheaper than Coal," Celsias, November 23, 2007, http://www.celsias.com/article/nanosolars-breakthrough-technology-solar-now-cheap.
37. "LED Holiday Lighting," US Department of Energy Office of Energy Efficiency and Renewable Energy, updated June 8, 2012, http://www.energysavers.gov/seasonal/led_lighting.html.
38. Brian Merchant, "Meet the Super-Efficient LED Bulb the Tea Party Wants You to Hate," March 12, 2012, Treehuger.com-energy-efficiency/introducing-super-efficient-led-light-bulb-right-wants-you-to-hate.html.
39. Andrew Tarantola, "Scientists Create 230-Percent Efficient LED Bulbs," Gizmodo, March 5, 2012, http://gizmodo.com/5890719/scientists-create-230+percent-efficient-led-bulbs.
40. Hans De Keulenaer, "The World's Largest Tidal Power Plant is Almost 40 Years Young," Global Community for Sustainable Energy Professionals, May 18, 2006, http://www.leonardo-energy.org/worlds-largest-tidal-power-plant-almost-40-years-young.
41. "Tidal Power," Alternative Energy News, http://www.alternative-energy-news.info/technology/hydro/tidal-power/.
42. ""Aqucadoura Generating Power for 1,500 Homes," Alternative Energy News, http://www.alternative-energy-news.info/agucadoura-generating-power-1500-homes/.
43. International Hydropower Association et al., *Hydropower and the World's Energy Future* (November 2000), http://www.ieahydro.org/reports/Hydrofut.pdf.
44. "AWEA First Quarter 2012 Market Report," American Wind Energy Association, http://www.awea.org/learnabout/industry_stats/index.cfm.
45. "Hydro Power," Orkustofnun, Iceland's National Energy Authority, (accessed June 26, 2012), http://www.nea.is/hydro.
46. Duncan Graham-Rowe, "Hydroelectric Power's Dirty Secret Revealed," *New Scientist*, February 24, 2005, http://www.newscientist.com/article/dn7046-hydroelectric-powers-dirty-secret-revealed.html.
47. Cutler Cleveland, "Three Gorges Dam and the Earth's Rotation," Energy Watch, June 18, 2010, http://www.theenergywatch.com/ 2010/06/18/three-gorges-dam-and-the-earths-rotation.
48. Amanda Chiu, "One Twelfth of Global Electricity Comes from Combined Heat and Power Systems," Worldwatch Institute, November 4, 2008, http://www.worldwatch.org/node/5924.
49. Elizabeth Svoboda, "The Truth about Energy," *Popular Mechanics*, July 2010, 70–78.
50. "Geothermal Energy Overview 2010," Our Energy, June 10, 2010, http://www.our-energy.com/news/geothermal_energy_overview_2010.html.
51. Massachusetts Institute of Technology, *The Future of Geothermal Energy* (Idaho Falls: Idaho National Laboratory, 2006), http://geothermal.inel.gov/publications/future_of_geothermal_energy.pdf.
52. Patrick Hughes, *Geothermal (Ground-Source) Heat Pumps: Market Status, Barriers to Adoption and Actions to Overcome Barriers*, no. ORNL/TM-2008/232 (Oak Ridge National Laboratory, December 2008), http://www1.eere.energy.gov/geothermal/pdfs/ornl_ghp_study.pdf.

53. Drake Bennett, "Nanotube-Filled Fabric Turns Heat into Power," *San Francisco Chronicle*, March 25, 2012.
54. European Renewable Energy Council and Greenpeace, *Working for the Climate: Renewable Energy and the Green Job [R]evolution* (Greenpeace International, August 2009), http://www.greenpeace.org/international/Global/international/planet-2/report/2009/9/working-for-the-climate.pdf.

CHAPTER 23

1. Ayesha Rascoe, "US Seeks to Spur Renewable Energy on Public Lands," Reuters, March 12, 2009, http://www.reuters.com/article/2009/03/12/us-usa-renewables-idUSTRE52A64Q20090312.
2. "Walmart Announces Goal to Eliminate 20 Million Metric Tons of Greenhouse Gas Emissions from Global Supply Chain," Walmart press release, February 25, 2010, http://www.walmartstores.com/pressroom/news/9668.aspx.
3. Gina Keating, "Disney Sets Plan to Cut Carbon Emissions to Zero," Reuters, March 10, 2009, http://www.reuters.com/article/2009/03/10/us-disney-idUSTRE52858G20090310.
4. Michael Mechanic, "Power Q&A: Amory Lovins," *Mother Jones*, May/June 2008, 17, http://motherjones.com/politics/2008/05/power-qa-amory-lovins.
5. Nicholas Stern, *Key Elements of a Global Deal on Climate Change* (London: London School of Economics and Political Science, 2008), 10.
6. *Impacts of Comprehensive Climate and Energy Policy Options on the US Economy* (Washington, D.C.: Johns Hopkins University Center for Climate Strategies, July 2010), See also: "Could New Climate Policy Create Jobs?" University of Southern California, Price School Newsletter, September 2010, http://www.usc.edu/schools/price/newsletter/september_2010/climate_policy_report.html
7. Van Jones, "Green-Collar Jobs: Energy Bill Includes Christmas Present for Nation's Job Seekers," Huffington Post (blog), December 21, 2007, http://www.huffingtonpost.com/van-jones/greencollar-jobs-energy-b_b_77934.html.
8. "Texas Meets Renewable Goals," *Power Engineering*, April 5, 2010, http://www.power-eng.com/articles/2010/04/texas-meets-renewable-goals.html.
9. "Renewable Energy: Pass SB 722," editorial, *Los Angeles Times*, June 26, 2010, http://articles.latimes.com/2010/jun/26/opinion/la-ed-renewable-20100626.
10. Dan Jones, "Europe: 100% Renewable by 2050?" *Power & Energy*, April 6, 2010, http://www.ngpowereu.com/news/europe-north-africa-renewable-energy.
11. "Striking a Blow for Wind Power," CNN Tech, March 3, 2010, http://www.cnn.com/2010/TECH/03/01/wind.power.denmark/index.html.
12. "Sweden Sets Its Objective to Become the First Oil-Free Country," Renewable Power News, April 7, 2010, http://www.renewablepowernews.com/archives/1223.
13. Karan Capoor and Philippe Ambrosi, *State and Trends of the Carbon Market 2007* (Washington, D.C.: World Bank, May 2007).
14. Phillip Inman, "Three Britons Charged over Carbon-trading 'Carousel Fraud,'" *Guardian*, January 11, 2010, http://www.guardian.co.uk/business/2010/jan/11/eu-carbon-trading-carousel-fraud.
15. Kevin Baumert, "Carbon Taxes vs. Emissions Trading," Global Policy Forum, April 17, 1998, http://www.globalpolicy.org/component/content/article/216-global-taxes/45883-carbon-taxes-vs-emissions-trading.html.
16. Brendan I. Koerner, "Power to the People: 7 Ways to Fix the Grid, Now," *Wired Magazine*, April 2009, 76, http://www.wired.com/science/discoveries/magazine/17-04/gp_intro.

17. Mariah Blake, "The Rooftop Revolution," *Washington Monthly*, March/April 2009, http://www.washingtonmonthly.com/features/2009/0903.blake.html.
18. Tyler Hamilton, "Everbrite Solar to Build 150MW Thin-Film Manufacturing Plant in Ontario," Clean Break, http://www.cleanbreak.ca/2009/03/26/everbrite-solar-to-build-150mw-thin-film-manufacturing-plant-in-ontario/.
19. John Farrell, "Gainesville, Florida, Uses CLEAN Contracts to Become a World Leader in Solar," Energy Self-Reliant States, Institute for Local Self-Reliance, December 1, 2011, http://energyselfreliantstates.org/content/gainesville-florida-uses-clean-contracts-aka-feed-tariffs-become-world-leader-solar.
20. More information about FeeBates is available at http://www.feebate.net.
21. Julia Levitt, "WorldChanging Interview: Amory Lovins," WorldChanging, March 23, 2009, http://www.worldchanging.com/archives/009560.html.
22. Daniel Snyder, "Pros and Cons of the Smart Grid—A National Power Grid for the US," Knoji, http://electrical-systems-lighting.knoji.com/pros-and-cons-of-the-smart-grid/.
23. Alex Spilius, "China and Russia Hack into US Power Grid," *The Telegraph*, April 8, 2009, http://www.telegraph.co.uk/news/worldnews/asia/china/5126584/China-and-Russia-hack-into-US-power-grid.html.
24. *A Systems View of the Modern Grid*, conducted by the National Energy Technology Laboratory for the US Department of Energy Office of Electricity Delivery and Energy Reliability (January 2007), http://www.netl.doe.gov/smartgrid/referenceshelf/whitepapers/ASystemsViewoftheModernGrid_Final_v2_0.pdf.
25. David R. Baker, "Jail Gets Its Own Grid," *San Francisco Chronicle*, March 16, 2012.
26. Jeanne Roberts, "NREL Gets Zero-Energy Office Building," EnergyBoom, July 18, 2010, http://www.energyboom.com/solar/nrel-gets-zero-energy-office-building.
27. "Top 10 Green US Cities," *Mother Earth News*, June 24, 2012, http://www.mnn.com/lifestyle/eco-tourism/photos/top-10-green-us-cities/the-greenness-of-a-city.
28. National Association of Realtors and Public Option Strategies, "Key Findings from a National Survey of 1,000 Adults Conducted October 5, 7, 9–10, 2007," October 2007, http://www.smartgrowthamerica.org/narsgareport2007/narslidesgraphics.pdf.
29. Meea Kang, "Creating More Livable Communities," *San Francisco Chronicle*, July 28, 2010, http://www.sfgate.com/cgi-bin/article.cgi?f=/c/a/2010/07/28/EDV91EKL1U.DTL.
30. Kalle Huebner, "2,000 Watt Society," Our World 2.0, United Nations University, June 2, 2009, http://ourworld.unu.edu/en/2000-watt-society/.
31. Jeffrey Ball, "The Homely Costs of Energy Conservation," *Wall Street Journal*, August 7, 2009, http://online.wsj.com/article/SB124959929532112633.html.
32. For more information on community aggregation, see One Block at a Time, http://1bog.org.
33. Students of Kinsale Further Education College, *Kinsale 2021:An Energy Descent Action Plan*, ed. Rob Hopkins (Kinsale, Ireland: Kinsale Further Education College, 2005), http://transitionculture.org/wp-content/uploads/kinsaleenergydescentactionplan.pdf.
34. More information about The Transition Network is available at www.transitionnetwork.org.
35. More information about the Post Carbon Institute is available at http://www.postcarbon.org/.
36. A Community Resilience Toolkit can be downloaded at http://www.baylocalize.org/toolkit.
37. Chris Meehan, "DOE Secures Funds for One of the World's Largest Solar Plants," Clean Energy Authority April 25, 2011, http://www.cleanenergyauthority.com/solar-energy-news/doe-provides-loan-guarantee-to-blythe-plant-042511/.
38. Gar Smith, "Privatizing Sunbeams," *Earth Island Journal*, Autumn 2011, http://www.earthisland.org/journal/index.php/eij/article/privatizing_sunbeams/.
39. David R. Baker, "Solar Startup BrightSource Calls Off IPO," *San Francisco Chronicle*, April 13, 2012.

40. David Myers, "Solar Belongs on Rooftops," *Earth Island Journal*, Autumn 2010, http://www.earthisland.org/journal/index.php/eij/article/rooftops/.
36. Ben Sills, Natalie Obiko Pearson, and Stefan Nicola, "Power Shifts as Off-grid Options Spread," *San Francisco Chronicle*, April 13, 2012.
41. "California Passes 1 Gigawatt Rooftop Solar Milestone: Study," US Department of Energy, Energy Efficiency and Renewable Energy News, November 16, 2011, http://apps1.eere.energy.gov/news/news_detail.cfm/news_id=17887.
42. Paul Gipe, "California's Solar Program Costs More than German Feed-in Tariffs," Wind-Works, July 6, 2011, http://www.wind-works.org/FeedLaws/USA/CaliforniasSolarProgramCostsMoreThanGermanFeed-inTariffs.html.
43. Ibid.
44. Ben Stills, Natalie Obiko Pearson, and Stefan Nicola, "Power Shifts."
45. Jeremy Rifkin, *The Third Industrial Revolution: How Lateral Power Is Transforming Energy, the Economy, and the World*, (Basingstoke: Palgrave Macmillan. 2011).
46. John Farrell, *Democratizing the Electricity System: A Vision for the 21st-Century Grid* (Minneapolis, Minn.: New Rules Project [now the Institute for Local Self-Reliance], June 2011).
47. Amory Lovins, "My Response to WSJ's, 'The Problem with Going Green,'" Rocky Mountain Institute, February 7, 2012, http://blog.rmi.org/my_response_to_wsjs_problem_with_going_green.
48. Sills, Pearson, and Nicola, "Power Shifts."

CHAPTER 24

1. Dan Chiras, "US Department of Energy Quiety Raises Insulation Standards," *Mother Earth News*, February 1, 2010, http://www.motherearthnews.com/Energy-Matters/US-Department-Energy-Quietly-Raises-Insulation-Standards.aspx, See also: "The Energy Savers: Tips on Saving Energy," DOE, December 2011, https://www1.eere.energy.gov/consumer/tips/insulation.html.
2. Center for the Advancement of the Steady State Economy, http://steadystate.org.
3. European Renewable Energy Council and Greenpeace, *Working for the Climate: Renewable Energy and the Green Job [R]evolution* (Greenpeace International, August 2009), http://www.greenpeace.org/international/Global/international/planet-2/report/2009/9/working-for-the-climate.pdf.
4. "The Uppsala Protocol," Uppsala Hydrocarbon Depletion Study Group, Uppsala University, Sweden, http://www.oilcrisis.com/Uppsala/.
5. Richard Heinberg, *Powerdown: Options and Actions for a Post-Carbon World* (Gabriola Island, B.C.: New Society Publishers, 2004).
6. Richard Heinberg, "Searching for a Miracle."
7. Richard Heinberg, *The Oil Depletion Protocol: A Plan to Avert Oil Wars, Terrorism, and Economic Collapse* (Gabriola Island, BC: New Society Publishers, September 1, 2006).
8. Richard Register, *Ecocities: Building Cities in Balance with Nature* (Berkeley, Calif.: Berkeley Hills Books, 2001).
9. "Top 15 Policies for Achieving a Steady State Economy," Center for the Advancement of Steady State Economics, http://steadystate.org/discover/policies/.
10. For more information about permaculture, visit the website of the Permaculture Institute at http://www.permaculture.org.
11. Ted Halstead, "GNP vs. GPI: A Better Way to Measure Economy," Third World Traveler, http://www.thirdworldtraveler.com/Reforming_ System/GNP_GPI.html.

12. Vaclav Smil, "Energy at the Crossroads," background note for a presentation at the Global Science Forum Conference on Scientific Challenges for Energy Research, May 17–18, 2006, http://www.oecd.org/dataoecd/52/25/36760950.pdf.
13. Richard Heinberg, *Searching for a Miracle: Net Energy Limits and the Fate of Industrial Society*, San Francisco: The International Forum on Globalization and the Post Carbon Institute, September 2009, page 69, http://www.ifg.org/pdf/Searching%20for%20a%20Miracle_web10nov09.pdf.

CONCLUSION

1. Yoichi Funabashi and Kay Kitazawa, "Fukushima in Review: A Complex Disaster, a Disastrous Result," *Bulletin of the Atomic Scientists* 68, no. 2 (March/April 2012), 9–21, http://bos.sagepub.com/content/68/2/9.
2. Funabashi and Kitazawa, op cit., p. 6. See also: "The Independent Investigation Commission on the Fukushima Nuclear Accident," February 28, 2012, http://rebuildjpn.org/en/project.
3. Dahr Jamail, "Fukushima Radiation Alarms Doctors."

Index

9/11 Commission, 158
2,000-watt societies, 214–15

Aboriginal peoples, 67
absorbed dose, defined, xxviii
accidents, 44–48, 130–36, 168–76. *See also* meltdowns; radiation; safety issues
Advanced Fuel Cycle Initiative, 15
advanced pressurized water reactor, defined, xxvii
Aeon, 124
aging facilities, 24–25, 49–55, 140–41, 145
air pollution, 57–61, 133–34
Akihito, Emperor, 128
Alexander, Donald, 21
Alexander, Gil, 161
Alexander, Lamar, 12
Alliance for Nuclear Responsibility, 163
alpha particles, defined, xxix
Alvarez, Robert, 24, 101, 132, 134
Amarillo site, 5
Ameren, 2
American Coalition for Clean Coal Electricity, 194
American Concrete Institute, 3
American Nuclear Society (ANS), 118, 130
American Wind Energy Association (AWEA), 196
ANS (American Nuclear Society), 118, 130
AP1000 reactors. *See* Toshiba-Westinghouse AP1000 reactors
AREVA company, 103, 119, 130
AREVA EPR reactors, 3, 6–8
Argentina, 45
Arkansas Nuclear, 72
Atomic Energy Commission, 35, 137
Atomic Energy of Canada, 16
Atoms for Peace program, 102–3
Atta, Mohamed, 158
Australia, 65, 67, 127
Austria, xxii
AWEA (American Wind Energy Association), 196
Axelrod, David, xxi

Babcock and Wilcox Co., 148
Balakovo plant, 45
Bataan nuclear facility, 16
Beaver Valley reactor, 3, 51
Bechtel National, 22
Becker, Rochelle, 163
becquerel, defined, xxviii
Bellefonte power station, 2
beta particles, defined, xxix
beyond-design-basis accidents, 143
Beyond Nuclear, 57–58, 141
Big Sol, 216–20
Bingaman, Jeff, 13
biodiesel, 205–6
biomass, 194
bird populations, xv–xvi
Birol, Fatih, 1
birth defects. *See* health effects
Blackburn, John O., 194
blackouts, 83, 85
Blair, Tony, 16
Blakeslee, Sam, 152, 154–55
Blanch, Paul, 142
Blayais plant, 46
Blue Ribbon Commission on America's Nuclear Future (BRC), 92–94
Bohunice plant, 45
boiling water reactors (BRWs), xxvii, 23
Boxer, Barbara, 145, 153
Boyd, Jim, 87
Bradford, Peter A., 31–32, 52
Braidwood plant, 47, 142, 172
BRC (Blue Ribbon Commission on America's Nuclear Future), 92–94
breeder reactors, 27
Bridenbaugh, Dale, xvii
British Petroleum, 216–17
Brower, David, 187
Brown, Lester, 204

Browns Ferry plant, 33, 34, 45, 55, 72, 142
Brunsbüttel reactor, 46
Brunswick site, 51, 55, 70, 82–83, 168–69, 172
Buffett, Warren, 10
Bulgaria, 47
Burnell, Scott, 161
Busche, Donna, 22
Bush, George H.W., 113
Bush, George W., 10, 14–15, 60–61, 107
Byron plant, 47, 140–41, 142

Calculation of Reactor Accident Consequences (CRAC-2) report, 35–36
Caldicott, Helen, 102, 134, 158–59
California Infill Builders Association, 214
Callaway site, 72, 170
Calvert Cliffs site, 10, 13, 83
Campbell, Colin, 222–23
Canada, 44, 65, 67, 192, 211
cancer. *See* health effects
cap-and-trade, 209–10
carbon offsets/cap-and-trade, 209–10
carbon taxes, 210
Carter, Jimmy, 107, 228
Cascadia Subduction Zone, 76
Catawba plant, 51, 72
CBS, 33
Centraco nuclear waste treatment facility, 47
Central and Eastern United States Seismic Source Characterization for Nuclear Facilities (CEUS) model, 71
cesium
- defined, xxix
- from Fukushima, 37, 39–40, 123–24, 126–27, 131, 133
- Simi Valley reactor and, 64
- spent fuel rods and, 101

Chalko, Tom, 70
Chalk River reactor, 44
Chapelcross reactor, 44
Cheney, Dick, 102
Chernobyl
- costs of, 9, 17
- deaths from, 32–33, 45, 57, 58–59
- long-term consequences of, xv–xvi
- as worst nuclear accident, 63

China, xxii–xxiii, 195
China National Nuclear Corp., 119
China syndrome, 41
Chu, Steven, 119, 217
Claude, André, xxiv
Clean Energy Deployment Administration (CEDA), 12–13
Clean Technica, 189
Clements, Tom, 119
climate change, 27–28, 80–86, 187
Clinch River site, 5
Clinton, Bill, 107
closed-loop recycling, 229
CMEs (coronal mass ejections), 84–86
CO_2 emissions, 1, 27–28, 80, 189, 191–92, 193, 208–10
coal, 184, 194, 196
cogeneration, 28, 190, 204
cold shutdowns, 41–42, 130
Collins, Elmo, 153
Columbia Generating Station, 19–20
Comanche Peak site, 5, 47, 172
combined heat and power (CHP) plants, 204
Commonwealth Radioactive Waste Management Act (Australia), 67
community aggregation, 215
concentrated solar, 182, 198–99
conservation, 186–87, 221–26
Constellation Energy, 7, 10
construction speed, 1–8
containment buildings, 51
coolant water, 28
Cooper reactor, 55, 83, 170
corium, 41
coronal mass ejections (CMEs), 84–86
corroded containment buildings, 51
corroded pipes, 51
costs. *See also* loan guarantees
- bailouts and, 10–14
- in Britain, 17
- of decommissioning, xxiv, 17–22
- of Fukushima, 125
- of new reactors, 2, 6, 9–11, 13–14, 26
- of refueling, 23–24
- of solar energy, 8

cover-ups, 130–36, 138–46
cracked tubes, 50–51, 140–41, 141–42
Creswell, James, 148
Crystal River site, 6, 12, 79

Cuba, 223
Cuomo, Andrew, 158, 159
curie, defined, xxviii

Dampier plant, 46
Davis-Besse reactor, 31, 33, 46, 52, 82, 147–52, 172
Deal, John "Grizz," 120
deaths, 32–33, 44–48, 57–59, 132, 159, 198
deceit, 130–36, 138–46
decommissioning, xxiv, 17–22, 95–101
decoupling and shared savings, 212
Deepwater Horizon disaster, 81
defense-in-depth, 169–70
demand for energy, 1
Democratizing the Electricity System (Farrell), 219
Denton, Harold, 29
Department of Energy (DOE), xviii, 20–22, 89–90, 132
Diablo Canyon reactors, 14, 47, 57, 70, 152–56, 169, 172–73
Diaz, Nils J., 169
distributed solar power, 199–200, 219, 228–29
Dominion Power, 73–74
Dow Chemical, 96–97
downsizing, 221–26
Dresden reactors, 55, 73, 75, 79, 97, 142, 175–76
Dricks, Victor, 161
dry-cask storage, 101
DTE Energy, 58
Duane Arnold reactor, 55, 75, 79

earthquakes
- Diablo Canyon and, 152–56
- Fukushima and, 76–77
- ground acceleration and, 74–75
- high risk US reactors, 79
- history of, 77–78
- Indian Point and, 156
- industry resistance, 75
- inland risks, 72–73
- New Madrid Fault and, 71–72
- North Anna reactors and, 73–74
- NRC and, 70–71
- potential for, 75–76
- risks, 69–70

eco-cities, 213–14, 225
Edano, Yukio, 38, 43
EDF (Électricité de France), xxiv, 10, 16, 130
The Effects of Nuclear Weapons, 110
efficiency, xxvii, 28, 188–92
Eisenhower, Dwight, 102
electrical cables, 51–52
Électricité de France (EDF), xxiv, 10, 16, 130
Electric Power Research Institute (EPRI), 142
electric pressurized water reactors (EPWR), 3
electric vehicles, 197, 214
Emanuel, Rahm, xxi
embrittlement, 50, 139–40
energy, defined, xxvii
energy efficiency. *See* efficiency
Energy Policy Act (2005), 10, 14
energy returned on energy invested (EROEI), xxvii, 198
Energy [R]evolution, 192
enhanced geothermal systems (EHS), 204–5
Enrico Fermi reactor, 44, 55, 58
Entergy Corp., 165–67, 170
Environmental Protection Agency (EPA), 132–33
EPA (Environmental Protection Agency), 132–33
EPRI (Electric Power Research Institute), 142
EROEI (energy returned on energy invested), xxvii, 198
European Renewable Energy Council, 192
evacuation plans, 136
Evolutionary Power Reactors, 7
Exelon Corporation, xxi, 49, 99–100, 142
exposure guidelines, 35, 39, 58, 60–61, 123–25, 131, 133–35

fallout. *See* exposure guidelines; radiation
Fanning, Thomas, xix
Farley reactors, 73, 79
Farrell, John, 219
fatalities. *See* deaths
feebates, 211

feed-in tariffs (FITs), 180–81, 210–11, 218
Feinstein, Dianne, 153
FENOC (FirstEnergy Nuclear Operating Company), 147–52
Fermi plutonium breeder reactor, 34, 82
Fertel, Marvin, 87
Fetter, Steve, 111–14
FirstEnergy Nuclear Operating Company (FENOC), 147–52
Fisher, Alison, xix
FITs (feed-in tariffs), 180–81, 210–11, 218
Fitzpatrick reactor, 55, 175
Flamanville plant, 6
floods, 82–84
Flowers, Brian, 111
FOE (Friends of the Earth), 162–63
Folkers, Cindy, 36
Ford, Gerald, 102, 107
Forsmark plant, 35, 47
Fort Calhoun reactor, 83–84, 142, 171
Fosmark facility, 90
fossil fuels, 1, 188
fracking, 74
France, xxiii–xxiv, xxv, 2, 16, 45, 46, 47, 60
French Nuclear Safety Authority (ASN), xxiii–xxiv
Friends of the Earth (FOE), 162–63
fuel cells, 205
fuel rods, 26, 39–40, 88–94. *See also* spent fuel rods
Fukushima 50, 38
Fukushima Daiichi
  accidents at, 45, 46, 47
  aftermath of, xv–xvii, 16–17, 37–43, 123–29, 227–28
  costs of, 17–18
  cover-ups and, 130–36
  earthquakes and, 69, 76–78
  fuel rod pool, 88–89
  NRC response, 143–45
  weather conditions and, 81
Fukushima Daini, 38
Full Body Burden: Growing Up in the Nuclear Shadow of Rocky Flats (Iverson), 97
Funabashi, Yoichi, 39, 227

gamma radiation, defined, xxix
Gandhi, Mahatma, 187
gas-cooled fast reactors, 117
Gates, Bill, 118
Gen4 Energy, Inc., 119
General Electric, xvii, 8, 33, 119
Generation IV nuclear reactors, 15, 116–20
genetic mutations. *See* health effects
genuine progress indicator (GPI), 225–26
GeoCap Study, 60
Geothermal: Getting Energy from the Earth (Brown), 204
geothermal power, 183, 194, 204–5
Germany, xxiii, xxv, 16, 46, 47, 60, 179–81, 210–11
gigawatt (GW), defined, xxvii
Gipe, Paul, 181
globalization, 224
Global Nuclear Energy Partnership (GNEP), 107–8
global warming, 27–28, 80–86, 187
Goldfinger, Chris, 75–76
Gould, Jay M., 59–60
GPI (genuine progress indicator), 225–26
Graham, Lindsey, 119
Grand Canyon National Park, 62, 68
Grand Gulf plant, 47–48
Gravelines plant, 47
gray (Gy), defined, xxviii
Great Charleston Quake, 72
green collar jobs, 180
Green Energy Act (Canada), 192
Green Jobs Act (2007), 208–9
Greenpeace International, 135–36, 192
Greifswald reactor, 46
Grim, Ryan, 146
Grossman, Karl, xx
ground acceleration, 74–75
Gundersen, Arnie, xx, 4–5, 9, 29, 83, 89, 131, 134
GW (gigawatt), xxvii

Hamaoka nuclear plant, 77
Hamm-Ueutrop plant, 46
Hanford Nuclear Reservation, 19–22
Hardebeck, Jeanne, 153–54
Hatch reactor, 50, 55, 173
H.B. Robinson plant, 72, 174
health effects. *See also* deaths; safety issues
  Chernobyl and, 32–33, 58–59

Fukushima Daiichi and, 123–29, 130–35
Indian Point and, 156–159
from nuclear plants and accidents, 44–48, 57–61
Simi Valley reactor, 62–64
Three Mile Island and, 32
Health Physics Society (HPS), 130
Hebert, Curtis, 165
Hedegaard, Connie, xxiv
Heinberg, Richard, 186, 188, 223, 226
high-temperature reactors (HTR), 117
Hirose, Kenkichi, 123
Hirsch, Dan, 64
Honor the Earth, 66
Hope Creek reactor, 55
Hopenfeld, Joram, 161
Hornung, Richard, 65
HPS (Health Physics Society), 130
HTR (high-temperature reactors), 117
Hubbard, Richard, xvii
human life, value of, 144–45
Humbolt Bay reactor, 97
Hungary, 46
hurricanes, 82–83
hydroelectric power, 183, 194, 203–4
hydrogen, 206

IAEA (International Atomic Energy Agency), 69, 105–6
ICRP (International Commission on Radiological Protection), 58
immune system damage. *See* health effects
independent spent fuel storage installations (ISFSI), 95
India, xxii, 15, 102, 223
Indian Point reactors, 33, 35–36, 49, 56, 71, 79, 97, 142, 156–60, 173
indigenous peoples, 65–68
infill building, 214
inland tsunamis, 83–84
Inslee, Jay, 13
International Atomic Energy Agency (IAEA), 69, 105–6, 119
International Commission on Radiological Protection (ICRP), 58
International Energy Agency (IEA), 1
International Nuclear and Radiological Event Scale, 34
iodine-131
defined, xxix
from Fukushima, 126, 133
permissible levels of, 60–61
releases of, 63
ionizing radiation, defined, xxviii
Iran, 102
Iraq Energy Institute, 119
ISFSI (independent spent fuel storage installations), 95
Ishihara, Shintaro, 105
Israel, 102
Italy, xxii, xxv
Iverson, Kristen, 97

Jacobson, Mark Z., 182
Jaczko, Gregory, xviii–xx, 75, 135, 145–46, 162
Japan, xxii–xxiv, 45–47, 77–78, 103–4, 123–29. *See also* Fukushima Daiichi
Johnson, Lyndon, 61
Johnson, Sam, 154
Jones, Evan, xxv
Jones, Thomas K., 113
joule, defined, xxvii

Kaku, Michio, 130
Kalpakkam nuclear power station, 78
Kaltofen, Marco, 133
Kamps, Kevin, xx, 54
Kan, Naoto, xxiv, 38, 39
Kashiwazaki-Kariwa plant, 37, 47, 77, 78
Kazakhstan, 65
Khan, Abdul Qadeer, 103
KI. *See* potassium iodide (KI) anti-radiation pills
kilowatt-hour (kWh), defined, xxvii
Kitazawa, Kay, 227
Kjaer, Christian, 196
Klea, Bonnie, 63
Kodama, Tatsuhiko, 125
Kosciusko-Morizet, Nathalie, xxiv
Kosloduy-5 reactor, 47
Krümmel reactor, 47
Kucinich, Dennis J., 150–52

LaDuke, Winona, 66
land pollution, 61–62
LaSalle reactor, 23, 142

lead-cooled fast reactors, 117
leakage, 50, 142
Le Havre facility, 90
Leningrad reactor, 47
leukemia. *See* health effects
Levy County site, 5, 6
light-emitting diodes, 201–2
Limerick reactors, 71, 79, 173
limited liability protection, 14
Lipsky, Jon, 96–97
Little, Mark M., 8
loan guarantees
  in the 2011 federal budget, xxiii
  for Calvert Cliffs, 10
  for new construction, 2–3
  Obama and, 12–13, 15, 155
  solar power and, 216
  Southern Company and, xviii
  Vogtle and, xxi
  wind power and, 196
lobbyists, xvii, xviii, xxi, 7, 10–12, 137, 174, 194, 217
local production, 224, 229
Lochbaum, Dave, 149
Lockbaum, David, 143
Lodge, Terry, 152
Loescher, Peter, xxv
Lovins, Amory, 28, 188–89, 215, 219
Lyman, Edwin, 143

Ma, John, 3
Makhijani, Arjun, 63
management problems, 169
Mangano, Joseph, 132
Marble Hill plant, 48
Mariotte, Michael, 10, 43
Mark 1 reactors
  Beyond Nuclear and, 166
  design of, xvi, 29–30, 40, 126
  earthquakes and, 75, 77
  names and locations of, 55, 69–70
  Oyster Creek, 35
Markey, Edward, xix, 36, 145
Master Curve, 139–41
Matsumura, Akio, 40
Mayak nuclear reprocessing plant, 44
McCain, John, 13
megawatt (MW), defined, xxvii
meltdowns
  Chalk River, 44
  Chernobyl, xxvi
  Fukushima Daiichi, xi, xvi–xvii, 18, 39, 47, 89, 124, 126–28, 130
  Lucens, Switzerland, 44
  Simi Valley, 44, 63–64
  Three Mile Island, xv, 9, 32
  in the United States, 34–35
Merkel, Angela, 16, 179–80
Merrifield, Jeffrey, xix–xx
microgrids, 213
micropower, 195
MidAmerican Energy Holdings Company, 10
Mihama reactor, 47
Millennium Ecosystem Assessment, 185
millirem, defined, xxviii
millisievert, defined, xxix
Millstone plant, 82, 100, 170, 173
mining, 61–62, 65–68
mini-nukes, 119
Minor, Gregory, xvii
Mironova, Natalia, xvii
Moglen, Damon, 145–46
molten salt reactors, 117
Monticello reactor, 55, 170
Mousseau, Timothy, xvi
MSNBC, 33
Mulley, George, 141
Murtha, Garvan, 166–67
Musegass, Phillip, 158
MW (megawatt), xxvii
Myers, David, 217

Nakayama, Shinichi, 125
Namibia, 65, 66, 67
National Energy Policy Act (2005), 15
National Renewable Energy Laboratory (NREL), 195–97
National Sacrifice Zones, 61
national security, 109–15
Native Americans, 65–68
natural gas, xxii
NBC, 33
negawatts, 191
net energy, 188–92
net metering programs, 218
neutron radiation, defined, xxix
New Madrid Fault, 71–72

Niger, 65, 66
Nine Mile Point plant, 48, 55
NIRS (Nuclear Information and Resource Service), 41
Nishina, Yoshio, 104
Nissim, Chaim, 114
Noda, Yoshihiko, 43, 128
North Anna site, 51, 71, 73–74, 79
Northeast Utility, 100
Northrop Grumman, 103
Norway, 27
Norwegian Institute for Air Research, 131
nozzles, 52
NPT (Nuclear Nonproliferation Treaty), 106–7
NRC. *See* Nuclear Regulatory Commission
NREL (National Renewable Energy Laboratory), 195–97
Nuclear and Industrial Safety Agency (Japan), 38, 77, 123
Nuclear Energy Act (Finland), 90
Nuclear Energy Institute, 87, 130
Nuclear Hydrogen Initiative, 15
Nuclear Information and Resource Service (NIRS), 41
Nuclear Nonproliferation Treaty (NPT), 106–7
Nuclear Regulatory Commission (NRC)
  Central and Eastern United States Seismic Source Characterization for Nuclear Facilities (CEUS) model, 71
  collusion with industry, 138–43
  cover-ups, 135
  Davis-Besse and, 147–52
  decommissioning and, 95–101
  Diablo Canyon and, 153–55
  earthquake hazards and, 70–72, 75
  Fukushima response, 143–45
  Indian Point reactors and, 156–60
  inspections and oversight, 169–76
  Jaczko resignation, 145–46
  licensing, xviii–xxii, 2, 11, 137–38
  operating extensions, 24–25, 49–55
  safety ratings, 33–34, 53–54
  San Onofre plant and, 160–64
  spent fuel rods and, 100–101
  Three Mile Island and, 31–32
  Vermont Yankee and, 164–67
  Vogtle plant and, 5–6
Nuclear Safety Commission (Japan), 123
nuclear samurai, 39
Nuclear Waste Fund, 90
Nuclear Waste Policy Act (NWPA), 92–94, 101
nuclear weapons, 102–8. *See also* national security
NuGen, 16
NWPA (Nuclear Waste Policy Act), 92–94, 101

Obama, Barack, xviii, xxi, 10, 15, 88, 131, 190, 195
Oconee plant, 72, 79, 170, 173
Office of Nuclear Reactor Regulation (ONRR), 98–99
Office of Technology Assessment (OTA), 110
off-the-grid movement, 218
Oi facility, 43
Oil Depletion Protocol, 223
Olkiluoto plant, 6, 16
Omaha Public Power District (OPPD), 84
Onagawa reactor, 77, 78
Onkalo site, 90, 91–92
ONRR (Office of Nuclear Reactor Regulation), 98–99
operating extensions, 24–25, 49–55, 138
organic agriculture, 225
Ortega, Rudy, 153
OTA (Office of Technology Assessment), 110
overheating reactors, 81
Oyster Creek reactor, 35, 49, 55, 83, 142

Pacala, Stephen, 1
Pacific Environment, 199–200
Pacific Ocean, 126
Pacific Plate, 76
Pakistan, 102
Paks reactor, 46
Palisades reactor, 51–52, 53–54, 170
Palo Verde reactors, 81, 142
passive solar, 200
passive thermal water heaters, 200
Pasternak, Judy, 65
PBMRs (pebble bed modular reactors), 117

Peach Bottom reactors, 24, 36, 55, 75, 79, 97, 142
Peacock, Marcus, 60
peak ground acceleration, 74
pebble bed modular reactors (PBMRs), 117
Pederson, Cyntha, 54
PEER (Public Employees for Environmental Responsibility), 134
pencil engineering, 139–41
performance records, 23–30
Perry reactor, 73, 74, 77, 79
Philippines, xxii, 15
Philippsburg reactor, 46
photovoltaic technology, 180–81, 182, 193, 198, 219–20
picocurie, defined, xxviii
Pilgrim reactor, 24, 55, 71, 88–89, 170
Point Beach reactor, 33, 173–174
policy reform
  2,000-watt societies, 214–15
  carbon offsets/cap-and-trade, 209–10
  carbon taxes, 210
  community aggregation, 215
  decoupling and shared savings, 212
  eco-cities, 213–14
  feebates, 211
  feed-in tariffs, 210–11
  microgrids, 213
  renewable offsets, 210
  renewables and, 208–9
  smart meters, 212–13
  Transition Towns, 215
  zero-energy buildings, 213
polluter pays principle, 225
population growth, 228
Possony, Stefan, 102
potassium iodide (KI) anti-radiation pills, 58
power, defined, xxvii
power-down economies, 221–26
Power Felt, 205
Prairie Island reactor, 83, 98, 142
precautionary principle, 223–24
pre-construction rate increases, xx, xxi, 2–3
pressurized water reactors, xxvii, 3, 6–8, 23, 50
Price-Anderson Act (1957), 14
Progress Energy, 6, 12
propaganda, 131
Protective Action Guides (PAGs), 60–61, 134
protests against nuclear energy, xxii–xxiii, 38, 128
Public Employees for Environmental Responsibility (PEER), 134

Quad Cities reactor, 55, 142

radiation. *See also* health effects
  exposure guidelines, 35, 39, 58, 60–61, 123–25, 131, 133–35
  from Fukushima, 37, 39, 42, 123–29, 131–36
  permissible doses of, 58
  from reactor bombings, 109–15
radioactive waste. *See* waste management
RADNET, 132–33
radon gas, 61–62
Ramberg, Bennett, 109
Rancho Seco station, 35, 45
Reagon, Ronald, 107
Rebuild Japan Initiative Foundation (RJIF), 39, 227
refueling costs, 23–24
regulatory agencies. *See* Nuclear Regulatory Commission
Reid, Harry, 101
rem, defined, xxviii
renewable energy, 179–84, 186, 193–206, 208–9, 213–14
Renewable Energy Sources Act (Germany), 180
renewable offsets, 210
reprocessing, 27, 107–8
RIA (risk-informed analysis), 140–41
Riccio, Jim, 25
Rifkin, Jeremy, 219
Ringhals reactor, 47
risk assessments, 35–36
risk-informed analysis (RIA), 140–41
River Bend reactor, 73, 79, 82
RJIF (Rebuild Japan Initiative Foundation), 39, 227
Rocard, Michel, xxv
Rocky Flats nuclear weapons facility, 96–97

Roisman, Anthony, 167
rooftop solar power, 199–200
Rothrock, Ray, 9
Röttgen, Norbert, 179
Rubin, Joe, 87
Runkle, John, 5
Runsfeld, Donald, 102
Russia, xxiv–xxv, 45, 46, 47

SACE (Southern Alliance for Clean Energy), xxi, 2
safety issues, 130–36, 138–46, 171–76
safety ratings, 33–34, 53–54
Saint Laurent reactor, 45
Saint Lucie reactors, 79
Salazar, Ken, 62, 68, 197
Salem Nuclear Generating Station, 56, 142
Sanders, Bernie, 145, 166, 200
San Onofre plant, 45, 57, 70, 98, 160–64
Sato, Yuhei, 125
Savannah River National Laboratory (SRNL), 119
Savannah River reprocessing plant, 44, 46
Schlesinger, James, 110
Scotland, 44
Seabrook plant, 48, 79
Searching for a Miracle (Heinberg), 186, 223
seismic risks. *See* earthquakes
Sellafield site, 27, 45, 90
Sendai tsunami, xv
Sequoia site, 72
Sequoyah power station, 24, 79, 174
Shearon Harris site, 5
Sherman, Janette D., 132
Sheron, Brian, 70
shield building design, 3
Shika reactor, 46
Shimizu, Masataka, 38
Shoreham site, 14, 48
Shoriki, Matsutaro, 103–4
Shumlin, Peter, 165
shutdowns, 23–30, 41–42
Siemens, xxiv–xxv, 180
sievert, defined, xxviii
Simi Valley reactor, 44, 62–64
Simpa Networks, 218
Slovakia, 45
small modular reactors (SMRs), 118–19
smart meters, 212–13
SOARCA (State-of-the-Art Reactor Consequences Analyses), 36
Socolow, Robert, 1
sodium-cooled fast reactors, 117
solar flares, 84–86
solar power
  Big Sol, 216–20
  capacity worldwide, 186
  community aggregation and, 215
  comparisons, 183
  concentrated, 182, 198–99
  cost of, 8
  distributed, 219, 228–29
  distributed and rooftop, 199–200
  in Germany, xxv, 181
  growth of, 193–94
  photovoltaics, 180–81, 182, 193, 198, 219–20
  superiority of, 26
  in Texas, 28
  thermal water heaters, 200
  thin-film, 200–201
solar thermal water tanks, 200
Solar Trust, 216–17
Solis, Hilda, 208–9
Solomon, Norman, 155
Solyndra solar energy, xxi
Sorrell, William, 165
Southern Alliance for Clean Energy (SACE), xxi, 2
Southern California Edison, 160, 162–64
Southern Company, xviii–xix, xxi
South Texas Project, 5, 12, 13, 28, 48, 70
Sovacool, Benjamin, 34, 114
Soviet Union, 44
Spain, 198
speed of construction, 1–8
spent fuel rods. *See also* fuel rods
  earthquakes and, 73, 75
  at Fukushima Daiichi, xi, xvi, 37–40, 42
  power outages and, 86
  reprocessing facilities, 27
  storage of, 88–94, 100–101, 144–45
  as targets in warfare, 111
SRNL (Savannah River National Laboratory), 119

State-of-the-Art Reactor Consequences Analyses (SOARCA), 36
Stern, Nicholas Lord, 208
strontium-90
defined, xxix
from Fukushima, 124, 126–27
permissible levels of, 60–61
Simi Valley reactor and, 64
subsidies, 10–11
SunPower, 216–17
super-critical water-cooled reactors, 117–18
Superphénix, 27
Surry reactor, 44, 46, 51, 169, 174
Sweden, 47
Switzerland, xxii, xxiv, 44

Taiwan, xxii, 77
taxpayer bailouts, 10–14
Teller, Edward, 113
tellurium, 133
Tennessee Valley Authority (TVA), xxii, 2
TEPCO, xvi, xx, 18, 29, 38–42, 123–29, 130, 136, 227
terawatt (TW), defined, xxvii
TerraPower, 118
terrorism, 114–15, 158
thin-film solar, 200–201
The Third Industrial Revolution (Rifkin), 219
Three Mile Island, 9, 31–34, 45, 59, 98, 174
tidal energy, 201
Tierney, John, 208
Tohoku quake and tsunami, 78
Tokaimura plant, 46
tombstone regulation, 142–43
Tomsk-7 reactor, 46
tornadoes, 82–83
Toshiba-Westinghouse AP1000 reactors, xviii–xix, xx, 2–6, 26, 57
Transition Towns, 215
traveling wave reactors (TWRs), 118
Tricastin reactor, 47
tritium, xxx, 57, 142, 165–66
Trojan plant, 46
Tsipis, Kosta, 111–14
tsunamis, xv
Turkey Point reactors, xxii, 5, 12, 13, 82
TW (terawatt), xxvii
TWRs (traveling wave reactors), 118

UCS (Union of Concerned Scientists), 143–45, 153, 170–71
UN Declaration on the Rights of Indigenous Peoples, 67, 68
undergrounding, 113
Union of Concerned Scientists (UCS), 143–45, 153, 170–71
UniStar Nuclear Energy, 7, 10
United Kingdom, xxii, xxiii, 16–17, 44, 45
United Nations Development Program, 187
United States
nuclear accidents, 44, 45, 46, 47–48
radiation from Fukushima, 126–27, 130–35
response to Fukushima disaster, xxii–xxiii
Uppsala Protocol, 222–23
uprates, 25
uranium
CO_2 emissions and, 80
defined, xxx
imports of, 26
indigenous peoples and, 65–68
land pollution and, 61–62

Venezuela, xxii
Vermont Yankee reactor, 55, 142, 164–67, 174
very-high temperature reactors (VHTR), 117
Vetter, Kai, 132–33
Virgil C. Summer Nuclear Generating Station, xx–xxi, 5, 72, 74, 79
Vogtle Electric Generating Plant, xviii–xxi, 3, 5–6, 74–75, 143, 174

Walmart, 208
Walt Disney Company, 208
Wasserman, Harvey, 6, 14
waste management, 7–8, 87–94. *See also* spent fuel rods
Waterford station, 70, 82
water pollution, 56–57
Watts Bar project, xxii, 1–2, 72, 79, 142
wave energy, 183, 202–3

Wellinghoff, Jon, 8
Westinghouse, 33, 119, 130
wind hydrogen, 206
wind power, xxv, 2, 26, 27–28, 180, 182, 193–98
Windscale facility, 44, 63
wind, water, solar, and geothermal energy (WWSG), 194–95
Wolf Creek reactor, 73, 79, 174–75
worker error, 169
World Nuclear Association, 7
Worldwatch Institute, 186
WPPSS plant, 48
Wright, Oliver, 16
WWSG (wind, water, solar, and geothermal energy), 194–95
Wyden, Ron, 40

Yankee Rowe reactor, 35, 59, 88
Yoshii, Hidekatsu, 77
Yucca Mountain, Nevada, 7, 66, 88, 91, 92–94

zero-energy buildings, 213
Zimmer plant, 48
Zion reactor, 95, 98, 99–100

False Solutions Publication Series

Previous publications in this series include:

Manifesto on Global Economic Transitions (2007). A joint project of 25 IFG associates and board members. Closely defines the global economic, political, and environmental crises of today and lists urgently needed economic and political transitions to reverse the problems.

The False Promise of Biofuels (2007) by Jack Santa Barbara (IFG board member). Argues that biofuels are no utopian cure for the global energy and climate crisis, and that they produce many more problems than they solve, including a vast conversion of food-producing lands to the production of fuel. Biofuels have proven to be less energy efficient than most other options, with a net energy ratio below zero.

The Rise and Fall of Global Agriculture (2007) by Debbie Barker (IFG board member). Predicts that globalized industrial agriculture—including export-orientated one-crop production, mechanization, and biotechnology—will bring severe food shortages, add to the global water crisis, push farmers off lands, diminish soil fertility, and increase the costs of food. Urges a rapid return to local sustainable food systems.

Searching for a Miracle (2009) by Richard Heinberg (IFG associate member and director of the Post Carbon Institute). Focuses on alternatives to fossil fuels, concluding that no single technology or even combination of technologies could sustain industrial society at its present level. Concludes that the truly optimistic alternative is to "power down."

For further information on International Forum on Globalization publications, please go to www.ifg.org.

About the Author

Gar Smith is editor emeritus of *Earth Island Journal*, a Project Censored award-winning investigative journalist, and cofounder of Environmentalists Against War. He has covered revolutions in Central America and has engaged in environmental campaigns on three continents. He lives a low-impact, solar-assisted lifestyle in Berkeley, California.

Chelsea Green Publishing is committed to preserving ancient forests and natural resources. We elected to print this title on 30-percent postconsumer recycled paper, processed chlorine-free. As a result, for this printing, we have saved:

27 Trees (40' tall and 6-8" diameter)
12 Million BTUs of Total Energy
2,324 Pounds of Greenhouse Gases
12,606 Gallons of Wastewater
844 Pounds of Solid Waste

Chelsea Green Publishing made this paper choice because we and our printer, Thomson-Shore, Inc., are members of the Green Press Initiative, a nonprofit program dedicated to supporting authors, publishers, and suppliers in their efforts to reduce their use of fiber obtained from endangered forests. For more information, visit: www.greenpressinitiative.org.

Environmental impact estimates were made using the Environmental Defense Paper Calculator. For more information visit: www.papercalculator.org.

the politics and practice of sustainable living

CHELSEA GREEN PUBLISHING

WIND POWER
Renewable Energy for Home, Farm, and Business
PAUL GIPE
9781931498142
Paperback • $50.00

WIND ENERGY BASICS
A Guide to Home- and Community-Scale Wind Energy Systems
PAUL GIPE
9781603580304
Paperback • $29.95

WHEN TECHNOLOGY FAILS
A Manual for Self-Reliance, Sustainability, and Surviving the Long Emergency
MATTHEW STEIN
9781933392455
Paperback • $35.00

WHEN DISASTER STRIKES
A Comprehensive Guide for Emergency Planning and Crisis Survival
MATTHEW STEIN
9781603583220
Paperback • $24.95

For more information or to request a catalog,
visit **www.chelseagreen.com** or
call toll-free **(802) 295-6300**.